POLYMER COLLOIDS:
A COMPREHENSIVE INTRODUCTION

Dedication

To Reta, David, Douglas and Christopher.
Their love and support, their joie de vivre,
and their encouragement have sustained
and uplifted me throughout this work.

Plate 1. Synthetic 'black opal' made from monodisperse PMMA particles in PS matrix, set into a gold mount.

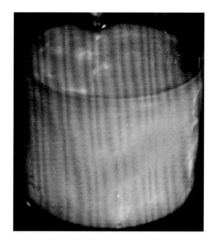

Plate 2. Monodisperse polyvinyl toluene latex. Courtesy of Eastman Kodak Co. from Bagchi, P., Birnbaum, S.M. and Gray, B.V. (1980). In Polymer Colloids (R.M. Fitch, ed.), Plenum Press, NewYork, USA.

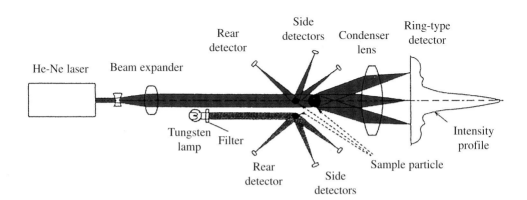

Plate 3. Fraunhofer-type apparatus for obtaining particle size distributions directly. Courtesy of Horiba Ltd, Irvine, CA, USA.

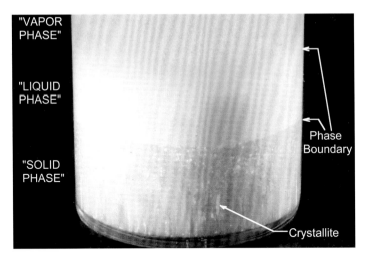

Plate 4. Monodisperse polymethyl acrylate colloid showing three phases at equilibrium.

Plate 5. PMMA/PHS colloid in decalin/CS$_2$ mixture with refractive indexes almost matched. Volume fraction decreases from left to right: A, glassy; B, glassy/crystalline; C, large colloidal crystals; D, small colloidal crystals, E, borderline disordered. Courtesy of Professor P.N. Pusey, Edinburgh University, Edinburgh, UK.

COLLOID SCIENCE

Editors
R. H. Ottewill and R. L. Rowell

In recent years colloid science has developed rapidly and now frequently involves both sophisticated mathematical theories and advanced experimental techniques. However, many of the applications in this field require simple ideas and simple measurements. The breadth and interdisciplinary nature of the subject have made it virtually impossible for a single individual to distil the subject for all to understand. The need for understanding suggests that the approach to an interdisciplinary subject should be through the perspectives gained by individuals.

The series consists of separate monographs, each written by a single author or by collaborating authors. It is the aim that each book will be written at a research level but will be readable by the average graduate student in chemistry, chemical engineering or physics. Theory, experiment and methodology, where necessary, are arranged to stress clarity so that the reader may gain in understanding, insight and predictive capability. It is hoped that this approach will also make the volumes useful for non-specialists and advanced undergraduates.

The author's role is regarded as paramount, and emphasis is placed on individual interpretation rather than on collecting together specialist articles.

The editors simply regard themselves as initiators and catalysts.

1. J. Mahanty and B. W. Ninham: Dispersion Forces.
2. R. J. Hunter: Zeta Potential in Colloid Science.
3. D. H. Napper: Polymeric Stabilization of Colloidal Dispersions.
4. T. G. M. van der Ven: Colloidal Hydrodynamics.
5. S. P. Stoylov: Colloid Electro-optics.
6. R. G. Laughlin: The Aqueous Phase Behavior of Surfactants.
7. R. G. Gilbert: Emulsion Polymerization.
8. R. M. Fitch: Polymer Colloids: A Comprehensive Introduction.

POLYMER COLLOIDS:
A COMPREHENSIVE INTRODUCTION

ROBERT M. FITCH Ph.D.

President
Fitch & Associates
Taos
New Mexico
USA

ACADEMIC PRESS

San Diego London Boston New York
Sydney Tokyo Toronto

This book is printed on acid-free paper

Copyright © 1997 by ACADEMIC PRESS

All Rights Reserved.

No part of this publication may be reproduced or transmitted in any form or by any means, electronic or mechanical, including photocopy, recording, or any information storage and retrieval system, without permission in writing from the publisher.

Academic Press, Inc.
525 B Street, Suite 1900, San Diego, California 92101-4495, USA
http://www.apnet.com

Academic Press Limited
24–28 Oval Road, London NW1 7DX, UK
http://www.hbuk.co.uk/ap/

ISBN 0-12-257745-0

A catalogue record for this book is available from the British Library

Typeset in Great Britain by Alden, Oxford, Didcot and Northampton

Printed in Great Britain by The University Press, Cambridge

97 98 99 00 01 02 EB 9 8 7 6 5 4 3 2 1

Contents

Preface   xv
Acknowledgements   xvii
List of symbols   xix

Chapter 1 Introduction   1
   1.1 Nomenclature   1
      1.1.1 Polymer colloids   1
      1.1.2 Latex   1
      1.1.3 Emulsion polymerization   1
   1.2 General descriptions   1
   1.3 Chemistry   3
   1.4 Physics   3
   1.5 Applications   4
      1.5.1 Industrial production   4
      1.5.2 Film formers   4
      1.5.3 Matrix materials   4
      Reference   5

Chapter 2 Emulsion polymerization   6
   2.1 Particle formation   6
      2.1.1 Homogeneous nucleation   7
         2.1.1.1 Fitch–Tsai theory   9
         2.1.1.2 Control of particle size and its distribution   12
         2.1.1.3 Hansen–Ugelstad theory   13
         2.1.1.4 Coagulation rates   16
         2.1.1.5 Absolute rates of radical capture   18
      2.1.2 Heterogeneous nucleation   21
         2.1.2.1 Micellar and microemulsion nucleation   21
         2.1.2.2 Monomer droplets   24
         2.1.2.3 Foreign particles   25
   2.2 Particle growth   26
      2.2.1 Monomer partition   26
         2.2.2.1 Swelling kinetics and equilibrium   26
      2.2.2 Coagulation   28
      2.2.3 Radical entry   29

	2.2.3.1 Initiation	29
	2.2.3.2 Propagation	29
	2.2.3.3 Termination	30
	2.2.4 Radical exit	31
	2.2.5 Aqueous phase reactions	31
	2.2.6 Smith–Ewart kinetics	31
	2.2.7 Ugelstad–Hansen theory	37
	2.2.8 Alexander–Napper–Gilbert theory	39
2.3	Particle morphologies in multiphase copolymers	41
	2.3.1 Thermodynamics of phase separation	41
	2.3.1.1 Torza and Mason theory and experiments	41
	2.3.1.2 Binodal phase separation and spinodal decomposition	42
	2.3.1.3 Sundberg theory and experiments	44
	2.3.2 Kinetics of phase separation and growth	44
2.4	Mechanical properties of derived films	45
	2.4.1 Polymer alloys	45
	References	46

Chapter 3 Practical applications of emulsion polymerization — 48
- 3.1 Applications — 48
  - 3.1.1 Laboratory — 48
  - 3.1.2 Batch emulsion polymerization — 49
  - 3.1.3 Semi-continuous Emulsion Polymerization — 50
    - 3.1.3.1 Multistage emulsion polymerization — 52
  - 3.1.4 Continuous emulsion polymerization — 53
    - 3.1.4.1 Tubular reactors — 54
    - 3.1.4.2 Continuous stirred tank reactors — 54
- References — 58

Chapter 4 Nonaqueous dispersions — 59
- 4.1 General characteristics of nonaqueous dispersions — 59
- 4.2 Nonaqueous emulsion polymerization — 60
  - 4.2.1 General characteristics — 60
  - 4.2.2 Particle formation — 60
  - 4.2.3 Particle growth — 61
  - 4.2.4 Nonradical polymerization — 61
  - 4.2.5 Methods of preparation of NADs — 63
    - 4.2.5.1 PMMA latex in aliphatic hydrocarbon — 63
    - 4.2.5.2 PET (polyethylene terephthalate) latex in aliphatic hydrocarbon — 63
- 4.3 Inverse emulsion polymerization — 64
  - 4.3.1 Polymerizing water-soluble monomers in dispersion — 64
  - 4.3.2 Inverse *micro*emulsion polymerization — 65
- 4.4 Emulsion polymerization in supercritical fluids — 68
- 4.5 Applications — 69
  - 4.5.1 Automotive finishes — 69

4.5.2 Industrial coatings	70
4.5.3 Flocculants	71
References	72

Chapter 5 Characterization of polymer colloids — 73

- 5.1 Light scattering — 73
  - 5.1.1 Rayleigh theory — 73
  - 5.1.2 Rayleigh–Debye theory — 76
  - 5.1.3 Mie theory — 78
  - 5.1.4 Higher order Tyndall spectra (HOTS) — 80
    - 5.1.4.1 Polydispersity effects — 81
  - 5.1.5 Experimental — 81
  - 5.1.6 Dynamic light scattering — 82
- 5.2 Neutron scattering — 85
  - 5.2.1 General considerations — 85
  - 5.2.2 Coherent scattering length — 85
  - 5.2.3 Coherent neutron scattering length density — 86
  - 5.2.4 Angular dependence of neutron scattering intensity — 87
  - 5.2.5 Neutron scattering experiments — 87
    - 5.2.5.1 Core/shell particle morphologies — 88
    - 5.2.5.2 Contrast matching technique — 89
  - 5.2.6 The structure factor in concentrated polymer colloids — 89
  - 5.2.7 Conclusion — 90
- 5.3 Microscopy — 91
  - 5.3.1 Optical microscopy — 91
    - 5.3.1.1 The ultramicroscope — 92
    - 5.3.1.2 Confocal laser scanning microscopy — 92
  - 5.3.2 Electron microscopy — 94
    - 5.3.2.1 Transmission electron microscopy (TEM) — 94
    - 5.3.2.2 Scanning electron microscopy (SEM) — 97
  - 5.3.3 Atomic force microscopy (AFM) — 98
- 5.4 Hydrodynamic exclusion chromatography (HEC) — 100
  - 5.4.1 Capillary hydrodynamic fractionation (CHDF) — 100
- 5.5 Sedimentation — 104
  - 5.5.1 Ultracentrifugation — 105
  - 5.5.2 Disk centrifugation — 108
  - 5.5.3 Sedimentation field flow fractionation — 110
    - 5.5.3.1 General principles — 110
    - 5.5.3.2 SdFFF apparatus — 111
    - 5.5.3.3 Other field flow fractionation methods — 112
    - 5.5.3.4 Adsorption of polymers by SdFFF — 113
  - 5.5.4 Conclusion — 114
- 5.6 Nuclear magnetic resonance spectroscopy (NMR) — 114
  - 5.6.1 Motional averaging and peak width — 114
  - 5.6.2 NMR spectra of aqueous polymer colloids — 115
    - 5.6.2.1 Morphology of hydrolyzed acrylate latex particles by NMR — 115
    - 5.6.2.2 NMR peak assignments — 117

CONTENTS

5.6.3 Films derived from two-stage emulsion polymerizations	118
5.7 Dielectric spectroscopy	120
5.7.1 Electrical double layer relaxations	120
5.7.2 Experimental	121
5.7.3 Ionic strength	124
5.7.4 Titration, ion-exchange and counterion condensation	125
5.7.5 Specific ion effects	126
5.7.6 Shear-dependence	126
5.8 Surface chemistry	128
5.8.1 Overview	128
5.8.2 Cleaning of polymer colloids	128
5.8.2.1 Ion-exchange	128
5.8.2.2 Dialysis and serum replacement	130
5.8.2.3 Centrifugation and washing	132
5.8.3 Surface group composition and concentration	133
5.8.3.1 Titration	133
5.8.3.2 X-ray photoelectron spectroscopy (XPS)	135
5.8.3.3 Ion scattering spectroscopy/secondary ion mass spectrometry (ISS/SIMS)	138
5.8.3.4 Nuclear magnetic resonance spectroscopy (NMR)	140
5.8.3.5 Surfactant adsorption	140
5.8.3.6 Dye partition	141
References	142
Chapter 6 Chemistry at the interface	**145**
6.1 Introduction	145
6.2 Origins of surface functional groups	145
6.2.1 Initiator-derived functional groups	146
6.2.1.1 Surfactant-free PS colloids	146
6.2.2 Comonomer-derived functional groups	151
6.2.2.1 Advantages and disadvantages of using comonomers	151
6.2.2.2 Independent control of particle size and surface group concentration	154
6.2.3 Chain transfer-derived surface groups	155
6.2.3.1 Coincidental transfer to surfactant	155
6.2.3.2 Using 'transurfs'	155
6.2.4 Chemical modification of surface groups	156
6.2.4.1 Reactions of persulfate	156
6.2.4.2 Hydrolysis of surface groups	156
6.2.4.3 Derivatization of benzyl chloride groups	158
6.2.4.4 Redox reactions at the particle surface	159
6.3 Heterogeneous catalysis	160
6.3.1 Acid-catalyzed hydrolyses of small molecules	160
6.3.2 Catalysis of other small molecule reactions	161
6.3.2.1 Decarboxylation	161
6.3.2.2 Hydrolysis	162

  6.3.3 Practical applications   163
 6.4 Biomedical applications   164
  6.4.1 Binding biomolecules to particle surfaces   164
  6.4.2 Immunospecific cell separation   166
   6.4.2.1 Immunomagnetic cell separation   166
   6.4.2.2 Immunofluorescent cell separation   166
   6.4.2.3 Cancer therapy   167
  6.4.3 Clinical diagnostics   168
   6.4.3.1 Detection of specific proteins   168
   6.4.3.2 Detection of specific DNA fragments   168
  6.4.4 Other applications involving DNA   169
  6.4.5 Chemotherapy: targeted drug delivery   170
   6.4.5.1 General considerations   170
   6.4.5.2 Targeted drug delivery   171
 References   172

Chapter 7 Latex stability   **173**
 7.1 Thermodynamic versus kinetic stability   173
 7.2 Kinetic stability of electrically charged hydrophobic colloids   174
  7.2.1 The van der Waals attraction   174
   7.2.1.1 Retardation   175
  7.2.2 The electrical double layer   176
   7.2.2.1 For flat plates   176
   7.2.2.2 For spheres   179
   7.2.2.3 Counterion condensation and the Stern layer   179
  7.2.3 Interactions between charged particles   180
   7.2.3.1 DLVO theory   180
   7.2.3.2 Lifshitz–Parsegian–Ninham (LPN) theory   185
   7.2.3.3 Langmuir–Ise–Sogami (LIS) theory   189
 7.3 Kinetic stability of electrically neutral hydrophobic colloids   192
  7.3.1 Polymeric stabilizers   192
  7.3.2 Interaction between two particles   196
   7.3.2.1 Smitham–Evans–Napper (SEN) theory   196
   7.3.2.2 Experimental results   198
   7.3.2.3 Practical considerations   200
 7.4 Electrosteric stabilization   201
 7.5 Kinetics of coagulation   202
  7.5.1 A second-order rate process   202
   7.5.1.1 Fast coagulation   204
   7.5.1.2 Slow coagulation   205
   7.5.1.3 Hydrodynamic interactions   207
   7.5.1.4 Specific ion effects   208
  7.5.2 Experimental methods   209
   7.5.2.1 Light scattering: integrated method   209
   7.5.2.2 Other methods   213
 7.6 Practical applications   216
  7.6.1 Latex paints   216
  7.6.2 Molding resins   217

7.6.3 Fibers	218
7.6.4 'Rubberized' concrete and stucco	218
7.6.5 Controlled heterocoagulation for novel particle morphologies	218
7.6.6 Reversible coagulation	219
7.7 Lyophilic polymer colloids	220
7.7.1 Synthesis of microgels	221
7.7.2 Transfer to good solvents	223
7.7.2.1 Destabilization and mixing with solvent	224
7.7.2.2 Gradual solvent transfer through semi-permeable membrane	224
7.7.3 Applications of lyophilic polymer colloids	225
References	225
Chapter 8 Electrokinetics	**228**
8.1 Flow past the electrical double layer	228
8.2 Electrophoresis	229
8.2.1 Basic concepts	229
8.2.2 Quantitative theory	230
8.2.2.1 Theory of O'Brien and White	232
8.2.2.2 The meaning of the zeta potential	232
8.2.3 Experimental results	235
8.2.3.1 Electrophoretic mobilities of low charge polymer colloids	235
8.2.3.2 Electrophoretic mobilities of highly charged polymer colloids	236
8.3 Dielectric spectroscopy	239
8.3.1 An ideal polymer colloid?	239
8.3.2 Hairy or porous particle surfaces?	240
8.4 Electrokinetic instrumentation	243
8.4.1 Electrophoretic light scattering (ELS)	243
8.4.2 Dielectric spectroscopy	245
8.4.2.1 The Solartron frequency response analyzer (FRA)	245
8.4.2.2 The four-electrode dielectric spectrometer	246
8.4.2.3 Electrodeless dielectric spectrometer	249
References	249
Chapter 9 Order–disorder phenomena	**250**
9.1 Origins of ordering in polymer colloids	250
9.1.1 Bragg diffraction	250
9.1.2 Ionic strength effects	252
9.1.3 Equilibrium three-phase systems	252
9.1.4 Hard sphere model	253
9.1.5 The Yoshino quantized oscillator model	255
9.2 Melting–freezing behavior of colloidal crystals	257
9.2.1 The role of Brownian motion	257

	9.2.1.1 Direct microscopic observation	257
	9.2.1.2 Neutron scattering	258
	9.2.1.3 Diffusing wave spectroscopy	260
	9.2.1.4 Crystalline and glassy states	262
	9.2.2 Reversible crystallization and melting	263
9.3	Bicomponent colloidal crystals	263
	9.3.1 Direct microscopic observation	264
	9.3.2 Binary phase diagram	265
9.4	Applications of colloidal crystal arrays	268
	9.4.1 Diffraction gratings	268
	9.4.1.1 Liquid arrays	268
	9.4.1.2 Gel arrays	269
	9.4.2 Photothermal nanosecond light-switching devices	270
	9.4.3 Electric field effects	271
	9.4.4 Elastic gel colloidal crystalline arrays	274
	References	275

Chapter 10 Rheology of polymer colloids

10.1	Introduction	277
10.2	Viscosity of hard sphere dispersions	278
	10.2.1 Principle of corresponding rheological states	280
	10.2.2 The Krieger–Dougherty equation	283
	10.2.3 The four limiting viscosities	285
	10.2.3.1 Low frequency, low shear regime	286
	10.2.3.2 High frequency oscillatory regime	286
	10.2.3.3 High shear limiting viscosities at $Pe \gg 1$ and $Pe = \infty$	287
10.3	Gelled dispersions	288
10.4	Electroviscous effects	291
10.5	Effects of ordering on rheology	293
	10.5.1 The shear modulus	293
	10.5.1.1 Buscall, Goodwin, Hawkins and Ottewill (BGHO) theory	293
	10.5.1.2 Comparison of experiments with theory	295
	10.5.1.3 Particle size effects	296
	10.5.2 The viscosity	297
	10.5.3 Creep compliance	298
10.6	Rheology of nonspherical particles	300
	10.6.1 Permanent aggregates and multilobed particles	300
	10.6.2 Shear-dependent aggregates	301
10.7	Rheological measurements	304
	10.7.1 Steady flow methods	304
	10.7.2 Oscillatory methods	305
	10.7.3 Creep compliance	306
	10.7.4 Pulse method	306
10.8	Practical applications	307
	References	312

Chapter 11	**Film formation**	**314**
11.1	Introduction	314
	11.1.1 The basic processes of film formation	314
11.2	Experimental results	315
	11.2.1 The minimum film forming temperature (MFT)	315
11.3	Electron microscopy	317
11.4	Polymer diffusion between particles	321
11.5	Theories of film formation	324
	11.5.1 The role of water	324
	11.5.2 The role of the polymer interface	325
	11.5.3 The role of the viscoelastic properties of the polymer	327
	11.5.4 Application to practical systems	331
	11.5.4.1 Wetting	332
	11.5.4.2 Adhesion	332
	11.5.4.3 Rheology	334
	11.5.5 Appearance	335
	11.5.5.1 Durability	337
	References	337
Index		**339**

Preface

There have been many articles and some books written about the various aspects of polymer colloids, emulsion polymerization, latex technology and the applications of functional latexes in catalysis, medical diagnostics and therapeutics. These are based, in turn, on a voluminous literature in scientific and technical journals and in patents. As a result, the scientist entering the field may well become bewildered as to where to start and what to believe, for the literature is also rife with controversy.

This text is an attempt to introduce the beginner to the field with a solid development of the fundamentals, without getting involved in details to the point where the overall picture is lost. It is an attempt at an holistic approach to a complex field. The author makes no excuses for taking a rather personal approach, based upon his more than 40 years of experience equally divided between industrial and academic research and development.

Generally speaking this text deals with subject matter that cannot be found in similar books elsewhere. The author assumes that the reader has a working knowledge of basic polymer science, for example kinetics of free radical polymerization, or can acquire the background on his own. The same would be true for colloid science. Some introduction to the appropriate material has been given, and the reader will be provided with references, especially those deemed to represent significant benchmarks or fundamental turning points in the development of the science and technology of polymer colloids.

There is a certain magic about this field which has appealed to chemists, physicists, engineers and formulators. It's easy to make a synthetic latex, but it can be incredibly difficult to understand all that went into the process and all that explains its subsequent behavior. It is something like having a baby, it's easy to do, it can bring great joy, but it can be impossibly difficult to understand in detail all that went into the process, and its subsequent behavior! Part of the appeal of polymer colloids is the great uniformity of particle size that can be achieved in many cases almost effortlessly. This in turn leads to beautiful optical phenomena such as the Higher Order Tyndall Spectra of dilute latexes and the opalescence of concentrated ones. This uniformity also makes polymer colloids superb standards for modelling molecular processes, from electromagnetic scattering to phase behavior to rheological properties. The ability also to control their surface chemistry has led to important developments in clinical diagnostics, cell separation and chemotherapy. These are the 'high tech' applications to a class of materials which have found huge volume utilization in latex paints, adhesives, varnishes, floor coatings, printing inks, plastics, fibers, and modified concrete among many others.

I have tried to cover at least the fundamentals at the level of a senior undergraduate or graduate student in chemistry or chemical engineering, or of a practising scientist or engineer who is assigned to a project involving polymer colloids. My desire is to provide enough of a working understanding to enable the reader to embark on his own to carry out research and/or development without stumbling too seriously. I welcome comments and criticisms which might lead to future improvements in this work.

Acknowledgements

Where the world ceases to be the scene of our personal hopes and wishes, where we face it as free beings admiring, asking, and observing, there we enter the realm of Art and Science.

Albert Einstein 1921

Einstein also said that, '... my inner and outer lives are based on the labors of other men ... and I must exert myself in order to give in the same measure as I have received and am still receiving'. Among those free beings who have contributed to this project are firstly the editors, Ronald Ottewill, who inspired me and gave me the confidence to undertake the effort in the first place, and Robert Rowell, who gave line-by-line editing help and unstinting encouragement throughout. Others who gave generously of their time and talents to review considerable parts of the text are Irvin Krieger, James Goodwin, Norio Ise and Stanislaw Slomkowski. Figures, photographs and micrographs, some unpublished, were generously donated by Peter Pusey, Robert Rowell, Ronald Ottewill, Norio Ise, Steven Siano, Mitchell Winnik, John Ugelstad, Arvid Berge, Mohammed El-Aasser, Horishi Yoshida and Didier Juhué. Direct assistance was also provided by former colleagues Lao So Su and Sunil Jayasuriya.

The International Polymer Colloids Group has been a source of invaluable help in a variety of ways, including the organization of Gordon Research Conferences and international symposia on Polymer Colloids, along with the publication of many books on the subject through third parties. Especially helpful has been the IPCG Newsletter, an informal means of correspondence, which has been generously edited by Donald Napper, and produced semi-annually for 25 years at no cost to the recipients through the beneficence of industrial colleagues.

The fine editing work of the people at Academic Press in London, especially that of Gioia Ghezzi, Joanna Craig and Roopa Baliga, has been invaluable in producing a professional book in spite of the author's limitations.

Finally I wish to acknowledge the many contributions of my graduate students who, over the years, taught me so much as we worked together discovering the mysteries of this ever-fascinating and engaging field.

List of Symbols

Notes: The order of symbols is lower case Roman, upper case Roman, lower case Greek, upper case Greek. Greek symbol order generally follows the current Greek alphabet with the following exceptions which are logical for most westerners: Φ, ϕ is considered as F, f; Z, ζ comes at the end of the alphabet; eta (H, η) comes after Roman H, h; and theta (Θ, θ) comes after the Greek H, η as in the Greek alphabet.

Although units of volume are strictly in m³ in the SI, we have acceded to convention in expressing concentrations in terms of dm³.

Symbol	Description	SI Units
a	particle radius	m
a	equilibrium partition coefficient of oligoradical	–
a_s	area occupied by a surfactant molecule	m²
a_h	hydrodynamic radius	m
$\bar{a}_w$	mass-average particle radius	m
$\bar{a}_n$	number-average particle radius	m
$\alpha_{lm(\tau,t)}$	deformation parameters	–
A	amplitude	–
A	Hamaker constant	J
A	area	m²
A	geometrical factor in light scattering	–
A'	second order coagulation frequency factor	m³/s (particles)
$A_k(\theta)$	geometrical scattering factor for a k-fold aggregate	–
α	Smith-Ewart partition coefficient for free radicals	–
α	Hansen-Ugelstad radical capture rate	–
α	Cole-Cole distribution constant	–
α'	Hansen-Ugelstad radical initiation rate	–
b	homogeneous nucleation effectivity	–
b	coherent electromagnetic scattering length	m
b	reduced viscosity fitting parameter	–
B	electrical field intensity	V
β_{lm}	resonance factor	–
c	concentration of particles	m⁻³
ccc	critical coagulation concentration	mol/dm³
C	capacitance	F or A s/V
C	concentration of atoms	m⁻³

LIST OF SYMBOLS

Symbol	Description	SI Units
C	extinction cross-section in turbidity measurement	–
d	half distance between two plates	m
d	distance between two surfaces	m
D	diffusion coefficient	m^2/s
D	diameter	m
D	interparticle spacing	m
δ	thickness (of, e.g., surface layer)	m
δ	thickness of the Stern layer	m
δ	thickness of overlap region in steric interaction	m
δ	hydrodynamic increment to particle radius	m
E	energy	J or $kg\, m^2/s^2$
E	electrical field	V/m or J/A s m
ε	permittivity	$A^2 s^4/kg\, m^3$
ε_0	permittivity of a vacuum	$A^2 s^4/kg\, m^3$
$\varepsilon\varepsilon_0$	dielectric constant	$A^2 s^4/kg\, m^3$
ε	strain in creep compliance	–
$\mathscr{E}$	dimensionless electrophoretic mobility	–
f	viscous resistance	kg/s
f	degree of ionization	–
F	capture efficiency	–
F	Helmholtz free energy	J or $kg\, m^2/s^2$
F_c	capillary force	N or $kg\, m/s^2$
F_{col}	colloidal force	N or $kg\, m/s^2$
F_{ext}	external force	N or $kg\, m/s^2$
F_{res}	viscous resistance force	N or $kg\, m/s^2$
F_{sed}	sedimentation force	N or $kg\, m/s^2$
ϕ	volume fraction	–
ϕ	viscous drag coefficient (see also f above)	kg/s
ϕ	angle	radian
ϕ_H	hydrodynamic volume fraction	–
g	acceleration due to gravity	m/s^2
g	autocorrelation function in light scattering	cd
g(r)	particle pair correlation function	–
g(τ)	autocorrelation function	–
g(κ, ζ)	relaxation effect function	–
g_v	free energy of condensation per unit volume	J/m^3
G	Gibbs free energy	J
G'	shear modulus	$kg/m\, s^2$
G_∞^{th}	theoretical high frequency, low amplitude modulus	$kg/m\, s^2$
ΔG^*	critical free energy for nucleation	J
ΔG_s	surface free energy increment	J
ΔG_v	condensation free energy increment	J
γ	surface tension (free energy)	mN/m
γ	gyromagnetic ratio	–
γ	exponential potential function	–
γ	surface potential function	–

LIST OF SYMBOLS

Symbol	Description	SI Units
$\dot{\gamma}$	strain rate	s^{-1}
$\hbar$	Planck's constant	J s
h_{ij}	center-to-center distance between ith and jth partls	m
H	distance of separation	m
H	enthalpy	J
η	viscosity	poise or kg/s m
θ	mean residence time	s
θ	scattering angle	radian
θ	phase angle	radian
i	number of segments per polymer chain	–
I	initiator molecule	–
i, I	light scattering intensity	cd
I	electric current	A
j	chain length of an oligoradical (no. of repeat units)	–
j_{cr}	critical chain length for self-nucleation	–
J	diffusive flux	(particles)/s m^2
J	creep compliance	m s^2/kg
$\langle J \rangle$	average hydrodynamic coefficient	–
k	diffusion-with-reaction constant	mol/dm^3 s
k	second order coagulation rate constant	dm^3/s (particles)
k or k_B	Boltzmann's constant	J/K
$\bar{k}_c$	overall average capture rate constant	dm^3/mol s
k_c	capture rate constant	dm^3/mol s
k_d	initiator decomposition rate constant	s^{-1}
k_{fast}	Smoluchowski fastest coagulation rate constant	dm^3/s (particles)
k_i	initiation rate constant	dm^3/mol s
$k_0 a$	Smith-Ewart radical exit frequency	s^{-1}
k_p	propagation rate constant	dm^3/mol s
k_{slow}	Fuchs slow coagulation rate constant	dm^3/s (particles)
k_t	termination rate constant	dm^3/mol s
k_t^*	termination rate constant in molecular units	dm^3/s
K	Smith-Ewart particle nucleation constant	–
κ	reciprocal Debye length	m^{-1}
l	distance from channel wall	m
L	luminosity	J/m^2
L	thickness of adsorbed layer	m
λ	wavelength	m
λ	ion conductivity	$A^4 s^5/kg^2 m^4$
m	radical exit rate	–
m	molecular weight	kg/mol
m	mass	kg
m	relative refractive index	–
m	Bragg diffraction order	–
M	monomer molecule	–
M	ion mass	kg
M	molecular weight	kg/mole

LIST OF SYMBOLS

Symbol	Description	SI Units
Mc	molecular weight between crosslinks	kg/mole
M•	free radical, monomer unit-terminated	–
M_j•	free radical, monomer unit-terminated, j units long	–
$\bar{M}_w$	weight-average molecular weight	kg/mol
μ	particle growth rate	dm^3/s
μ	chemical potential	J/mol
μ	ionic strength	mol/dm^3
n	number of free radicals in a particle	–
n	refractive index	–
n	surface atom concentration	m^{-2}
n	concentration of ions	dm^{-3}
n	particle number density	dm^{-3}
$\bar{n}$	time-average number of free radicals per particle	–
N	number concentration of particles	mol/dm^3 or dm^{-3}
N_1	number concentration of primary particles	mol/dm^3 or dm^3
N_A	Avogadro's number	$mole^{-1}$
N_n	concentration of particles containing n radicals	dm^{-3}
NOE	nuclear Overhauser effect	–
N_p	number of polymer molecules per particle	–
N_s	number concentration of seed particles	dm^{-3}
ν	number of chains per unit volume	(chains) m^{-3}
p	packing fraction	–
p_2^∞	pair distribution function at high shear, steady flow	–
p_2^{hyd}	pair distribution function at extremely high shear	–
P	number concentration of terminated radicals	dm^{-3}
P	pressure	Pa or N/m^2
P(Q)	electromagnetic scattering form factor	–
P_c	capillary pressure due to surface tension	Pa or N/m^2
P_e	Peclet number	–
Π	osmotic pressure	Pa
q_m	swelling ratio	–
Q	electromagnetic scattering vector	m^{-1}
r	radius	m
r^*	critical radius for stable nucleus	m
r_p	radius of a particle	m
r_s	radius of seed particle	m
r_{pq}	radial collision distance between particles p and q	m
R	interparticle distance	m
R	Reynold's number	–
R	radial distance	m
R	electrical resistance	ohm or kg $m^2/s^3 A^2$
R_{90}	Rayleigh ratio at 90°	m^{-1}
R_c	overall rate of capture of oligoradicals by particles	dm^{-3}/s

LIST OF SYMBOLS

Symbol	Description	SI Units
R_f	overall rate of coagulation	dm^{-3}/s
R_i	rate of initiation	$mol/dm^{-3}\ s$
R_i	internal Reynold's number	–
R_p	rate of propagation (polymerization)	$mol/dm^{-3}\ s$
$R^\bullet$	initiator free radical	–
ρ	coherent scattering length density	m^{-2}
ρ	density	kg/m^3
ρ	ionic charge density	C/dm^3
$\hat{\rho}_d$	normalized segment density distribution function	(segments) dm^{-3}
ρ_d	segment density distribution function	(segments) dm^{-3}
S	total number of surfactant molecules	dm^{-3}
S	spreading coefficient	J/m^2
S	sedimentation coefficient	s
S	entropy	J/K
$S(Q)$	electromagnetic scattering structure factor	–
S^B	Brownian pair interaction	–
S^H	hydrodynamic dipolar interaction	–
σ	interfacial tension	J/m^2 or mN/m
σ	electromagnetic scattering cross-section	m^2
σ	photoelectron cross-section	m^2
σ	stress	$kg/m\ s^2$
σ_0	surface charge density	C/m^2 (or $\mu C/cm^2$)
σ_ζ	charge density at slipping plane	C/m^2 (or $\mu C/cm^2$)
σ_r	reduced shear stress	–
σ_y	yield stress	$kg/m\ s^2$
t	time	s
$t_{1/2}$	half life	s
t^0	minimum elution time	s
t_r	retention time	s
Δt_n	time interval for formation of stable particles	s
T	absolute temperature	K
T_1	nuclear spin-lattice relaxation time	s
$\bar{\tau}$	average lifetime of a free radical in a particle	s
τ	half life for coagulation	s
τ	turbidity	m^{-1}
τ	time delay in autocorrelation	s
τ	yield stress	$kg/m\ s^2$
τ_r	nmr rotational correlation time	s
u	fluid velocity	m/s
u	electrophoretic mobility	$\mu m\ s^{-1}/V\ cm^{-1}$
U	adiabatic pair potential	J
$\bar{v}$	specific volume	dm^3/g
v	velocity	m/s
v	valence (charge) of an ion	–

LIST OF SYMBOLS

Symbol	Description	SI Units
v	velocity	m/s
v, v_p	volume of a particle	m^3
v_t	terminal velocity	m/s
V	molecular, or chain segment, volume	m^3
V	electric voltage or potential difference	V or $kg\ m^2/s^3\ A$
V_1	molar volume of solvent	dm^3/mole
V_m	molar volume	dm^3/mole
V_p	volume fraction of monomer-swollen particle	–
V_T	total interaction pair potential	J
w	channel thickness	m
w_0	reduction in particle radius due to coalescence	m
W	weight	kg
W	Fuchs stability ratio	–
$\langle W \rangle$	average Brownian coefficient	–
$[x]$	concentration of substance x	mol/dm^3
χ	Flory-Huggins interaction parameter	–
Y	Hansen-Ugelstad 'aqueous termination' rate	–
Y	electrical admittance	mho or $s^3\ A^2/km\ m^2$
$Y^*_{lm(\theta,\phi)}$	spherical harmonics	–
ψ	electrical potential	V or J/A s
ω	frequency, or frequency shift	Hz or s^{-1}
ω	rotational velocity	(radians) s^{-1}
ω^*	characteristic relaxation frequency	Hz or s^{-1}
z	valence of ions	–
Z	electrical impedance	ohm or $kg\ m^2/s^3\ A^2$
Z	number of surface charges	–
$\tilde{\zeta}$	dimensionless zeta potential	–
ζ	zeta potential	V or J/A s

Chapter 1

Introduction

1.1 Nomenclature

1.1.1 Polymer colloids

The field of polymer colloids involves the intersection of the science of polymers and that of colloids, 'polymer ∩ colloids' in Boolean terms. We may think of a Venn diagram of two circles, one labeled 'polymers' and the other 'colloids'. The two circles overlap, and the region of overlap is the subject of this book. A colloid – strictly speaking a fluid, lyophobic colloid – is a dispersion of fine particles in a fluid medium; and in our case the particles are composed of polymers, nearly always synthetic, and nearly always formed by free radical polymerization.

1.1.2 Latex

'Latex' refers to the milky appearing sap of certain plants such as milkweed, lettuce and rubber trees (*Hevea braziliensis*). These turn out to be colloids of natural rubbers suspended in an aqueous medium containing proteins and other substances which act as stabilizers. Polymer colloids came to be known as 'synthetic latexes' or simply 'latexes'. The terms often can be employed interchangeably.

1.1.3 Emulsion polymerization

This is a somewhat confusing term. It based upon the historical fact that when water-insoluble monomers were emulsified in water with a suitable surfactant and a water-soluble, free radical initiator was added, polymerization could ensue to produce a polymer colloid. As will be seen, however, the monomer need not be water-insoluble, and an emulsion need not be formed in order to produce by free radical polymerization a polymer colloid.

1.2 General descriptions

Polymer colloids generally have a milky appearance, but can have a bluish, translucent appearance when the particles are small enough. When the particles are very uniform in size, the latexes can exhibit a beautiful opalescence, as shown in Plate 2 [1].

Latexes usually have a low viscosity – almost like that of water, although they may be very viscous under some conditions. The colloidal particles range in size from about 10 nm to 1000 nm in diameter, and are generally spherical, although almost every conceivable shape has been observed. Examples of some particle morphologies are shown in Figs 1a, b, c & d. The surface tension of these systems can range from about $20\,\mathrm{mN\,m^{-1}}$ to that of pure water, $73\,\mathrm{mN\,m^{-1}}$.

Typical of any colloid, the internal interfacial area is large, with a concomitant free energy. For example, a liter of latex containing 10 nm spherical particles will have a particle/liquid interfacial area of about $3 \times 10^9\,\mathrm{cm}^2$, and a corresponding internal free energy of approximately 3×10^{10} ergs (3×10^3 J).

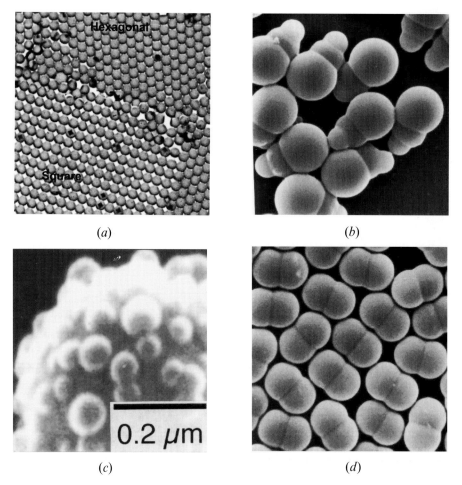

Fig. 1. Particle morphologies: (a) uniform particles; (b) 'ice cream cone' particles; (c) 'framboidal' particles; (d) dumb-bell particles. Micrographs (b) and (d) courtesy of Emulsion Polymers Institute, Lehigh University, Bethlehem, PA; (c) courtesy of M. Okubo, Faculty of Engineering, Kobe University, Kobe, Japan.

1.3 Chemistry

The chemistry of polymer colloids involves both their formation by emulsion polymerization and reactions at the particle–fluid interface in the final latex. The polymerization is almost always by a chain-growth mechanism involving free radical propagation. However, ionic mechanisms, both in aqueous and nonaqueous media, have been shown to occur under appropriate conditions. There is a class of polymer colloids comprised of relatively low molecular weight polymers which are emulsified into water *after* polymerization, e.g. alkyd resins, polyurethanes and silicones. Some post-application curing reaction to crosslink the derived film is required to achieve appropriate mechanical properties. An example is an alkyd resin containing copolymerized acid-containing monomer which self-emulsifies upon adding base, such as an amine. These emulsions are used in electrocoating metal for corrosion protection of automobile bodies.

Chemistry at the interface of polymer colloids includes surface reactions involving catalytic functional groups placed at the interface either during emulsion polymerization or afterwards by some post-polymerization reactions. The surface concentrations of such groups can be very high, e.g. 10 M, and this, combined with a hydrophobic attraction of a substrate molecule, can lead to enhanced rates of reaction. Antibody molecules attached to the surface of latex particles can be used to attach the particles to cell surfaces, thereby allowing the particles to serve as highly immunospecific markers for the corresponding antigen. If the particles are fluorescently labeled, they become extremely sensitive detectors for proteins, and have found uses in medical diagnostics, cell separation, cancer chemotherapy and targeted drug delivery systems.

1.4 Physics

It turns out that it is relatively easy to make polymer colloids with extremely narrow particle size distributions, so-called 'monodisperse' systems (Fig. 1a). The particles usually carry an electrostatic charge, and therefore repel each other. Otherwise, because of their van der Waals attractions, the particles would tend to aggregate. When they are all the same size and carry the same amount of charge, all particles position themselves away from each other to the same average distance, thereby forming regular three-dimensional arrays. As a result, these systems are ideal for studying order–disorder phenomena. For example, they can form macroscopic analogs of crystals with face-centered cubic (fcc) or body-centered cubic (bcc) symmetry, and can exhibit Bragg diffraction (which is responsible for their opalescence), and reversible melting upon heating.

Because it is possible to vary the particle size and surface charge, and because they can be transferred from one fluid medium to another, latexes can serve as model colloids for rheological and dielectric studies, as well as neutron- and light-scattering investigations.

1.5 Applications

1.5.1 Industrial production

Billions of pounds of synthetic latexes are produced worldwide every year for an enormous variety of applications. Most production is of aqueous latexes conducted at atmospheric pressure and temperatures from around 40 to 70° C, so that it is a relatively safe process and easy to conduct in standard stirred tank reactors. Simple batch-mode polymerization, in which all ingredients are mixed together at the beginning, is often the method of choice, but more sophisticated modes are also routinely employed today, involving semi-continuous monomer feed into a batch reactor, or a fully continuous process in which all ingredients are fed over long times either in a series of continuous stirred tank reactors (CSTRs) or, more rarely, in tubular reactors. With the same 'recipe' very different results, in terms of particle size, molecular weight, particle morphology and latex rheology are achieved, depending upon which mode of operation is employed.

1.5.2 Film formers

Latex paints were considered almost magical when they first appeared on the market in the 1950s because, although they were water-borne, the derived paint films were water-proof. The secret was in the fact that the polymer particles had a sufficiently low glass transition temperature, T_g, that when the water evaporated from the applied paint, the particles coalesced to form a continuous film. The paint was easily formulated from an aqueous latex and a dispersion of pigment in water.

Film-forming latexes are now employed not only in paints, but also, for example, in floor coatings ('polishes'), printing inks, adhesives, paper overprint varnishes, carpet backing and paper making.

1.5.3 Matrix materials

In some applications a polymer colloid is used in admixture with another material to enhance its properties. For example, rubberized concrete is formed by mixing cement and sand with a rubbery latex (such as polystyrene/butadiene, which of course contains water) instead of with water to form a material in which the cement particles are held together in part by the synthetic rubber. As a result the concrete has improved flexibility and thermal expansion properties, such that roads or airport runways made from it do not require the usual asphalt expansion strips (which cause the thumping of tires on highways).

The process of emulsion polymerization is often preferred over solution methods because of its simplicity and the fact that high molecular weight polymer can be produced at high reaction rates and low viscosities. The polymer is then isolated by coagulating the latex and filtering off the aqueous medium and washing the derived 'crumb.' Certain kinds of polymers can only be made by emulsion polymerization, e.g. acrylonitrile-butadiene-styrene (ABS) and impact polystyrene. The dried crumb may be used as a molding resin, or in some cases it is dissolved into a good solvent and spun into fibers (e.g. Orlon® polyacrylonitrile).

Reference

1. Bagchi, P., Birnbaum, S.M. and Gray, B.V. (1980). In *Polymer Colloids* (R.M. Fitch ed), Plenum Press, New York. Frontispiece (courtesy of Eastman Kodak Co.).

Chapter 2

Emulsion Polymerization

2.1 Particle formation

Colloids generally, and polymer colloids in particular, are formed in either of two ways: (a) by disintegration of larger particles or (b) by condensation of smaller particles, usually molecules. In the first case one has to emulsify a polymer into the diluent. This requires large amounts of emulsifying agent and mechanical energy. Polymers above a certain molecular weight are so viscous that it is essentially impossible to shear bulk material, or even a solution of the polymer, into small particles. Low molecular weight polymers, especially those containing ionic groups along the chain, can be quite readily emulsified to form stable dispersions. For example, alkyd resins containing some copolymerized maleic acid groups, when neutralized with an amine such as morpholine, will produce charge-stabilized latexes. The particle size of such dispersions tends to be large and the distribution of sizes quite broad. These systems have many industrial applications, one of the most important being automotive primer paints. In this case, the dispersion is applied to the metal by electrodeposition, and is subsequently baked in an oven to remove water and to crosslink the polymer in order to develop strong film properties.

This text will deal primarily with water-borne polymer colloids formed by condensation of small molecules, i.e. by emulsion polymerization, because of the enormous variety of industrial and academic applications of the process and its versatility. Particle formation in this case occurs by a process of nucleation, either homogeneous or heterogeneous. We shall also confine the discussion primarily to free radical polymerization. Such reactions are comprised of four fundamental processes, namely initiation, propagation, termination and chain transfer. Propagation is the only one which produces the polymer. Over the major part of the process, steady state conditions must obtain, i.e. where the rate of generation of free radicals (initiation) is equal to the rate at which they disappear (termination). This implies a constant overall concentration of propagating free radicals, [M·]. Otherwise, if [M·] decreases, the reaction ultimately dies, and if it increases, the reaction heads towards an explosion!

In an ordinary batch-mode emulsion polymerization all of the ingredients are mixed together and heated to reaction temperature. The reaction mixture contains one or more monomers containing double bonds capable of free radical polymerization, water, emulsifier (surfactant) and initiator. At reaction temperature, the initiator forms free radicals which attack the monomer molecules and form new radicals with added monomer units.

2.1.1 Homogeneous nucleation [2–7]

In emulsion polymerization if one of the monomers has some finite solubility in the diluent, and if the free radical initiator, I, is dissolved in the diluent (as it must be in order to avoid polymerization in the monomer phase), then initiation will most likely take place in the diluent phase. The reaction can be represented as:

$$I \xrightarrow{k_d} 2R \cdot \qquad (1)$$

The kinetics of this reaction, in terms of the rate of appearance of free radicals would then be:

$$R_i = \frac{d[R\cdot]}{dt} = 2k_d[I] \qquad (2)$$

Whenever these events are taking place in the water phase, the subscript w will be used. For example:

$$R_{iw} = \frac{d[R\cdot]_w}{dt} = 2k_{dw}[I]_w \qquad (3)$$

The initiator radicals attack monomer molecules to initiate chain growth:

$$R_w\cdot + M_w \xrightarrow{k_{iw}} M_w\cdot \qquad (4)$$

The corresponding rate equation is:

$$R_{iw} = \frac{d[M\cdot]_w}{dt} = k_{iw}[R\cdot]_w[M]_w \qquad (5)$$

Because the rate of initiator decomposition is much slower than that of Reaction (4), it is the rate-determining step. Once a chain is started it propagates at a rate determined by the rate constant k_{pw} and the concentrations of the two reacting species:

$$M_{jw}\cdot + M_w \xrightarrow{k_{pw}} M_{(j+1)w}\cdot \qquad (6)$$

The subscript j indicates the degree of polymerization (number of monomer units) in the growing chain. Typically the initiator free radical, R·, is highly water-soluble and ionic, such as ·SO_3^- or ·OSO_3^-, which introduce sulfonate or

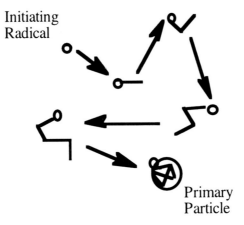

Fig. 2. Homogeneous nucleation. o depicts initiator radical, — depicts a monomeric repeat unit.

sulfate end groups, respectively. As monomer units are added, ordinarily the growing chain molecule becomes increasingly less water-soluble until it reaches a critical chain length, j_{cr}, at which phase separation occurs, i.e. the chain comes out of solution and forms a primary particle, as shown in Fig. 2.

The well-known theory of homogeneous nucleation can be applied to such a process. When a new interface is created, by the formation of a polymer particle in this case, there is a corresponding interfacial free energy required:

$$\Delta G_s = 4\pi r^2 \gamma_{pw} \tag{7}$$

where γ_{pw} is the interfacial tension between polymer and water and r is the radius of the incipient particle. For the nucleation process to be spontaneous this energy must be supplied from some internal source, which in this case derives from the aggregation of the hydrophobic parts of the growing chain:

$$\Delta G_c = -\tfrac{4}{3}\pi r^3 g_v \tag{8}$$

where g_v is the free energy of condensation per unit volume of the primary particle or nucleus. The overall free energy of a growing chain is then simply the sum of Eqs 7 and 8:

$$\Delta G = \Delta G_s + \Delta G_c \tag{9}$$

At very small sizes, r, the first term in Eq. 9 dominates, preventing formation of a particle, but as the 'embryo' grows there becomes enough energy of condensation to overcome the surface energy requirement, and a stable 'nucleus,' or primary particle, can form. This is shown schematically in Fig. 3.

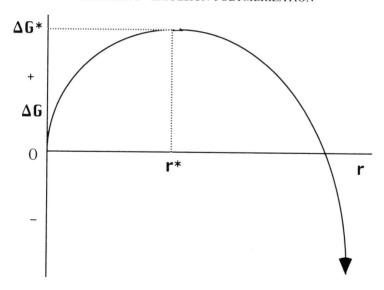

Fig. 3. Energetics of homogeneous nucleation.

For many colloids the interfacial tension (free energy), γ, is quite high and they require high degrees of supersaturation before nucleation can take place. In polymer colloids, however, the polymer/water interface usually has a low interfacial tension, especially in the presence of surfactants, and thus homogeneous nucleation occurs quite readily.

2.1.1.1 Fitch–Tsai theory

In practical terms this means that when the propagating chain in solution reaches the critical chain length, j_{cr}, it forms a particle by collapse of the repeat units upon themselves, and a new organic phase is introduced into the system. Fitch and Tsai proposed that the rate of appearance of primary particles would thus initially be equal to the rate of generation of free radicals, R_{iw}, since according to Fig. 2, every effective radical generated leads to the formation of a new particle. Some radicals will terminate in the aqueous phase prior to nucleation, and others may be annihilated by wall reactions, etc. and still others may aggregate, so that a factor, b, is introduced to account for the reduction in rate due to these side reactions:

$$\left(\frac{dN}{dt}\right)_0 = bR_{iw} \qquad (10)$$

where N is the number of particles per liter (dm^{-3}) and the subscript 0 indicates an initial rate. The rate of generation of free radicals is on the order of 10^{15} dm^{-3} s^{-1}, and therefore new particles appear at somewhat the same rate, so that within a few seconds time, as j approaches j_{cr} for each radical, the number concentration of particles becomes very large, on the order of 10^{15} to 10^{17} dm^{-3}. At this point another process becomes highly probable: the growing chains in solution are surface-active by virtue of the fact that they possess an ionic, polar head group and a hydrophobic tail. These oligomeric radicals, now presented with a huge internal surface area due to the new particles, may adsorb, reversibly, on the surface of primary particles before they reach the critical chain length, j_{cr}. For each of these events, which is irreversible (more probable at higher j), the rate of generation of new particles is reduced by one:

$$\left(\frac{dN}{dt}\right) = bR_{iw} - R_c \qquad (11)$$

where R_c is the overall average rate of oligoradical capture by particles. Usually the changeover of the rate from that in Eq. 10 to that governed by Eq. 11 occurs over a short period of time, an example of which is shown in Fig. 4 [8].

The steep initial slope shows the rapid proliferation of primary particles, and the rapid transition to a plateau indicates how effectively capture takes over. Note that in the plateau region $dN/dt = 0$ because $bR_{iw} = R_c$, i.e. a steady state has been reached in which essentially all of the radicals formed are captured by particles before they can nucleate. The experimental conditions employed by Krieger and Juang [8] in this reaction involved a large amount of surface-active agent.

Fitch and Tsai [9] had earlier performed similar experiments with methyl methacrylate (MMA) emulsion polymerization, both in the presence and total absence of surfactant, with very different results, as shown in Fig. 5. The initial, steep, positive slope is expected, but then there is a sharp drop in N, indicating the disappearance of particles! This can only happen by the aggregation of particles by coagulation, the rate of which is given the symbol

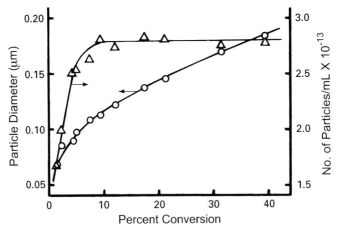

Fig. 4. Particle formation in styrene emulsion polymerization. High surfactant concentration [8].

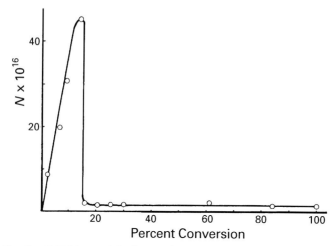

Fig. 5. MMA particle formation. Low surfactant concentration.

R_f. The process is represented in Fig. 6 in which two charged particles undergo a Brownian collision, stick together and fuse. They stick because their van der Waals force of attraction is greater than the electrical force of repulsion.

The particles fuse because they are softened by the presence of monomer which migrates to this new organic phase as soon as it is formed (but only if the monomer is soluble in the polymer). The new, larger particle is spherical because the interfacial tension tends to minimize the interfacial area. The surface ionic groups arise from adsorbed emulsifier and initiator-derived polymer end groups. As coagulation progresses and as more oligo-radicals are adsorbed, the surface charge density of the resulting particles increases, which in turn increases their surface electrical potential, leading to a greater mutual repulsion and reduction in the overall average rate of coagulation, R_f.

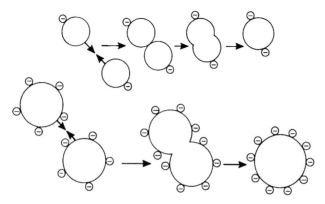

Fig. 6. Brownian collision of charged particles leads to coagulation, fusion and increased surface charge density.

All of this leads to a third term in the equation for the rate of formation of particles:

$$\left(\frac{dN}{dt}\right) = bR_{iw} - R_c - R_f \tag{12}$$

This is the Fitch–Tsai Equation, which can be a useful tool in developing a qualitative understanding of the processes which govern the formation of particles in emulsion polymerization, and for building strategies for regulating particle size and its distribution in practical applications. Some examples (below) will help to illustrate, and then subsequent discussion will elaborate upon the absolute values of the rates of initiation, R_i; capture, R_c; and coagulation, R_f, in Eq. 12.

2.1.1.2 Control of particle size and its distribution

It is often very easy to obtain extremely narrow particle size distributions – so-called 'monodisperse systems' – in emulsion polymerization. La Mer and Dinegar in 1950 were the first to formulate the prerequisites for monodispersity as a result of their work with sulfur sols: all of the particles must be nucleated in a short period of time, and all must grow at the same rate. They will then end up with the same size. We shall define the initial period, Δt_n, as the time for stable particle formation, which for our purposes may include a stage during which N is changing rapidly, either up or down. This is shown schematically in Fig. 7 where the two curves from Figs 4 and 5 are superimposed.

So even though a great deal of coagulation occurs, as long as it stops early enough in the reaction, i.e. Δt_n is sufficiently small, the latex can be monodisperse. In the Krieger–Juang case the primary particles were highly stabilized against coagulation with large amounts of surfactant, whereas in the

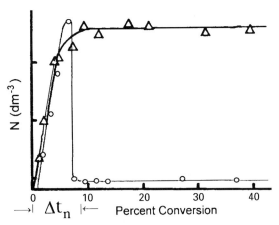

Fig. 7. Extreme cases in emulsion polymerization in which Δt_n is small.

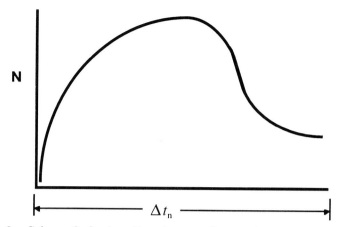

Fig. 8. Schematic for low R_i or intermediate surfactant concentration.

Fitch–Tsai case, low amounts (well below the critical micelle concentration) were employed. In the former case the final number of particles was large, so that their size was small; in the latter case the opposite was true: there were relatively few and therefore large particles. These are extreme cases, but are well documented experimentally.

Intermediate situations can also be envisaged in which the initial slope of $(dN/dt)_0$ is lower, either because R_i is less or R_f is greater, as shown in Fig. 8. In these cases it can be seen that Δt_n is much longer than that in Fig. 7, so that some particles are just formed at later times when others have been growing for longer times, with the result that the particle size distribution will be broad, and the average size will be intermediate between the two described above. Again, many experimental examples of this behavior are described in the literature.

Bimodal size distributions are also well known. For example, if surfactant is injected into the reaction mixture after some time, a second crop of particles may be formed, as shown in Fig. 9. The fact that new, stable particles appear when more surfactant is introduced suggests that this phenomenon is related to the stability of particles already formed.

This means that R_f must be positive throughout the reaction (certainly where N is decreasing), and that the added surfactant simply stabilizes particles that are constantly being nucleated, i.e. R_f is reduced. This results in an increase in dN/dt as seen in Fig. 9.

2.1.1.3 Hansen–Ugelstad theory

Ugelstad and Hansen made significant improvements on the Fitch–Tsai theory, both theoretically and experimentally [3–7]. Their first contribution was to show that particle nucleation could be viewed in terms of the polymerization process, i.e. that the rate of appearance of primary particles is equal to the rate at which oligomers of the critical chain length are formed by

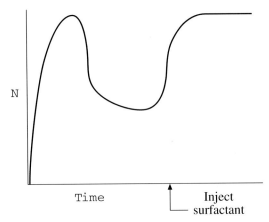

Fig. 9. Schematic for injection of surfactant during emulsion polymerization.

propagation from chains with one monomer unit less in length:

$$\frac{dN_1}{dt} = k_{pw}[M]_w[M\cdot]_{(j_{cr}-1)w} \qquad (13)$$

The subscript 1 in N_1 indicates that these are primary particles. There will be a series of steps forming oligoradicals of increasing chain length starting with initiator radicals. The intermediate radicals can undergo other reactions, e.g. propagation, termination and capture by particles along the way. The first of these steps involves initiator radicals, $R\cdot$, which are formed by initiator decomposition and disappear by propagation (to larger chain radicals), termination and capture:

$$\frac{d[R\cdot]}{dt} = bR_i - k_{iw}[R\cdot]_w[M]_w - k_{tw}[R\cdot]_w[M_j\cdot]_w - k_c[R\cdot]_w N \qquad (14)$$

Intermediate oligomeric free radicals of any chain length j, dissolved in the aqueous phase, undergo the same kinds of reactions:

$$\frac{d[M_j\cdot]}{dt} = k_{pw}[M_{j-1}\cdot][M]_w - k_p[M\cdot][M]_w - k_{tw}[M_j\cdot][M\cdot]_w - k_{cj}[M_j\cdot]N \qquad (15)$$

For styrene, experimental evidence [3–7, 10] indicates that j_{cr} is 5, whereas for MMA it is around 10 according to Gilbert [11], although earlier experimental molecular weight determinations [9] indicated it to be about 65. These equations can be combined, along with a few simplifying assumptions (k_c is averaged over all values of j to give $\bar{k}_c$, steady state in $[M_j\cdot]$ and termination of primary radicals, $R\cdot$, is negligible), to give the following simple equation for the rate of formation of primary particles:

$$\frac{dN_1}{dt} = bR_i\left(1 + \frac{k_{tw}[M\cdot]_w}{k_p[M]_w} + \frac{k_c N}{k_p[M]_w}\right)^{1-j_{cr}} \qquad (16)$$

If this is integrated over the time period of interest, one obtains the total number of particles:

$$N_1(t) = \left(\frac{k_{pw}[M]_w}{\bar{k}_c}\right)\left\{\left(\frac{\bar{k}_c}{k_{pw}[M]_w}j_{cr}bR_i t + 1\right)^{1/j_{cr}} - 1\right\} \quad (17)$$

This is an excellent equation to do 'what if' calculations on the effects of various experimental parameters on the final number of particles, noting importantly that the rate of coagulation, R_f, is assumed to be zero. The overall rate of capture, R_c, is given by $k_c N[M·]_w$; hence the term $k_c N$ in Eq. 16. More will be said about capture later (Eqs 24 and following). Some examples are given below in Fig. 10 for MMA using $j_{cr} = 60$, and other imaginary alkyl acrylates with values of j_{cr} as shown.

Figure 10, which is plotted using Eq. 17, shows the importance of the critical chain length, j_{cr}, in determining the final number concentration of particles (and therefore of particle size). Monomer concentration in the aqueous phase, $[M]_w$, is also important [12]. The size of the particles is also

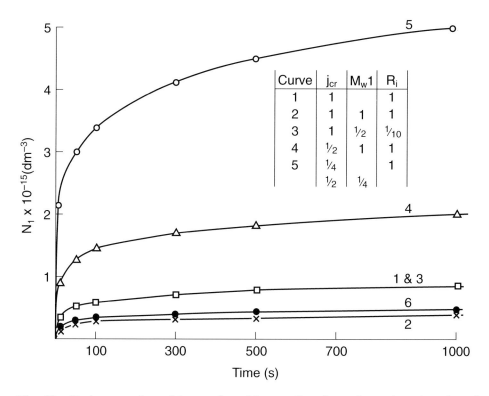

Fig. 10. Estimates of particle number, N, as a function of reaction time for alkyl methacrylates in emulsion polymerization: homogeneous nucleation and no coagulation, using Eq. 17. Curve 1 is for MMA using literature values of parameters. Parameters for other curves are relative to those for MMA.

determined by the rate at which they grow and by the overall time of growth by polymerization. This will be dealt with in a later section.

Gilbert has pointed out that the capture process may not occur for oligo-radicals below a critical size, j_z, which is smaller than the critical size for self-nucleation, j_{cr}. This would tend to increase somewhat the concentration of radicals in solution, leading ultimately to slightly larger calculated values for the number of particles at any given time. More importantly, Gilbert has pointed out that the rate constants for small species may be larger than those for larger homologs, so that radicals in solution reach j_{cr} more rapidly than calculated in Fig. 10. This will increase N_1 considerably [11].

The rate of effective radical generation, bR_i, can be independently controlled by choice of initiator, its concentration and the reaction temperature. The radical effectivity factor, b, can be estimated as follows:

The rate of termination of free radicals in the aqueous phase is

$$R_{tw} = 2k_{tw}[M\cdot]_w^2 \tag{18}$$

so that the total number of radicals terminated, P, during the time interval Δt is

$$P = \int_t^{t+\Delta t} R_{tw} \, dt \tag{19}$$

If the total number of radicals generated in Δt is $[M\cdot]_w$, which is

$$[M\cdot]_w = \int_t^{t+\Delta t} R_i \, dt \tag{20}$$

then

$$b \equiv 1 - \frac{P}{[M\cdot]_w} \tag{21}$$

When b represents the aggregation number of a 'micelle' formed by oligoradicals, it is more difficult to calculate, and probably will have to be determined experimentally or estimated.

2.1.1.4 Coagulation rates

The question now is: how important is coagulation in determining the number of particles? Fitch, Palmgren, Aoyagi and Zuikov looked at the emulsion polymerization of various acrylate monomers and of methyl methacrylate during the first few tens of seconds of the reaction [13]. Typical results are shown in Fig. 11, in which the intensity of light scattered at 90° is plotted as a function of time. Because the particles first formed are so small, the Rayleigh theory of light scattering can be applied, and it tells us that the smaller the

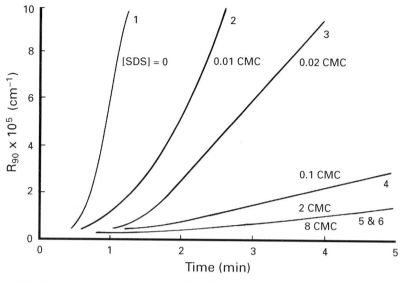

Fig. 11. Nucleation and growth rates for MMA emulsion polymerization at various SDS concentrations. Rayleigh ratio, R_{90}, in units of cm^{-1}.

particles, the lower is the scattered intensity: $R_{90} \propto r^6$ for a fixed number of particles, where r is the particle radius.

It can be seen from Fig. 11 that the higher the concentration of sodium dodecyl sulfate (SDS), but still well below the critical micelle concentration (CMC),[*] the lower is R_{90} and therefore the smaller are the particles at a given time. This was interpreted as strong evidence that the SDS surfactant was stabilizing primary particles, slowing or inhibiting their coagulation, and that coagulation was important from the outset of the reaction even when N was increasing rapidly. Thus, R_f in Eq. 12 is positive from the outset, except when the SDS – or presumably other surfactant – concentration is at or above its CMC.

The value of R_f is calculated from the so-called DLVO theory [14], which in turn is based upon the theories of Smoluchowski and Fuchs:

$$R_f = \left(\frac{4\pi D_{pq} r_{pq}}{W_{pq}}\right) N^2 \qquad (22)$$

Here D represents the mutual diffusion coefficient for particles of size p and q, r is the collision radius, i.e. the sum of the radii of p and q particles, and W is the so-called Fuchs stability factor. W is the ratio of the collision rate based on purely Brownian motion, $R_{f(fast)}$, to the collision rate when there is a repulsive potential due to charges (or, in the case of steric stabilization, due to adsorbed polymer) on the particles, $R_{f(slow)}$:

$$W \equiv \frac{R_{f(fast)}}{R_{f(slow)}}$$

[*] Surfactant molecules in solution will aggregate to form 'micelles' above a critical concentration.

The Brownian collision rate, often called the 'Smoluchowski fast coagulation rate' after the man who developed the theory, may be characterized by a half-life which turns out to be independent of particle size:

$$\tau = \frac{3\eta}{4Nk_\mathrm{B}T} \tag{23}$$

where η is the viscosity, k_B is the Boltzmann constant, and T is the temperature. If the particle concentration, N, is $10^{16}\,\mathrm{dm}^{-3}$, the half-life will be 12 ms, which is to say that in the absence of stabilizer coagulation proceeds very rapidly even in fairly dilute dispersions. Thus it is likely that in many – if not most – emulsion polymerizations coagulation is the principal factor determining particle size and its distribution, rather than the rate of nucleation.

> In many – if not most – emulsion polymerizations, coagulation is the principal factor determining particle size and its distribution.

It is conceivable also that nucleation occurs continuously and that R_f is positive throughout the entire reaction, even when the total number of particles is unchanging, such that a steady state exists in which $bR_i = R_c + R_f$. This condition is known as 'heterocoagulation', in which small particles preferentially coagulate with large ones. Thus the principal mechanism for particle growth can be by this means rather than by swelling of the original particles by monomer and polymerization within them. A more detailed discussion of latex stability and the DLVO and other theories will be dealt with in Chapter 7. It suffices to point out here that not only is the amount and nature of the surfactant important in determining particle size, but also the nature of the particle (swollen by monomer) surface. For example, experimental evidence exists that SDS is less effective as a stabilizer for polyvinyl acetate (PVAc) than it is for polymethyl methacrylate (PMMA). This may be because of the difference in adsorption isotherms of SDS on the two polymers: the area per adsorbed molecule is 1.1 nm² on PVAc and 0.79 nm² on PMMA [15]. This in turn would mean a higher surface charge density on the PMMA which should lead to a larger Fuchs stability ratio, W_{pq}, a lower R_f as given in Eq. 22 and as suggested in Fig. 6.

2.1.1.5 Absolute rates of radical capture

The process of capture of radicals by particles may be represented by:

$$P + M_{jw}\cdot \xrightarrow{k_{cj}} (PM_{jp}\cdot) \tag{24}$$

in which P is a particle, the total number of which will be N per liter. This means that the overall rate of capture will be given by:

$$R_c = \bar{k}_c[\text{M·}]_w N \tag{25}$$

The rate constant for the capture of oligoradicals by particles, as mentioned earlier, is an average over all values of j:

$$\bar{k}_c = \sum_{j=1}^{j_{(cr-1)}} \frac{k_{cj}[\text{M}_j\text{·}]_w}{[\text{M·}]_w} \tag{26}$$

where $[\text{M·}]_w$ represents the total concentration of all radicals in the aqueous phase. The electrostatic and diffusional factors which affect $\bar{k}_c$ are similar to those for coagulation, but in addition there are terms for diffusion into the particle (and back out again) along with possible reaction (by propagation, termination or chain transfer) and partition between the two phases. These were taken into account by Hansen and Ugelstad in the following expression:

$$k_{cj} = 4\pi D_{wj} r F_j \tag{27}$$

where F is the rate-reduction factor. These are expressed mathematically as follows:

$$\frac{1}{F} = \left(\frac{D_w}{aD_p}\right)(X \coth X - 1)^{-1} + W' \tag{28}$$

where $X \equiv r_p(k/D_p)^{1/2}$ and where $k \equiv k_p[\text{M}]_p + (nk_{tp}/v_p)$.

This k represents 'diffusion with reaction' in which the radical progresses in space (within the monomer-swollen polymer particle) by propagation or else is terminated. The Ds represent physical diffusion coefficients of the radicals, with subscripts w or p to denote either water or polymer phases. The a in Eq. 28 is the equilibrium partition coefficient of a radical between the organic and water phases, which in turn will be governed by the value of j of any oligoradical ($a_j \equiv [\text{M·}]_{jp}/[\text{M·}]_{jw}$); as j increases, the value of a increases exponentially. W' is the Fuchs stability factor for a singly charged radical colliding with a particle or a micelle, and r_p and v_p are the particle radius and its volume, respectively. When the particles are very small, they can only accommodate one radical at a time, since upon entry of a second radical, termination occurs instantaneously. Under these conditions n, the number of radicals in a particle, is either 1 or 0.

Hansen and Ugelstad showed that there are three limiting cases under which simple solutions for Eq. 25 can be obtained:

(1) $F = 1/W'$ in which capture is governed solely by electrostatic interactions between particle and oligoradical, and is therefore irreversible. In this case

$$k_c = \frac{4\pi D_w r_p}{W'} \quad \text{and} \quad R_c \propto r_p N \tag{29}$$

(2) When capture is reversible and depends on oligoradical solubility in the polymer phase, and diffusion into the particles with possible reaction therein, then $F = ar_p(kD_p)^{1/2}/D_w$ and the capture rate constant becomes:

$$k_c = 4\pi r_p^2 (kD_p)^{1/2} \quad \text{and} \quad R_c \propto r_p^2 N \qquad (30)$$

(3) When reversible capture is determining, $F = ar_p^2 k/3D_w$, and

$$k_c = av_p k \quad \text{and} \quad R_c \propto r_p^3 N \qquad (31)$$

When the particles and the radicals are both small, the efficiency of capture, F, is very low, on the order of 10^{-6} to 10^{-4} because the j-mers will be very water-soluble and because the particles will contain very little monomer (Kelvin effect, see Eq. 34), so that D_p will be low. If the particle contains a radical already, the capture of another from the water phase is much more probable. At the outset, all primary particles contain radicals, but this rapidly changes until in a few seconds or minutes the average number of radicals per particle, $\bar{n}$, has declined to 1/2 (the reasons for this will be discussed later in Section 2.2.6). At any instant there will be in the continuous medium a mixture of oligoradicals of all values of j. The largest of these, with $j \to j_{cr}$, will be captured irreversibly regardless of the nature of the particle. The strong dependence of F on particle radius, r, the diffusion coefficient in the polymer phase, D_p, and upon a is graphically shown in the theoretical curves [7] reproduced in Fig. 12.

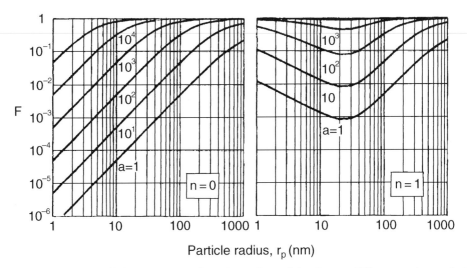

Fig. 12. Capture efficiency F as a function of particle size at different values of the radical partition coefficient, a, and the number of radicals in a polymer particle, $n = 0$ and 1. For styrene polymerization: $j_{cr} = 5$, $W' = 1$, and $k = 760 \text{ sec}^{-1}$, $D_w = 1 \times 10^{-7}$, $D_p = 5 \times 10^{-8} \text{ dm}^2 \text{ s}^{-1}$.

As we shall see later in this chapter, the rate of capture has important implications for the rate of emulsion polymerization as well as for particle formation.

2.1.2 Heterogeneous nucleation

Nucleation theory tells us that heterogeneous nucleation is energetically favored over homogeneous nucleation simply because some kinds of particles (not necessarily polymer) are already present, so that the surface tension barrier, $\Delta G_s = 4\pi r^2 \gamma_{pw}$, does not exist.

2.1.2.1 Micellar and microemulsion nucleation

Surfactant micelles can play the role of heterogeneous nuclei. Sütterlin and coworkers have studied the influence of monomer solubility in a homologous series of acrylate and methacrylate monomers [16]. For example, in the acrylate series the value of the ratio of the rate constants for propagation and termination, k_p/k_t, is relatively independent of the number of carbon atoms in the alkyl group, whereas monomer (and thus oligomer) solubility vary greatly. Sütterlin's group studied the effect of SDS concentration on N at constant R_i. These were all batch polymerizations in which all ingredients were added at the beginning of the reaction. Above the CMC micelles are present and they are swollen with monomer. Typical results are shown in Fig. 13, in which the dependence of particle number on SDS concentration over four orders of magnitude is given for a series of acrylate monomers, ethyl (EA), propyl (PA), butyl (BA) and 2-ethylhexyl (2-EHA) acrylate.

The curves are characteristic for monomers which differ in water-solubility. In contrast to the soluble EA, for example, which readily forms many polymer particles at submicellar surfactant concentrations, the poorly soluble monomer 2-EHA exhibits relatively low N values until the CMC is passed, after which it has more particles than the EA system. Interestingly, N increases more rapidly with SDS concentration above the CMC in the case of EHA. The more hydrophobic monomer will be solublilized in surfactant micelles above the CMC to a greater extent because of more favorable interactions with the hydrocarbon core of the micelles.

On the other hand, the solubility of EA is relatively independent of surfactant concentration. For this monomer N increases with SDS concentration primarily because of the progressive reduction in R_f, as observed for MMA in Fig. 11. Just below the CMC the more water-soluble monomer produces almost 1000 times as many particles (approx. 10^{18} vs 10^{15} L^{-1}) because of its greater propagation rate in the aqueous phase ($k_p[M]_w$), as predicted in Curves 1 and 6 in Fig. 10. Above the CMC there is apparently more of the hydrophobic monomer available for nucleation in the micelles because of its greater degree of solubilization. Thus EHA produces more particles above the CMC than does EA. This effect may be enhanced by the fact that SDS is adsorbed more strongly on the more hydrophobic polymer, giving it a higher

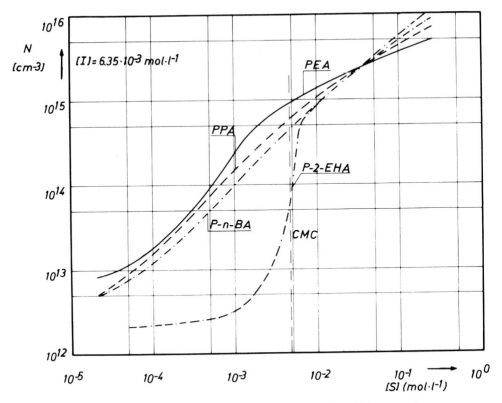

Fig. 13. Dependence of log N on surfactant concentration [S] for various acrylates [16].

surface electrical potential, Ψ_0, above the CMC and consequently a lower R_f. Apparently below the CMC this effect is insufficient to overcome the vast differences in $[M]_w$ between the two monomers (EA is 340 times more soluble than 2-EHA).

> Just below the CMC the more water-soluble monomer produces almost 1000 times as many particles.

The steep rise in N when micelles are present with a poorly soluble monomer like that shown in Fig. 13 for 2-EHA, led Harkins and subsequently Smith and Ewart [17] to develop their famous theory of particle formation based upon radical capture by surfactant micelles. Subsequently Roe [18] showed that the same equations could be derived without invoking micelles, allowing for the possibility that homogeneous nucleation occurs, at least until j becomes large enough so that oligoradicals are irreversibly absorbed by micelles. The basis of their theories is that particle formation stops when the interfacial area (polymer/water) stabilized by the surfactant equals the area which can be covered by the amount of surfactant added, $a_s S$, where a_s is the area stabilized per surfactant molecule and S is the total number of surfactant molecules in the system. This surface coverage may be distributed over a few large particles or many small particles, depending upon the competition between new particle formation, represented by the rate of free radical generation, bR_i, and the rate of particle volume growth, μ (assumed to be constant during this period of particle formation). The Smith–Ewart–Roe equation resulting from these considerations is:

$$N = K\left(\frac{bR_i}{\mu}\right)^{0.4} \times (a_s S)^{0.6} \qquad (32)$$

where K is a constant with the value of 0.53 for purely micellar capture, and 0.37 if capture by the new particles is also taken into account. The equation has been experimentally verified for a very limited number of systems in which relatively water-insoluble monomers, e.g. styrene, were used with surfactant at concentrations above the CMC [19]. Smith and Ewart did not consider coagulation, since they used high concentrations of surfactant. Figure 14 shows schematically what the curves for N as a function of time should look like.

Microemulsions represent a special case of micellar nucleation, in that they are concentrated systems of (monomer-swollen) micelles. Incidentally, there are both normal and 'inverse' microemulsions. The latter are composed of a continuous organic phase, e.g. toluene, and micelles which have their surfactant molecules with their polar heads on the inside, and hydrophobes on the outside. For inverse microemulsion polymerization the micelles contain an aqueous solution of a water-soluble monomer such as acrylic acid, acrylamide, or sodium styrene sulfonate.

Microemulsions are thermodynamically stable systems, so that the ingredients can be mixed in any order, unlike regular emulsions. Otherwise the principles for heterogeneous nucleation cited above apply. Because microemulsions ordinarily contain large amounts of surfactant plus a co-surfactant,

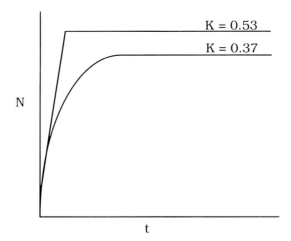

Fig. 14. Schematic representation of particle formation by heterogeneous nucleation in surfactant micelles, with no coagulation, according to Smith and Ewart [17].

e.g. an alcohol or the monomer itself, there usually are micelles present during the entire polymerization. Thus the probability of radical capture by micelles remains very high throughout and the rate of coagulation is negligible. The net result is that every oligoradical enters an 'unstung' micelle, initiating a new particle. Thus new particles are nucleated continuously throughout the polymerization, and the particles on the average contain only one polymer molecule! Reaction rates can be very fast and molecular weights, very high (acrylamide, for example, can go to completion in 10 min with average molecular weights of around 1×10^6–10×10^6 Da) [20].

2.1.2.2 Monomer droplets

In ordinary emulsion polymerizations the monomer is emulsified into small droplets in the reactor, but the free radicals, generated in the continuous phase, do not seem to cause polymerization in these droplets. If the initiator were oil-soluble, however, then free radicals would be generated in the monomer phase, and most of the polymerization could take place there. Indeed, this is well-known as 'suspension,' or 'bead' polymerization. In this case the particles are on the order of the size of the original emulsified monomer droplets, i.e. a few micrometers to millimeters in diameter. In true 'emulsion polymerization' oligoradicals of course can be captured by monomer droplets and start polymerization therein, but ordinarily this is negligible because of statistical considerations: there will be on the order of 10^{11} monomer droplets and 10^{17} polymer particles per liter, 1 000 000 times as many. Thus a free radical in the aqueous phase is essentially completely surrounded by polymer particles and can hardly 'see' any monomer drops. But there is clearly a finite probability that some polymerization will occur in the monomer phase, and this can cause problems in practical manufacturing situations. For example, as the monomer droplets are consumed by diffusion to polymer particles and polymerization therein, the polymer residue in the monomer droplets will remain as insoluble, amorphous material often considered erroneously as 'coagulum.'

> There is a finite probability that some polymerization will occur in the monomer phase, and this can cause problems in practical manufacturing.

As the size of the monomer globules is made smaller (e.g. by stronger agitation and/or more or better surfactant), their number increases to the point where they come to be the most probable sites for heterogeneous nucleation [21]. This has come to be known as 'miniemulsion polymerization.' Generally this results in particles which are larger than those obtained by homogeneous nucleation, as a result of which the kinetics of the reaction, and thus the molecular weights, can be quite different in the two cases. Smith–Ewart Case 3 kinetics usually are observed in miniemulsion polymerization, which ordinarily results in lower molecular weights than in the 'normal' situation.

2.1.2.3 Foreign particles

Other materials, such as clays, pigments, silica and even colloidal gold, can serve as heterogeneous nuclei for the formation of particles, although not much work in this area has been done, and even less has been successful. There have been various technical needs for polymer-coated particles, e.g. for paints, printing inks, etc. Most successful attempts have employed dispersions of inorganic particles which have been functionalized with polymerizable groups on their surface. For example, a titanium ester containing a methacrylate group chemically bonded onto the surface of TiO_2 pigment can in a sense copolymerize with methyl methacrylate under emulsion polymerization conditions to obtain PMMA-coated pigment particles [22]. The concentration of pigment in the initial dispersion must be high enough so that the oligoradicals are adsorbed onto the pigment before they can self-nucleate to form new particles [23]. When such polymer-coated pigment particles are used in latex paint films, there is not much evidence to date that the results are sufficiently superior to justify the added expense of this process.

Heterogeneous nucleation is most frequently practised by so-called 'seeded' emulsion polymerizations in which a pre-formed latex is introduced to the reaction mixture at the beginning. There must be enough seed particles to avoid new particles being subsequently nucleated. But how much is enough? This depends upon the capture efficiency discussed in Section 2.1.1.5 above. Hansen and Ugelstad applied the combined HUFT theory [Hansen–Ugelstad–Fitch–Tsai] to the surfactant-free, seeded emulsion polymerization of styrene. They varied the size of the seed latex particles, r_s, their number concentration, N_s, the surface charge on the seed particles, σ_0, and the ionic strength, μ, of the medium, and then they measured the total number of new particles, N, in the final reaction mixture. Some of their results are shown in Fig. 15.

In the absence of seed and surfactant ($N_s r_s = 0$), only homogeneous nucleation could occur, and a relatively small number of particles was formed, on the order of $10^{14}\,dm^{-3}$. No doubt considerable coagulation (large R_f) was involved.

As seed particles were added in separate experiments, N decreased because of two effects, very similar to each other: capture of oligoradicals by the seed particles (largely unaffected by the charge density on the seed, according to calculations) and coagulation of tiny primary particles onto seed particles, heterocoagulation strongly affected by σ_0. The theoretical curves could only be made to fit the experimental points when coagulation among primary particles also occurred and was calculated to be very fast ($W_{11} = 0.1$), radical adsorption was irreversible ($F = F_s = 1$), and coagulation of primary particles with seed particles was more or less rapid ($W_{1s} = 1$ or 10, depending upon the value of σ_0).

In the presence of surfactant, however, both radical capture and coagulation rates are reduced so that the tendency towards nucleation is greatly enhanced, i.e. both R_c and R_f are reduced in Eq. 12, leading to greater values of N.

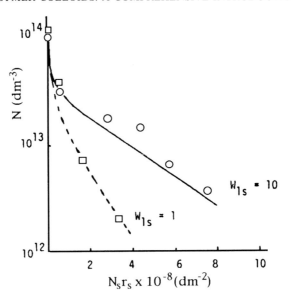

Fig. 15. Particle formation in presence of polystyrene seed latex. Total number of particles, N, as function of number, N_s, and radius, r_s, of seed. Points are experimental. ○: $r_s = 81$ nm, $\sigma_0 = 9.5\,\mu C\,cm^{-2}$; □: $r_s = 40$ nm, $\sigma_0 = 1\,\mu C\,cm^{-2}$. Curves are theoretical: $W_{1s} = 1$, $W_{1s} = 10$, and in both $F = Fs = 1$ and $W_{11} = 0.1$.

2.2 Particle growth

2.2.1 Monomer partition

2.2.1.1 Swelling kinetics and equilibrium

Once the particles are formed how do they grow? The appearance of the new polymer phase means that the monomer has a new place to go, so that it now partitions itself among the monomer droplets, the aqueous phase and the polymer particles, assuming of course that the monomer is a good solvent for its own polymer (acrylonitrile is not, and vinyl chloride is only a poor solvent, but most monomers are). The driving force for the swelling of a polymer by a solvent has been developed by Flory and Huggins. The free energy of mixing has both entropic and enthalpic components which have been published in many books on polymer chemistry and physics. The basic equation, from Gibbs and Helmholtz, is:

$$\Delta G_{mix} = \Delta H_{mix} - T\Delta S_{mix} \tag{33}$$

which for polymer–solvent mixing in small particles becomes:

$$\ln \phi_m = \left(\frac{1}{j} - 1\right)\phi_p - \chi\phi_p^2 - \frac{2\bar{V}_m \gamma}{RTr_p} \tag{34}$$

where ϕ_p and ϕ_m are volume fractions of polymer and monomer, χ is the Flory–Huggins interaction parameter between polymer and monomer, $\bar{V}_m$ is the partial molar volume of the monomer, and γ is the interfacial tension between the particle and the surrounding medium. This is often referred to as the Morton Equation [24]. The last term, due originally to Lord Kelvin, represents the resistance because of the creation of new surface area upon swelling, important only when the surface/volume ratio is large, i.e. in small particles. The smaller r_p is, the less will be ϕ_m, the degree of swelling. Also, when the interfacial tension of the particles is higher, the swelling will be less. Gardon has calculated particle swelling for just about any situation and has provided graphs [25] which can be of practical assistance to practitioners of emulsion polymerization (Fig. 16).

During polymerization the particles are swollen with monomer at a rate which ordinarily is fast compared with the rate of consumption of monomer by polymerization, especially when the particles are small, so that the swelling

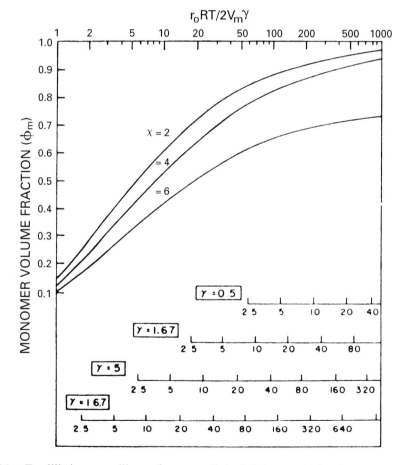

Fig. 16. Equilibrium swelling of uncrosslinked latex particles by monomer or a solvent, according to Eq. 34 [25]. The lower scales in the figure give r_0 in nm at indicated γ values in mN m^{-1} for $V_m = 100$ mL mol^{-1} and $T = 298$ K.

is essentially at equilibrium at all times. The rate of swelling depends upon diffusion of monomer from the emulsified monomer droplets through the aqueous phase into the polymer particles. When the solubility of the monomer is extremely low, somewhere less than about 10^{-5} M, then the assumption of equilibrium swelling will no longer hold.

In practical terms, a common, slightly water-soluble monomer such as styrene will have a concentration of about 10^{-3} M in the aqueous phase and about 5 M within the particles, as long as there are pure monomer droplets present in equilibrium with the other two phases. At some stage of the reaction, all the monomer droplets will have been consumed, and there still will be at the moment of their disappearance the equilibrium concentration of monomer in the particles. Thereafter the concentration of monomer in the polymer particles decreases, and this is accompanied by a decrease in the rate of polymerization.

Ordinarily there are thus three distinct intervals during an emulsion polymerization:

- Interval I: Particle formation. This ends when $dN/dt \to 0$.
- Interval II: Constant rate period during which $d[M]_p/dt = 0$.
- Interval III: Declining $[M]_p$, which usually is accompanied by declining rate of polymerization, but may initially involve acceleration due to the Norrish–Tromsdorff effect [26].

Thus one readily may calculate from the conversion at which the rate changes what the equilibrium value of $[M]_p$ was during Interval II [27]. Incidentally, in Interval III as one approaches the end of the reaction, the concentration of polymer becomes very high within the particles, so that the internal viscosity rises. This in turn leads to reduced rate constants for termination and perhaps also propagation, but also to reduced radical capture efficiency (Eqs 27 and 28), which may lead to new particle nucleation late in the reaction, with a consequent bimodal particle size distribution. It can also lead to considerable chain transfer to polymer, which in turn results in chain branching and crosslinking.

2.2.2 Coagulation

We have seen that coagulation can be a major factor in the initial formation of particles. It can also be the principal mechanism of particle growth. For example, even though $dN/dt = 0$, R_f may be positive throughout the entire reaction under conditions in which surfactant is limited, so that the primary particles are colloidally unstable and heterocoagulate with larger particles formed earlier in the polymerization. Evidence for this comes from electron micrographs of two kinds: one in which dried particles have been subjected to selective etching by an oxygen plasma such that the internal structure of the particles is revealed [28], and the other in which careful examination of the background in micrographs of large particles shows extremely small particles in great numbers. In the latter case negative staining technique [29] is recommended because such small particles have almost no contrast relative to the support material under the electron beam (see Section 5.3.3.2 below). Unfortunately, this mechanism of particle growth has been largely overlooked because of the experimental difficulties involved, but it does provide one means of regulating particle morphology.

> Coagulation can be the principal mechanism of particle growth.

2.2.3 Radical entry

After the polymer particles are formed and are swollen with monomer, they become the principal locus of polymerization, with few exceptions (e.g. in the case of very water-soluble monomers). The free radicals are generated continuously in the aqueous phase. They diffuse randomly about until they collide with, and enter, a particle. Thereupon all three of the major processes in radical polymerization can occur, i.e. initiation, propagation and termination (chain transfer reactions will be dealt with separately). Radical entry continues to be governed by Eqs 24–31 throughout the reaction, depending upon the specific conditions of monomer concentration, particle size, radical concentration within the particles, etc.

> Free radicals are generated continuously in the aqueous phase. They diffuse about until they collide with, and enter, a particle.

2.2.3.1 Initiation

When the particles are very small – say, less than $c.\ 0.1\ \mu m$ – and contain no free radicals, an entering (oligo)radical will initiate polymerization therein. If a propagating radical already exists in the particle, then the entering radical immediately terminates it and polymerization stops in that particle until a third radical enters. (Note that 'entry' in this context means irreversible entry. A radical conceivably could enter, add one or more monomer units and exit again if j were still less than j_{cr}.) Larger particles, especially those whose viscosity is high because of low monomer swelling (such as at the approach to the end of the polymerization, or where the Norrish–Tromsdorff effect exists), can accommodate more than one free radical at a time. Thus the distribution of radicals among the particles is a stochastic process, which is dealt with below.

2.2.3.2 Propagation

Polymerization, the formation of long molecules by the addition of monomer units can be represented by:

$$M_{jp}^{\cdot} + M_p \xrightarrow{k_p} M_{jp+1}^{\cdot} \tag{35}$$

where the subscript p designates the polymer particle phase, in analogy to Eq. 6 for propagation in the water phase. The corresponding rate of polymerization is then:

$$R_p \equiv -\frac{d[M]}{dt} = k_p[M]_p[M\cdot]_p \quad (36)$$

assuming that essentially all of the monomer is consumed by polymerization in the particles. The rate constant k_p is that for homogeneous kinetics because the propagating radical 'sees' a homogeneous medium in its immediate environment. The monomer concentration $[M]_p$ can be determined experimentally or calculated using Eq. 34, but the value of the radical concentration $[M\cdot]_p$ is much more difficult to obtain. It can be represented by:

$$[M\cdot]_p = \frac{\bar{n}N}{N_A} \quad (37)$$

where $\bar{n}$ is the average number of free radicals per particle, N is the number of particles per dm^3 and N_A is Avogadro's number. Thus the overall rate of any emulsion polymerization is:

$$R_p = k_p \left(\frac{\bar{n}N}{N_A}\right)[M]_p \quad (38)$$

where N_A is used to convert numbers of radicals into moles. The quantity $\bar{n}$ is thus an all-important parameter in determining the kinetics of emulsion polymerization. It can be determined by obtaining experimentally the overall rate of polymerization, R_p, the number concentration of particles, N, and the monomer concentration within the particles, $[M]_p$. The value of k_p for common monomers can be obtained from the literature. Then from Eq. 38:

$$\bar{n} = \frac{R_p N_A}{k_p N [M]_p} \quad (39)$$

2.2.3.3 Termination

Because of the very small volume of latex particles and the very high values for the rate constants for termination ($k_t \approx 10^7$ dm^3 mol^{-1} s^{-1}), two free radicals may coexist for only extremely short times before mutually terminating. This means that a particle either contains one radical or none, and on a time-average $\bar{n} = 1/2$.

There are, however, a number of exceptions to this general rule: (1) When the particle viscosity is high (e.g. at low $[M]_p$), then k_t will be lower. For example, this starts at c. 35% and 75% conversion in MMA and styrene polymerizations, respectively, and becomes more pronounced as $[M]_p$ decreases until k_t may go to zero at very high conversions. (2) When the particle size is large enough, two radicals may not 'see' each other for longer times, if at all. Under these circumstances $\bar{n}$ can be much greater than one half.

Thus, with free radicals originating in the continuous medium and entering the particles in a random fashion, initiating or terminating polymerization depending upon a number of factors, the situation in most cases is quite complicated. In the following sections we shall see that solutions can be obtained in many situations.

2.2.4 Radical exit

There is, however, one more important consideration, namely the escape of free radicals from the particles. Clearly if $j \geq j_{cr}$, then the radical cannot depart from the particle. But chain transfer to monomer is considerable with some important monomers such as vinyl acetate, vinyl chloride and ethylene. This occurs when the propagating free radical abstracts a hydrogen atom from the monomer, terminating its own growth and producing a monomeric radical:

$$M_p\cdot + MH_p \xrightarrow{k_{tr(M)}} P_p + M_{1p}\cdot \qquad (40)$$

The product $M_{1p}\cdot$ can easily diffuse out of the particle into the aqueous phase, where it can undergo termination, propagation or re-entry into another particle. The kinetics under these circumstances ordinarily are characterized by $\bar{n} \ll 1/2$.

2.2.5 Aqueous phase reactions

The continuous medium is not usually considered important in emulsion polymerization, but there are many exceptions to this rule, especially with more water-soluble monomers, e.g. acrylic acid, acrylamide, methyl acrylate and the like. It is even possible for the major part of the polymerization to take place in the aqueous phase, or for much polyelectrolyte or nonionic polymer to be formed which is totally dissolved in the water, or which becomes adsorbed onto the particles. Often this is desired, as it contributes to the colloidal stability of the system, especially against freezing and thawing. More subtle reactions, such as termination in the aqueous phase, can also have a pronounced effect on the overall polymerization kinetics. These will be dealt with in the following sections.

2.2.6 Smith–Ewart kinetics

The first to develop a viable, quantitative theory for emulsion polymerization kinetics were Wendell Smith and Roswell Ewart [17] in 1948. During World War II Smith and Ewart were involved in developing synthetic rubber by means of the emulsion copolymerization of styrene and butadiene: styrene/butadiene rubber, or SBR. They were smart enough to choose styrene for their fundamental studies, as it behaves in a regular and reproducible manner. Smith and Ewart looked at two parts of the process: (1) particle formation by heterogeneous nucleation in surfactant micelles (discussed in Section 2.1.2.1 above) and (2) emulsion polymerization kinetics under steady state conditions where N is a constant.

> The first to develop a quantitative theory for emulsion polymerization kinetics were Wendell Smith and Roswell Ewart in 1948.

They looked at the distribution of the radicals among particles by assuming a steady state in the appearance and disappearance of radicals within any class of particles containing n radicals:

$$\left(\frac{R_c}{N}\right) N_{(n-1)} + k_o a \left(\frac{n+1}{v}\right) N_{(n+1)} + k_t \left[\frac{(n+2)(n+1)}{v}\right] N_{(n+2)}$$
$$= N_n \left[\left(\frac{R_c}{N}\right) + k_o a \left(\frac{n}{v}\right) + k_t n \frac{(n-1)}{v}\right] \quad (41)$$

i.e. the rate of capture, exit and termination to form N_n type particles is equal to the sum of the rates which lead to their disappearance, where n is the number of free radicals in a particle, N_n is the number concentration of particles containing n radicals, k_o is the rate constant (frequency) of radical exit from particles, a is the particle surface area (all particles have the same size), v is the volume of a particle, and k_t is the rate constant for termination (in molecular units).

This is the famous 'recursion equation' and it is perfectly general (if we take $k_o a$ to represent a rate constant for exit which may or may not be proportional to particle radius). What complicates matters is the details of the capture of oligoradicals, exit kinetics and the viscosity dependence of the termination rate, k_t. Smith and Ewart went on to show that there were three limiting cases in which rather simple solutions to Eq. 41 could be obtained. Let us take the simplest first, their Case 2:

As noted in Section 2.2.3.3 above, small latex particles cannot accommodate more than one radical at a time, since capture of a second radical results in essentially instantaneous termination. This can be represented schematically in Fig. 17 in which the number of radicals, n, in an *average* particle is plotted as a function of time during which entry and termination repeatedly occur. Radicals are being generated continuously in the aqueous phase. In this case chain transfer is negligible. If the rate of generation of free radicals in the aqueous phase is slower than that in Fig. 17, the situation might be represented by Fig. 18 in which the time between entries (capture frequency) is longer. The average value of n is seen to remain unchanged!

Since the overall rate of polymerization is given by Eq. 38:

$$R_p = k_p \left(\frac{\bar{n}N}{N_A}\right) [M]_p$$

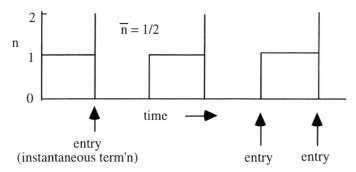

Fig. 17. Smith–Ewart Case 2 in which the time-average value of $n = 1/2$.

for Smith–Ewart Case 2 we have the extremely simple formula:

$$\text{Case 2 } (\bar{n} = 1/2): \quad R_p \equiv -\frac{d[M]}{dt} = k_p \left(\frac{N}{2N_A}\right)[M]_p \quad (42)$$

Because the average particle in Figs 17 and 18 contains essentially either 1 or no radical, these are known as 'zero-one' type kinetics. Thus the overall rate of polymerization is independent of the initiator concentration, but dependent upon particle size, i.e. $R_p \propto 1/r^3$, since for a given amount of polymer, the smaller the particles, the greater their number.

Since the polymer chains are growing when a radical is present in a particle ($n = 1$), the chain length, and thus the molecular weight, will be directly proportional to average radical lifetime, $\bar{\tau}$, which is equal to the time between entries: $\bar{\tau} = N/R_c$. The average degree of polymerization of the polymer being formed will always be equal to the frequency of addition of monomer units, $k_p[M]_p$ per second, multiplied by the radical lifetime:

$$\overline{DP} = k_p[M]_p\bar{\tau} = k_p[M]_p\left(\frac{N}{R_c}\right) \quad (43)$$

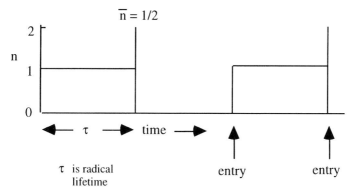

Fig. 18. Smith–Ewart Case 2 in which the time-average value of $n = 1/2$ (slower R_c).

Clearly, in this case, the molecular weight will be a function not only of particle size, but also of initiator concentration. If all the effective free radicals generated are captured, then

$$R_c = bR_i = 2bk_d[I] \tag{44}$$

where b is the radical effectivity, k_d is the initiator decomposition rate constant and $[I]$ is the initiator concentration.

> If there is chain transfer to a small molecule, the resulting small free radical can diffuse out of the particle.

If there is significant chain transfer to a small molecule, typically monomer (cf. Section 2.2.4 above), the resulting small free radical can diffuse out of the particle before it adds many or any monomer units. This can also result in zero–one kinetics, but the time in which a particle is in the 'zero state' may be much longer than that during which it contains a radical. This would occur when the $k_o a$ term in Eq. 41 – which now clearly must include also a frequency of chain transfer to monomer – is quite large relative to the frequency of radical entry, i.e. $k_o a \gg R_c/N$. This is called the Smith–Ewart Case 1 and is shown schematically in Fig. 19. The overall rate of polymerization then depends upon whether termination occurs predominantly in the particles or in the continuous phase:

Case 1 ($\bar{n} \ll 1/2$):   where $k_o a \gg R_c/N$
  Case 1A: termination in the continuous phase

$$R_p = k_p[M]_p V_p \alpha \left(\frac{R_c}{2k_t}\right)^{1/2} \tag{45}$$

where α is the partition coefficient of free radicals between the two phases: $\alpha \equiv c_p/c_w$, and the cs represent concentrations of free radicals in the polymer and water phases; V_p is the volume fraction of the (monomer-swollen) polymer phase.

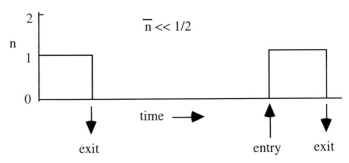

Fig. 19. Smith–Ewart Case 1 in which the time-average value of $n \ll 1/2$.

The average radical lifetime in this case will be:

$$\bar{\tau}_p = \frac{V_p \alpha}{(2k_t b R_i)^{1/2}} \qquad (46)$$

Case 1B: termination in the particle phase

$$R_p = k_p[M]_p \left(\frac{V_p R_c}{2k_o a}\right)^{1/2} \qquad (47)$$

The average radical lifetime here becomes

$$\bar{\tau}_p = \left(\frac{V_p}{k_o a b R_i}\right)^{1/2} \qquad (48)$$

Thus in Case 1 (A or B) the rate of polymerization depends on initiator concentration (through R_c), but not on particle size, just contrary to Case 2! However, in this case where many radicals are escaping the particles, many may re-enter other particles, which means that R_c can be considerably larger than R_i, so that we do not know *a priori* how the rate of polymerization depends on initiator concentration.

> Where many radicals are escaping particles, many may re-enter others, which means that R_c can be larger than R_i.

Again, the particle size must be relatively small in order for the radicals formed by transfer to exit readily. If the particles are large, and if the internal viscosity is high, diffusion out may be so slow that the radical polymerizes to a size larger than j_{cr} and cannot escape.

When the particles are large enough and conditions are such that they can accommodate more than one radical at a time, $\bar{n}$ can become greater than unity, Smith–Ewart Case 3. A 'chair diagram' might look something like that in Fig. 20. As soon as $\bar{n}$ gets to be more than about 2, the kinetics become identical to solution polymerization kinetics, as given in Eq. 49:

Case 3 ($\bar{n} \gg 1/2$):

$$R_p = k_p[M]_p \left(\frac{V_p R_c}{2k_t}\right)^{1/2} \qquad (49)$$

in which the term V_p merely corrects for the fact that polymerization is occurring only in the organic phase. Again, equating the rate of capture to bR_i is probably a good bet; and b is usually around 0.5. So for Case 3 the rate of polymerization is independent of particle size and generally is proportional to the square root of initiator concentration.

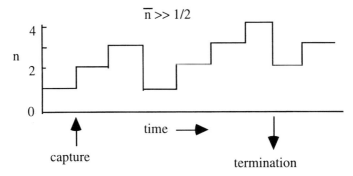

Fig. 20. Smith–Ewart Case 3 in which $\bar{n} \gg 1/2$.

As $\bar{n}$ gets to be more than about 2, the kinetics become identical to solution polymerization kinetics.

The radical lifetime for Case 3 will be

$$\bar{\tau}_p = \left(\frac{V_p}{2k_t R_c}\right)^{1/2} \tag{50}$$

The three Smith–Ewart limiting cases apply to many systems, but the theory, as we have seen, leaves a number of questions unanswered for many others, such as

(1) What about intermediate cases where $\bar{n}$ is just a little bit more or less than 1/2? (one possible example is given in Fig. 21).
(2) How does the rate of chain transfer enter into the theory?
(3) Is it possible to relate the rate of polymerization and the molecular weight to experimentally determined factors, rather than the unknown rate of capture, R_c?
(4) How is the rate of capture of the oligoradicals affected by their solubility, i.e. the value of j, and whether they are charged or not?

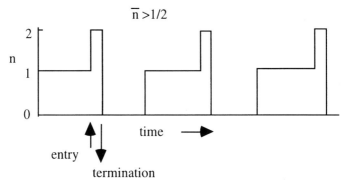

Fig. 21. Case in which $\bar{n}$ is slightly larger than 1/2.

2.2.7 Ugelstad–Hansen theory

Hansen and Ugelstad addressed many of these issues in their landmark paper [30] published in 1976. They built upon the earlier work of Stockmayer [31] and O'Toole [32] who had developed solutions for the Smith–Ewart recursion equation for certain conditions using Bessel functions. To relate the rate of capture of radicals to the rate of generation of radicals, Ugelstad and Mørk earlier had observed that:

rate of capture = effective rate of initiation + overall rate of exit

− termination in the water phase

because all of the radicals which are generated from the initiator along with all those which exit must either be captured or undergo termination in the aqueous phase. Expressed mathematically, it is:

$$R_c = bR_i + \sum_n k_d N_n n - 2k_{tw}^*[M_j \cdot]_w^2 \qquad (51)$$

where k_d is their 'desorption' constant, equivalent to the Smith–Ewart $k_o a$, and the asterisk on the k_{tw}^* is to remind us that this is in molecular units rather than in moles. The rate of capture is also given by Eq. 25: $R_c = \bar{k}_c[M \cdot]N$. Each of the four terms in Eqs 51 and 25 is rendered dimensionless by dividing by the termination rate (actually frequency), to give an extremely simple equation which encompasses all of the concerns mentioned in the previous section:

$$\alpha = \alpha' + m\bar{n} - Y\alpha^2 \qquad (52)$$

with the following definitions:

Capture: $\quad \alpha \equiv \left(\dfrac{R_c/N}{k_t^*/v} \right)$

Initiation: $\quad \alpha' \equiv \left(\dfrac{R_i/N}{k_t^*/v} \right)$

Exit: $\quad m \equiv \dfrac{k_d}{k_t^*/v}$

'Aqueous termination': $\quad Y \equiv \left(\dfrac{2Nk_{tw}^*}{\bar{k}_c^2} \right) \left(\dfrac{k_t^*}{v} \right)$

Y is taken by Ugelstad and coworkers as a measure of the amount of termination taking place in the aqueous phase. The solution to Eq. 52 was taken from the work of O'Toole, who used Bessel functions to solve the Smith–Ewart formula for certain conditions. But Ugelstad et al. found that a simple procedure could be employed involving continued fractions. The idea is to solve for $\bar{n}$ in terms of α', which in turn is experimentally obtainable from initiator decomposition rates and solution termination rate constants, both of which are abundantly available from the literature. The O'Toole equation is:

$$\bar{n} = \left(\frac{a}{4}\right)\left(\frac{I_m(a)}{I_{m-1}(a)}\right) \quad (53)$$

where $a \equiv \sqrt{8\alpha}$. The ratio of Bessel functions in Eq. 53 is given by an infinite series which is in this case a continued fraction:

$$\frac{I_m(a)}{I_{m-1}(a)} = \cfrac{a/2}{m + \cfrac{a^2/4}{m+1+\cfrac{a^2/4}{m+2+\cfrac{a^2/4}{m+3+\cdots}}}} \quad (54)$$

From the definition of a above and from Eq. 53, we can now write in mathematical shorthand:

$$\bar{n} = \left(\frac{1}{2}\right)\frac{2\alpha}{m} + \frac{2\alpha}{m+1} + \frac{2\alpha}{m+2} + \cdots \quad (55)$$

And finally $\bar{n}$ can be obtained as a function of α' by choosing values of α and m, calculating $\bar{n}$ using Eq. 55, and then computing corresponding values of α' from Eq. 52, also choosing arbitrary values of Y. Typical results from the paper by Ugelstad and Hansen are shown in Figs 22 and 23.

Several things can be observed from these figures. Firstly, there are three regions in which $\log \bar{n}$ is a linear (or almost linear) function of $\log \alpha'$. These correspond to the three Smith–Ewart limiting cases, as shown schematically in Fig. 24.

The transitional regions in which there is curvature are relatively small, so that for many, if not most, emulsion polymerizations the dependence of $\bar{n}$ on the initiator concentration is a simple one, and no Bessel functions are required.

At values of $m \leq 1$ the region of Case 1 is independent of the rate of aqueous phase termination, as given by Y. At values of $m \geq 0.001$ there is really no distinct Case 2 region. At the higher values of m the value of $\bar{n}$ becomes very sensitive to the amount of termination in the water phase.

Experimentally, if one wishes to determine the kinetics of emulsion polymerization of a particular system, the first thing to do is to obtain the value of $\bar{n}$. This can be done by determining the overall rate of polymerization and measuring the particle size (without monomer) and the concentration of monomer in the particles. Rearranging Eq. 38 gives:

$$\bar{n} = \frac{R_p N_A}{k_p N[M]_p} \quad (56)$$

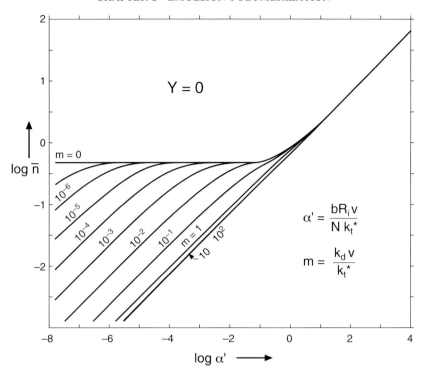

Fig. 22. Calculated values of $\bar{n}$ as a function of α' for $Y = 0$.

From the particle size and the total volume of polymer (without monomer) produced at the time of the rate measurement, one can calculate the number concentration of particles, N. One needs a literature value for k_p to then calculate the desired result. One caveat: a glance at the literature indicates how unreliable some data are because of the spread of values reported. One may wish to look at the source and the methods used before choosing! Experimental methods for obtaining rates of polymerization are dealt with in Section 3.1.

Once the value of $\bar{n}$ is known, the experimentalist has knowledge of where approximately on the Ugelstad–Hansen plots he is, and how the rate of reaction and the molecular weight of the polymer formed will be affected by changes in initiator concentration, for example. By varying the value of α', one can quickly determine exactly which curve he is on, and thereafter make predictions about other conditions. This becomes especially powerful if the temperature dependences of the rate constants involved are known.

2.2.8 Alexander–Napper–Gilbert theory

Napper and Alexander, and more recently Napper and Gilbert, at the University of Sydney, Australia, have contributed greatly to our detailed, quantitative understanding of the processes taking place in emulsion polymerization. For

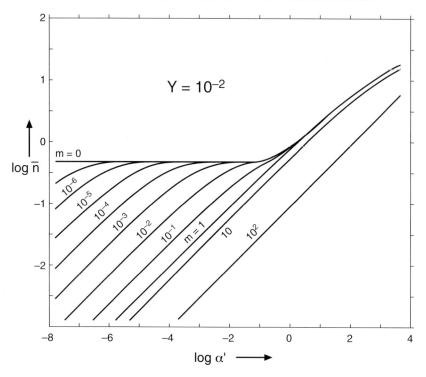

Fig. 23. Calculated values of $\bar{n}$ as a function of α' for $Y = 0.01$.

example, they showed that particle size distributions during particle formation (Interval I) were positively skewed, confirming the role of coagulation. They chose the unfortunate term 'coagulative nucleation' to describe the phenomenon, which is a contradiction in terms. They also have provided evidence that there can be coagulation even when the *initial* surfactant concentration is above the CMC. It is possible that so many particles could be formed, and their

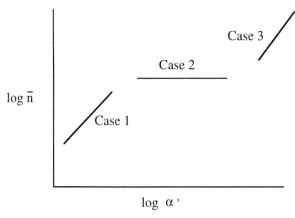

Fig. 24. The three Smith–Ewart cases.

surface area could increase so much during growth, that they 'run out' of sufficient stabilizer. This would result in coagulation.

The Sydney group has also pointed out the importance of the chain length-dependence of the rate constants for termination and propagation. It is well known that small molecules and free radicals have higher rates of reaction than larger ones in a homologous series (such as our j-mers). As soon as the chain length reaches about 10 carbon atoms ($j \approx 5$), the rate becomes independent of the chain size. Gilbert and Napper indicate that their experimental determinations of absolute reaction rates confirm this, and that it can greatly affect the amount of chain termination in the aqueous phase [33]. They also find j_{cr} for styrene to be equal to 5, confirming what others have found. An excellent overview of emulsion polymerization is given in the book by Gilbert [34].

2.3 Particle morphologies in multiphase copolymers

If a seed latex of a polymer P1 is used to nucleate the polymerization of a second monomer M2 to form polymer P2, then it generally can be expected that P1 and P2 will be incompatible, so that the resulting latex particle will contain two polymer phases. Under some circumstances this is a highly desirable result. For example, in ABS (acrylonitrile/butadiene/styrene) plastics there are rubbery domains of polybutadiene (PB) dispersed in a glassy matrix of polystyrene/acrylonitrile (PS/AN) copolymer. The rubbery domains impart impact resistance to an otherwise brittle material, but only if they are within a certain size range.

The question is what experimental factors determine the morphology of the two-phase particles? Are they like dumb-bell shaped, 'ice-cream cone' shaped, 'framboidal' (raspberry) shaped (Fig. 1), or core/shell? And if core/shell, what determines which polymer is core and which is shell? Let us take the case where M2 is a good solvent for P1, so that the seed particles are swollen with the second monomer. This is the case for ABS, in which PB seed latex has a mixture of styrene and acrylonitrile added in the second stage of the emulsion polymerization. Stabenow and Haaf found that the resulting particle morphologies depended on the size of the seed particles, the degree of crosslinking of the seed latex and the amount of M2 added [35]. Small seed particles with low crosslinking tended to give core/shell morphology; large seeds with low crosslinking gave 'warty' or framboidal shapes; whereas higher degrees of crosslinking, regardless of size, tended to give particles in which PS/AN was finely dispersed within a matrix of PB.

2.3.1 Thermodynamics of phase separation

2.3.1.1 Torza and Mason theory and experiments

When drops of two immiscible liquids, 1 and 2, suspended in a third liquid, 3, are brought together, the drops will interact by one completely engulfing the

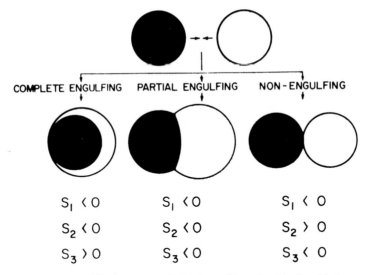

Fig. 25. Equilibrium morphologies of immiscible liquid drops.

other ('core/shell' 1/2 or 2/1), forming a dumb-bell shape ('partial engulfing'), or by remaining separate depending upon the three spreading coefficients involved, S_1, S_2, and S_3, where $S_i = \sigma_{jk} - (\sigma_{ij} + \sigma_{ik})$ (see Fig. 25) and σ represents the interfacial tension or free energy [36]. These morphologies are completely determined by thermodynamics and will manifest themselves when all the phases involved are sufficiently mobile. Thus two-phase latex particles can be expected to exhibit similar morphologies under conditions of nucleation and growth when there is sufficient monomer swelling them that adequate mobility is assured, i.e. when they will act as liquids. In a seeded emulsion polymerization, after the swelling of P1 by M2, there are only two phases of interest, namely the swollen particles and the water phase. As M2 polymerizes to form P2, at some point the P2 oligomers will no longer be soluble in the P1/M2 and the new P2 phase nucleates.

2.3.1.2 Binodal phase separation and spinodal decomposition

If thermodynamic equilibrium is maintained, then the situation just described can be represented by a classical three-component phase diagram, such as that shown in Fig. 26. If we start with a 90/10 M2/P1 mixture (point X in Fig. 26), and start polymerization, the composition will follow the reaction vector b. The particle contents remain homogeneous until it arrives at the binodal curve, whereupon phase separation occurs. Tiny particles of P2 will nucleate and be immediately swollen with M2. These particles grow with further polymerization, and they will tend to coagulate and coalesce to form one of the structures shown in Fig. 25, which would constitute a single latex particle, except in the last case depicted in that figure.

For pure liquids, under conditions of sufficient supercooling, nucleation is diffusion-controlled, such that the separating phases do not have time to

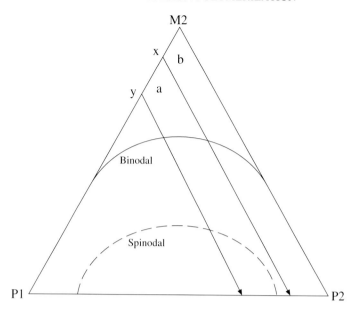

Fig. 26. Three-component phase diagram for two polymers and one monomer. ——, Binodal and – – – –, spinodal curve. Arrows are reaction vectors, a + b.

organize in fully separated domains. This leads to a bicontinuous phase structure [37], a phenomenon known as 'spinodal decomposition.' Similarly, in our case if we were to start with a 70/30 M2/P1 mixture (point Y, Fig. 26), the reaction vector passes across the spinodal curve at which spinodal decomposition sets in. The internal morphology of the latex particle in this case would consist of two polymer phases randomly intermingled as shown in Fig. 27.

Some of the earliest experimental and theoretical work on this was done by Johnston, who studied the morphologies which developed as a result of phase separation accompanying the polymerization of styrene containing dissolved PMMA [38]. He related these to the three-component phase diagram of the system styrene/polystyrene/polymethylmethacrylate, and was in fact the first to propose this approach. Since he worked in bulk, however, he could not

Fig. 27. Bicontinuous phase structure of two incompatible polymers.

observe what would happen as a result of the interfacial tensions against water.

2.3.1.3 Sundberg theory and experiments

When polymers and surfactants are present along with liquid monomers in an emulsion polymerization, then the interfacial free energies and spreading coefficients are affected in more complex ways. Sundberg and coworkers have calculated predicted equilibrium morphologies as a function of conversion and the presence of ionic polymer end groups, which have been confirmed by model experiments [39]. They show that in some cases, although rarely, it is possible for the system to switch from one structure to another as reaction conditions change during the polymerization. For example as new surface area is generated by the formation and growth of latex particles, surfactant in aqueous solution becomes depleted with a consequent increase in interfacial tension, σ_{pw}.

2.3.2 Kinetics of phase separation and growth

In real life things are not so simple. Every conceivable biphasic morphology has been observed, from particles containing many smaller particles within them, as in Fig. 28, to those with 'warts,' or framboidal shapes, to ones with hollows on the surface like golf balls!

The formation of many nuclei within a single particle results from nucleation and growth, as we have seen. When the internal viscosity of the particles is high enough, these nuclei diffuse and coagulate together more or less incompletely, so that they remain separate to an extent which depends on monomer content, temperature, molecular weight and the degree of crosslinking of the polymer. Even the latex particle size becomes a determinant of the internal heterophase particle size because of the distances over which the newly formed nuclei must travel to coagulate. Thus we see more and smaller

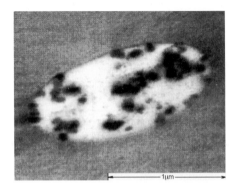

Fig. 28. PS microdomains within PMMA latex particles. Microtomed cross-section, RuO_4 stained [40].

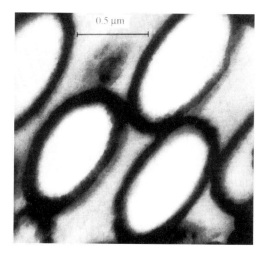

Fig. 29. PMMA/PS core/shell particles [40].

internal particles as the amount of crosslinking is increased, and more of them when the particle size is larger. This has been confirmed by experiment, for example in early work on ABS [41].

Another way to regulate morpholgy by ensuring that it is diffusion-controlled is through monomer-starved conditions by adding monomer more slowly than it is polymerized during the emulsion polymerization. This is ordinarily done to effect core/shell morphology by forcing the second polymer phase to nucleate and grow on the surface of the seed particles (Fig. 29 – note distortion due to microtome slicing of the sample).

2.4 Mechanical properties of derived films

Phase-separated polymer systems are so important primarily because of the unique dynamic mechanical properties which they exhibit.

2.4.1 Polymer alloys

These systems may be regarded as the analogs of metal alloys, which long ago were seen to be phase-separated systems whose properties and morphologies were understood in terms of phase diagrams and kinetics (whether a mixture of elements was cooled slowly or quenched to give wrought iron or case-hardened steel, for example).

In the case of copolymers, the most famous are glassy polymers containing dispersed particles of polybutadiene rubber, such as ABS (acrylonitrile/butadiene/styrene) and high-impact polystyrene. On impact the microvoids generated in the glassy plastic propagate until they reach a rubbery domain in which the energy is dissipated as heat, thus preserving the integrity of the material.

To make objects from such a latex, the colloid ordinarily is coagulated by the addition of salt or acid. The coagulum, which looks like cottage cheese (and is in some ways a synthetic analog to cheese), is filtered off, and dried. The resulting 'crumb' is then ready for the molding machine, where it is melted and extruded or compressed into – a boat hull, telephone case, shower stall, etc. Unless the rubber phase has the same refractive index as that of the continuous phase, the material will scatter light, and look cloudy and opaque. But the plastics made this way – and there is no other way as cheap and effective – are incredibly tough, and yet still may retain some flexibility.

References

2. Fitch, R.M. (1973). *Br. Polym. J.* **5**, 467.
3. Hansen, F.K. and Ugelstad, J. (1978). *J. Polym. Sci., Polym. Chem. Ed.* **16**, 1953.
4. Hansen, F.K. and Ugelstad, J. (1979). *J. Polym. Sci., Polym. Chem. Ed.* **17**, 3033.
5. Hansen, F.K. and Ugelstad, J. (1979). *J. Polym. Sci., Polym. Chem. Ed.* **17**, 3047.
6. Hansen, F.K. and Ugelstad, J. (1979). *J. Polym. Sci., Polym. Chem. Ed.* **17**, 3069.
7. Hansen, F.K. (1992). In *Polymer Latexes* (E.S. Daniels, D. Sudol and M.S. El-Aasser eds), ACS Symposium Series 492, pp. 13 ff., American Chemical Society, Washington, DC.
8. Krieger, I.M. and Juang, M.S. (1976). *J. Polym. Sci., Polym. Chem. Ed.* **14**, 2089.
9. Fitch, R.M. and Tsai, C.H. (1973). In *Polymer Colloids* (R.M. Fitch ed), p. 73, Plenum Press, New York.
10. Goodall, A.R., Wilkinson, M.C. and Hearn, J. (1975). *Prog. Colloid Interface Sci.* **53**, 327.
11. Gilbert, R.G. (1995). *Emulsion Polymerization*, Academic Press, London.
12. Fitch, R.M. (1982). In *IUPAC Macromolecules* (H. Benoit and P. Rempp eds), p. 39, Pergamon Press, Oxford.
13. Fitch, R.M., Palmgren, T.H., Aoyagi, T. and Zuikov, A. (1984). *Angew. Makromol. Chem.* **123/124**, 261.
14. Verwey, E.J.W. and Overbeek, J.Th.G. (1948). *The Theory of the Stability of Lyophobic Colloids*, Elsevier, Amsterdam.
15. Vijayendran, B.R. (1980). In *Polymer Colloids II* (R.M. Fitch ed), p. 209, Plenum Press, New York.
16. Sütterlin, N. (1980). In *Polymer Colloids II* (R.M. Fitch ed), p. 583, Plenum Press, New York.
17. Smith, W.V. and Ewart, R.H. (1948). *J. Chem. Phys.* **16**, 592.
18. Roe, C.P. (1968). *Ind. Eng. Chem.*, Sept. 20, p. 20.
19. Gerrens, H. (1959). *Fortsch. Hochpolym. Forsch.* **1**, 234.
20. Carver, M.T., Dreyer, U., Knoesel, R., Candau, F. and Fitch, R.M. (1989). *J. Polym. Sci. A, Polym. Chem.* **27**, 2161.
21. Ugelstad, J., El-Aasser, M. and Vanderhoff, J.W. (1973). *J. Polym. Sci., Polym. Lett. Ed.* **11**, 505.
22. Caris, C.H.M. (1990). *Polymer Encapsulation of Inorganic Submicron Particles in Aqueous Dispersion*, Ph.D. Thesis, Technical University, Eindhoven, Netherlands.
23. Fitch, R.M. (1993). *Proceedings of the Taniguchi Conference on Polymer Emulsions* **1**.
24. Morton, M., Kaizerman, S. and Altier, M.W. (1954). *J. Colloid Sci.* **9**, 300.
25. Gardon, J.L. (1968). *J. Polym. Sci., Part A-1* **6**, 2859.
26. Flory, P.J. (1953). *Principles of Polymer Chemistry*, pp. 125–129, Cornell University Press, Ithaca, New York.

27. Verdurmen, E. (1992). Ph.D. Thesis, Technical University, Eindhoven, Netherlands.
28. Eliseyeva, V.I., Ivanchev, S.S., Kuchanov, S.I. and Lebedev, A.V. (1981), Emulsion Polymerization and its Applications in Industry, pp. 96–100, Consultants Bureau, New York.
29. Scholsky, K.M. and Fitch, R.M. (1985). *J. Colloid Interface Sci.* **104**, 595.
30. Ugelstad, J. and Hansen, F.K. (1976). *Rubber Chem. Tech.* **49**(3), 536.
31. Stockmayer, W.H. (1957). *J. Polym. Sci.* **24**, 314.
32. O'Toole, J.T. (1965). *J. Appl. Polym. Sci.* **9**, 1291.
33. Maxwell, I.A., Morrison, B.R., Napper, D.H. and Gilbert, R. (1991). *Macromolecules* **24**, 1629.
34. Gilbert, R.G. (1995). *Emulsion Polymerization*, Academic Press, London.
35. Stabenow, J. and Haaf, F. (1973). *Angew. Makromol. Chem.* **29/30**, 359, 1.
36. Torza, S. and Mason, S.G. (1970). *J. Colloid Interface Sci.* **33**, 67.
37. Hammel, J.J. (1969). In *Nucleation* (A.C. Zettlemoyer ed), p. 519, Marcel Dekker, New York.
38. Johnston, G.J. (1972). *Polymerization of a Vinyl Monomer Containing a Second Dissolved Polymer*, M.A.Sc. Thesis, University of Waterloo.
39. Winzor, C.L. and Sundberg, D.C. (1992). *Polymer* **33**, 4269; Sundberg, E.J. and Sundberg, D.C. (1993). *J. Appl. Polym. Sci.* **47**, 1277.
40. Lee, S. and Rudin, A. (1992). In *Polymer Latexes, Preparation, Characterization and Applications* (E.S. Daniels, E.D. Sudol and M.S. El-Aasser eds) p. 234, ACS Symposium Series 492, American Chemical Society, Washington, DC.
41. Stabenow, J. and Haaf, F. (1973). *Angew. Makromol. Chem.* **29/30**, 359.

Chapter 3

Practical Applications of Emulsion Polymerization

3.1 Applications

This chapter deals with the various ways polymer colloids are actually made in the laboratory and in manufacturing. In each case the program of addition of ingredients is important, whether it is in batch, semi-continuous, multistage or continuous mode. From the same set of materials, especially when more than one monomer is copolymerized, the results can be very different. This was already made clear in the section on particle morphology in Chapter 2.

> From the same set of materials, especially when more than one monomer is copolymerized, the results can be very different.

3.1.1 Laboratory

Laboratory apparatus for emulsion polymerization can be assembled from commercially available components. A typical laboratory set-up is shown in Fig. 30. Typically a wide-mouthed flask (L) with a flanged top containing several openings with ground glass joints (J) is used so that removal of the product and cleaning up after the reaction can be undertaken easily. The reaction is exothermic, and with some monomers, such as vinyl acetate, they can be dangerously so. Thus, means for efficient heat transfer are required, both its removal during the major part of the reaction, and its provision during the latter stages when the monomer has been largely consumed, by good stirring (F) and a bath surrounding the reaction flask (M). For most systems it is necessary to exclude oxygen because it is an effective free radical inhibitor. This is done by providing a blanket of inert gas, usually nitrogen. For all but batch reactions, the controlled addition of monomer(s) from some reservoir (Q) during the polymerization is accomplished by means of a metering pump (O) or dropping funnel. Often other ingredients are also added, such as surfactant and initiator solutions. If the temperature is high enough and the boiling point of monomers is low enough, a reflux condenser (C) must be provided as well. Samples are customarily taken at various times

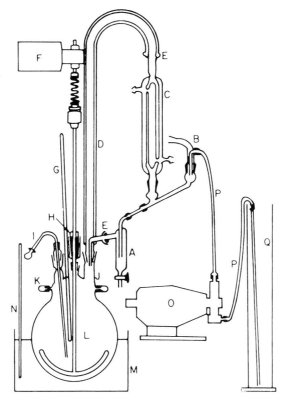

Fig. 30. Laboratory apparatus for emulsion polymerization [42].

to measure the progress of the reaction, so that a sampling device must be installed (I). All this is required, along with some means of measuring the temperature (G)!

3.1.2 Batch emulsion polymerization

In a batch reaction all the ingredients are charged to the flask at the beginning, the contents are heated to reaction temperature and kept there by heating or cooling, as needed, until the samples removed indicate the desired conversion of monomer to polymer. Although in most cases it is desirable to take the reaction to as high a conversion as possible, there are occasions when it is terminated (e.g. by rapid cooling or adding a free radical inhibitor) in order to control copolymer composition, degree of crosslinking or molecular weight distribution.

A typical 'recipe' (among many thousands of possibilities) for a commercially useful, film-forming polymer, e.g. for latex paint, might be as follows:

- Methyl methacrylate   60 g
- Butyl acrylate   38 g
- Methacrylic acid   2 g

- Sodium dodecyl sulfate   1 g
- Ammonium persulfate   0.5 g
- Distilled water   100 g

When this mixture is heated to about 60–80° C under a nitrogen atmosphere, the ammonium persulfate decomposes slowly to form sulfate ion radicals, $\cdot OSO_3^-$ which initiate the polymerization. In about 2 h the polymerization is complete. Particles of c. 200 nm diameter are formed, and the average molecular weight of the terpolymer is on the order of 250 000 Da. The methacrylic acid is added to enhance the stability of the latex particles, since it tends to be incorporated at the particle surface. It has the disadvantage, however, of increasing the water-sensitivity of the derived films.

Another disadvantage of this method is that, because the rate constants for propagation (the k_p values) and the water-solubilities of the three monomers are different from each other, there is considerable compositional drift during course of the reaction. This can be obviated by adding the monomers gradually during the polymerization according to a calculated regime which takes into account not only the k_p values and solubilities, but also the partition of the monomers among the aqueous and organic phases. This is known as the 'semi-continuous' or 'semi-batch' method.

> Because the k_p values and the water-solubilities of the monomers are different from each other, there is compositional drift during the reaction.

3.1.3 Semi-continuous emulsion polymerization

In this technique, the same ingredients as above could be used, but now a portion is started just like a smaller batch reaction. The monomer composition, however, has been calculated to ensure the desired polymer composition according to the Copolymer Equation, but with the added considerations of drift mentioned above [43, 44]. After the reaction is under way the remaining monomer mixture is programmed in so as to maintain a constant polymer composition throughout. Dramatic differences from batch polymers are obtained in this manner in polymer microstructure and therefore T_g.

The work of Guyot, Pichot, Guillot and Rios Guererro with styrene/acrylonitrile (S/AN) emulsion copolymerization will serve as an illustration [45]. Styrene is very slightly soluble in water, whereas acrylonitrile is soluble. In a *batch* reactor the initial condition was [AN]/[S] = 4 molar ratio; $K_2S_2O_8$ initiator; $C_{12}H_{25}SH$ chain transfer agent in trace amounts; SDS as surfactant; and monomer/water ratio = 0.10. The reactions were run at 50°C, samples were removed periodically, and the polymer separated in an ultracentrifuge and then analyzed. The amounts of the two monomers in the particles are not constant, but rise until the monomer droplets disappear. Then they fall, as shown by curve number 3 and 7 in Fig. 31. In these curves the total number of grams of each monomer in the swollen particle phase is plotted as a function of conversion.

CHAPTER 3 PRACTICAL APPLICATIONS OF EMULSION POLYMERIZATION 51

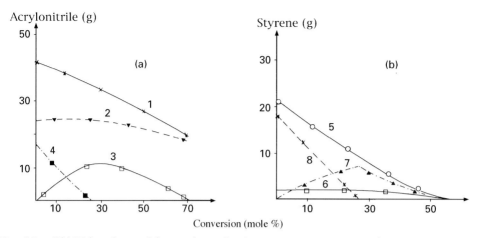

Fig. 31. S/AN batch emulsion polymerization monomer contents in: 2,6 Water; 3,7 Particles; 4,8 Monomer Droplets; 1,5 Overall [45].

In a *semi-continuous* process under otherwise identical conditions, 25% of the monomers were charged to the flask initially, and the remainder was programmed in. There is greater compositional uniformity in this case, as shown in Fig. 32. The calculations are based upon the thermodynamics of swelling as expressed in the Morton Equation, Eq. 34 (see Section 2.2.1.1), but with the added requirement that with each increment of the polymerization, the amount of each monomer in the particles changes, so that there is re-equilibration thoughout the system. Equation 34 is thus extended to take two monomers into account. In this equation the chemical potential is given by μ, subscripts represent monomers 1, and polymer 2, ϕ represents volume fraction, m is molecular weight, and χ is the Flory–Huggins interaction parameter. V_m is the molar volume of liquid inside the particles, γ is the interfacial tension of the particles, and r_0 is the radius of the

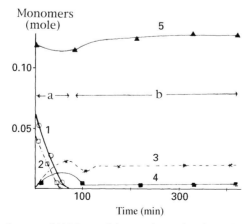

Fig. 32. Semi-continuous S/AN emulsion polymerization. Curve labels as in Fig. 31.

unswollen particle:

$$\mu_1^p = \mu_{01}^p + RT\left[\ln\phi_1 + (1-\phi_1) - \phi_2\frac{m_1}{m_2} - \phi_p\frac{m_1}{m_2}\right.$$
$$\left. + (\chi_{12}\phi_2 + \chi_{1p}\phi_p)(\phi_2 + \phi_p) - \chi_2\frac{m_1}{m_2}\phi_2\phi_p\right] + 2\frac{V_m\gamma}{r_0}\phi_p^{1/3} \quad (57)$$

The full calculation requires a similar consideration of the chemical potential of monomer 2, along with appropriate differential equations for the incremental changes in the conversion of monomer to polymer with time [43].

Incidentally, the other components may also be added during the course of the reaction. One may wish to withhold some of the sodium dodecyl sulfate (in this case) at the beginning of the reaction to obtain larger particles (recalling that surfactant concentration at the beginning determines particle size by its effect on R_f). Then the remainder of the SDS is added in order to maintain stability of the particles as their surface area increases with growth. The rate of polymerization, the molecular weight, latex stability, rate of film formation on drying, and the rheological properties all depend upon particle size and size distribution. It may be desirable in other circumstances to employ an initiator which introduces different end groups on the polymer chains; or one may wish to operate at a temperature at which the initiator half-life is shorter than the time required for 100% conversion. In such cases it may be desirable to add initiator continuously.

3.1.3.1 Multistage emulsion polymerization

In Section 2.3, the consequences of adding different monomer(s) to a seed latex in a two-stage reaction were described, namely that phase separation was very likely to occur, since very few polymers are compatible with each other. As a result, and because the products can be commercially useful in many ways, multistage emulsion polymerization has become a highly developed art. Beyond what already has been said in the earlier section, it should be pointed out here that core/shell type morphologies may be imposed upon a system even in the absence of polymer incompatibility by 'monomer-starved' addition in subsequent stage(s). The second stage monomer is simply added at a rate which is less than that at which it is being polymerized.

Hoy and Bassett have described a technique in which there are, in a sense, an infinite number of stages. This is achieved by continuously changing monomer feed composition, by means of what they call 'power feed' [46]. They start with two (or more) monomer feed tanks, and during reaction they run the contents of one into the other, which in turn feeds the reactor. The apparatus is illustrated in Fig. 33. These authors define the term X as the ratio of the weight of monomer in the far tank to that in the near tank. The rate of change in co*monomer* (not copolymer) composition is determined by the value of X, as shown in Fig. 34.

CHAPTER 3 PRACTICAL APPLICATIONS OF EMULSION POLYMERIZATION 53

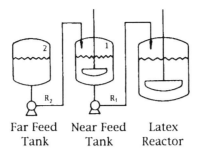

Fig. 33. Monomer tanks arranged for power feed [46].

What results is an 'onion skin' structure of the latex particles, with the outer layers richer in that monomer which is added mostly towards the end. When this is monomer rich in acrylic- or methacrylic acid, then the surface of the particles is hydrophilic and swollen more or less, depending upon the pH. Such a surface condition leads to enhanced film formation when the latexes are used in coatings.

3.1.4 Continuous emulsion polymerization

When product latex is being removed from the reactor at the same rate that all the reactants are being added, then one has a truly continuous reactor. There are two basic types: the continuous stirred tank reactor (CSTR) and the tubular reactor. Tubular reactors are generally unsatisfactory for industrial production because of the very long tube lengths required and the necessity

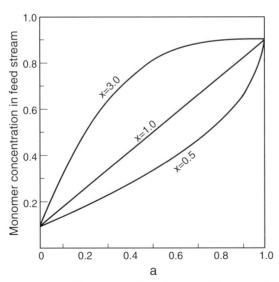

Fig. 34. Monomer composition as a function of dimensionless time for various values of X.

for 'plug flow.' A styrene/butadiene emulsion polymerization, for example may require 8–10 h reaction time, necessitating perhaps hundreds of feet of tube, all of which must be thermostatted. There is one notable exception, mentioned below, where a short tube feeds a CSTR.

3.1.4.1 Tubular reactors

In a tube there are two kinds of flow, laminar and turbulent, depending upon the velocity and viscosity of the fluid and the diameter of the tube. In laminar flow, fluid in the center of the tube flows faster than that near the walls, so that residence times in the reactor will be widely different, leading to an extremely broad distribution of conversions and other properties of the product. In turbulent flow, there is good mixing across the diameter of the tube, resulting in essentially plug flow, where every small 'plug' or volume of the reaction mixture spends the same amount of time in the reactor as all others. But turbulence occurs only at high flow velocities, as measured by the Reynold's number, $R = ru\rho/\eta$, where r is the tube radius, u is the velocity of flow, ρ is the fluid density and η is the viscosity. Turbulence usually occurs at $R > 2000$. It is possible to overcome the limitations of laminar flow by introducing gas bubbles at regular intervals, thereby isolating plugs of the reaction mixture. This can only be done, though, on laboratory scale where the diameter is sufficiently small so that the bubbles fill the entire cross-section of the tube. In either case the plugs can be considered as minute batch reactors, and batch reactor theory should be applied.

> In a tubular reactor introduce gas bubbles to isolate 'plugs' of reaction mixture.

3.1.4.2 Continuous stirred tank reactors

CSTRs are very different in the nature of the product they produce, even when starting with identical recipes. It should be noted, incidentally, that one may have a single CSTR or several in series, one feeding into the next. A typical laboratory set-up is shown in Fig. 35.

In order to understand what is going on in such a reactor, certain simplifying assumptions are made: (1) the contents are 'perfectly mixed' such that the inflowing stream is instantly mixed throughout the reactor; (2) all particles in the tank have the same probablility of leaving in the effluent within a time increment Δt. A consequence of this model, which fits reality reasonably well (although less well with very large reactors), is that the distribution of ages of the particles is very different from that in a batch reactor. In the latter case, all of the particles can be formed within a short time period, after which they all grow together, and end up about the same age, as shown in Figs 4 and 5.

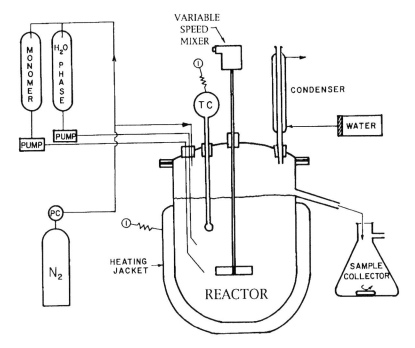

Fig. 35. A continuous stirred tank reactor.

> All properties of the latex will exhibit a double distribution due to stochastic effects of random entry and exit of radicals, and to the distribution of residence times.

In the CSTR it turns out that the particle age distribution (and the residence time distribution) is given by [47]

$$f(t) = \frac{1}{\theta} e^{-t/\theta} \tag{58}$$

where θ is the mean residence time, given by V/F in which V is the working volume of the reactor, F is the volumetric feed and effluent rate, and t is the time or age of a particle. This means that all properties of the latex, such as particle size, molecular weight, copolymer composition, degree of crosslinking and particle morphology will exhibit a double distribution – that due to the stochastic effects of random entry and exit of radicals, and that due to the distribution of residence times. If there is more than one reactor in series the distribution of residence times becomes narrower, as shown in Fig. 36.

Upon start-up, it will take some time, on the order of 3–5 residence times, before the system arrives at a steady state in which the composition of the effluent is time-invariant. In some situations this never occurs, but rather the system undergoes limit-cycle oscillations, as shown in Fig. 37.

This phenomenon is exhibited by more water-soluble monomers, such as vinyl acetate and methyl methacrylate [48]. It arises from the fact that particle

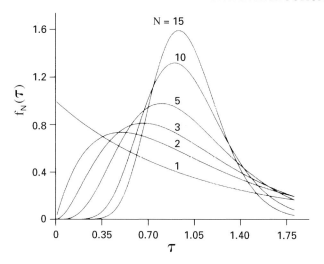

Fig. 36. Residence time distributions – CSTR series with N reactors [47].

nucleation and growth in the early stages can produce a large internal surface area which adsorbs the surfactant faster than it is being fed in. Then any new particles nucleated are unstable and heterocoagulate onto the pre-existing ones. The rate of surface area production then slows (since the surface/volume ratio of large particles is smaller than that of smaller ones), with the result that surfactant concentration in the medium, $[S]_w$ starts to increase. In the meantime particles are leaving the system and no new ones are being formed, with the result that the rate of polymerization decreases. When $[S]_w$ rises sufficiently, new primary particles are stabilized, and there is a burst of new particles formed; the rate of reaction increases – and so on.

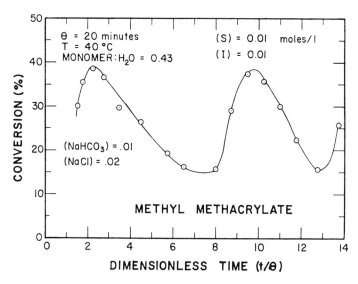

Fig. 37. Oscillations in MMA emulsion polymerization in a CSTR [47].

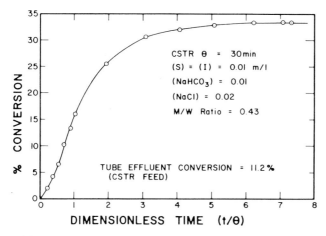

Fig. 38. Emulsion polymerization of MMA. Approach to steady state in tube–CSTR combined reactor system [47].

A simple solution to this problem is to provide a steady supply of pre-formed seed particles, either by adding seed latex as one of the feed streams or by hooking up a relatively short tube reactor to supply the seed continuously on line. Under these conditions the surface area is relatively constant, and the rate of surfactant supply is the same as that of its adsorption onto the particle surfaces. The result is illustrated in Fig. 38.

Besides the obvious advantages of continuous reactors for large-scale production, there are some disadvantages peculiar to polymer colloids. The most important is that colloids are intrinsically unstable systems, with a tendency to coagulate. Any coagulum which builds up on the reactor walls, stirrer blades or cooling coils will lead to clogging of the vessel and decreased heat transfer rates and ultimately to product which cannot meet specifications. Then the reactor must be shut down, cleaned, and upon start-up, all material formed prior to the steady state must be discarded because it, too, is off-specifications. The clean-up can be especially problematical in tubular reactors.

> It is especially desirable in continuous reactors to employ latex recipes which produce very clean product.

For this reason it is especially desirable in continuous reactors to employ latex recipes which produce very clean product. Glass-lined reactors are preferred over metal-lined ones, since there is a tendency for some latexes to deposit onto metal, presumably because of a sufficient concentration of polyvalent metal ions at the surface. Steric stabilizers (to be discussed later in this book) are especially good for producing clean, stable colloids. High shear rates of the stirrer may induce coagulation, especially at the periphery of the stirrer blades where the velocity is maximal. Good reactor design, including the use of baffles can overcome some of these problems, and still ensure good mixing.

References

42. Barrett, K.E.J. and Thompson, M.W. (1975). In *Dispersion Polymerization in Organic Media* (K.E.J. Barrett ed), p. 236, John Wiley, London.
43. Guillot, J. (1985). *Makromol. Chem.*, Suppl. **10/10**, 235.
44. Delgado, J., El-Aasser, M.S., Silebi, C.A., Vanderhoff, J.W. and Guillot, J. (1987). In *Future Directions in Polymer Colloids* (M.S. El-Aasser and R.M. Fitch eds), NATO ASI Series E, No. 138, p. 79, Martinus Nijhoff, Dordrecht, Netherlands.
45. Guyot, A., Guillot, J., Pichot, C. and Rios Guererro, L. (1981). In *Emulsion Polymers and Emulsion Polymerization* (D.R. Bassett and A.E. Hamielec eds), ACS Symposium Series 165, p. 415, American Chemical Society, Washington, DC.
46. Bassett, D.R. and Hoy, K.L. (1981). In *Emulsion Polymers and Emulsion Polymerization* (D.R. Bassett and A.E. Hamielec eds), ACS Symposium Series 165, p. 371, Americal Chemical Society, Washington, DC.
47. Poehlein, G.W. (1983). In *Science and Technology of Polymer Colloids* (G.W. Poehlein, R.H. Ottewill and J.W. Goodwin eds), NATO ASI Series E, No. 67, p. 112, Martinus Nijhoff, Dordrecht, Netherlands.
48. Greene, R.K., Gonzalez, R.A. and Poehlein, G.W. (1976). In *Emulsion Polymerization* (I. Piirma and J.L. Gardon eds), ACS Symposium Series 24, p. 341, American Chemical Society, Washington, DC.

Chapter 4

Nonaqueous Dispersions

4.1 General characteristics of nonaqueous dispersions

Nonaqueous dispersions (NADs) are simply polymer colloids in which the continuous medium is an organic liquid. The polymer, of course, must be insoluble in the medium, e.g. polystyrene/acrylic acid (87/13) in aliphatic hydrocarbon or polyoctyl methacrylate in methanol. In some cases the particles, instead of being solid polymer, are comprised of an aqueous solution of a water-soluble polymer. These generally are called 'inverse emulsions,' and provide a means for capitalizing on the advantages of polymer colloids and yet still be able to employ water-soluble polymers. A specific example is high-molecular weight polyacrylamide (PAM), which is used for flocculation of sewage solids in water treatment plants.

> Some dispersions are thermodynamically stable, with no tendency to coagulate, and therefore need no stabilizer.

Because organic liquids generally have low dielectric constants, most dissolved salts do not ionize well in them, with the consequence that ionogenic surfactants are not very effective in organic media. Nonionic surface active compounds, especially polymeric ones – so-called steric stabilizers – are excellent, however. These are discussed in Chapter 7. In some cases the dispersions are thermodynamically stable, with no tendency to coagulate, and therefore need no stabilizer. An example would be the well-known polyvinyl chloride (PVC) organosols, in which the polymer is semi-crystalline, and where the amorphous regions are soluble in the medium. Another example is the 'microgels' that are soluble in the medium, but because each particle is crosslinked, it retains its identity and spherical shape. In the PVC case above, the 'crosslinks' are the crystalline domains, which are insoluble in the medium at room temperature, but which can be rendered soluble by heating. If the medium is a plasticizer for PVC, and the colloid is poured into a mold and heated, upon cooling the system recrystallizes in the shape of the mold, to form 'rubber' dolls, pipes or wall coverings, to cite a few examples.

4.2 Nonaqueous emulsion polymerization

4.2.1 General characteristics

> Emulsion polymerization in organic media follows essentially the same rules as in water.

Emulsion polymerization in organic media follows essentially the same rules as in water, with some exceptions. Initiators should be organic-soluble, e.g. benzoyl peroxide or azobisisobutyronitrile (AIBN). Steric stabilizers are ordinarily amphiphilic block- or graft-copolymers in which one moiety is oil-soluble and the other will adsorb or 'anchor' well onto the particle surface, either by lyophobic forces or by chemical bonding. These may be pre-formed, purchased commercially, or they may be formed *in situ* during the early stages of the reaction. A more complete discussion of steric stabilization of colloids is given in Section 7.3. An excellent monograph on the subject of NADs is the book edited by Barrett [49].

4.2.2 Particle formation

Free radicals are generated and grow in the medium until they reach a critical size, j_{cr}, at which self-nucleation occurs or when oligomers are large enough to form particle nuclei by aggregation. Coagulation of primary particles was considered by Barrett to be unimportant generally in nonaqueous emulsion polymerization, although Kamath and Fitch found it to be the main mechanism of particle formation [50]. Final numbers of particles are strongly influenced by the initial concentration of stabilizer, as in aqueous systems. The basic relationship governing this behavior is just as in aqueous systems (see Eq. 12):

$$\left(\frac{dN}{dt}\right) = bR_{im} - R_c - R_f$$

where R_{im} signifies the rate of radical generation in the nonaqueous medium.

> The nature of the medium changes continuously throughout the reaction.

Generally the monomers will be totally soluble in the continuous medium, so that there are only two phases present during the emulsion polymerization: the medium and the monomer-swollen particles. This means, however, that the nature of the medium changes continuously throughout the reaction. Take for example the polymerization of methyl methacrylate in hexane. If all of the monomer is added at the beginning, the 'medium' may be a 50/50 solution of the ester MMA and the hydrocarbon, a fairly good solvent for the oligomers being formed, with the result that j_{cr} should be fairly large. This will lead to the formation of relatively few particles, and therefore large particle sizes. At high conversion, most of the MMA has been consumed and the medium becomes almost pure hexane, j_{cr} will be much smaller, and there will thus be an increasing tendency towards the nucleation of new particles at higher conversions.

4.2.3 Particle growth

> The ratio of rate constants varies almost continuously with the conversion.

Although the general theory of emulsion polymerization can be applied to NADs, as exemplified in the Smith–Ewart recursion formula, Eq. 41, there are differences from aqueous systems:

- The degree of swelling of the particles is considerably smaller because there is no separate monomer phase.
- As a result, the internal viscosity of the particles is high, with a consequently larger average number of radicals per particle, i.e. generally the kinetics will be in Smith–Ewart Case 3.
- This means that bulk kinetics can usually be applied, taking into account that the reaction is occurring only in the disperse phase.
- High internal viscosity also means that the ratio of rate constants, $k_p/k_t^{1/2}$, varies almost continuously with the conversion, as shown in Fig. 39, in which the 'factor' on the ordinate scale is this ratio relative to its initial value.

Monomers which are more polar than MMA, such as vinyl acetate and methyl acrylate, will have greater partition coefficients and thus greater swelling into the particles with consequently less pronounced acceleration.

The acceleration shown in Fig. 39 arises from the well-known Norrish–Tromsdorff effect in which the value of k_t is diffusion-controlled almost from the very start because of the high internal viscosity of the particles. Polymeric radicals within the particles diffuse extremely slowly, so that mutual termination by two such radicals becomes highly unlikely and $\bar{n}$ rises to values far beyond 1/2. At higher conversions the acceleration factor decreases as even the diffusion of monomer to the reactive sites becomes difficult, i.e. when k_p becomes diffusion-controlled. The beauty of emulsion polymerization is that even with such high rates of polymerization and high molecular weights, one can readily remove the large heat of reaction because of the low viscosity of the *medium*, unlike bulk or solution polymerizations.

4.2.4 Nonradical polymerization

> Ionic and step-growth condensation polymerizations can be carried out in NADs.

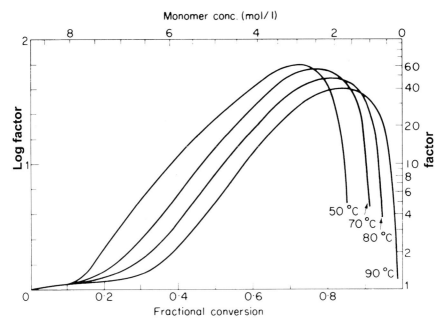

Fig. 39. Variation in relative rate factor with conversion for MMA bulk polymerization at various temperatures [51].

Because water can be totally excluded from NAD polymerizations, ionic and step-growth condensation polymerizations can also be carried out. For example, a rigorously dried solution of styrene in *n*-heptane when treated with butyl lithium undergoes 'living' anionic polymerization to give stereospecific polystyrene. The polymer is insoluble in the medium, so that again, at some critical degree of polymerization, j_{cr}, particles nucleate and need to be stabilized. This can be accomplished by the addition of an appropriate amphiphilic steric stabilizer which does not react with the highly reactive initiating or propagating ions. The stabilizer is often a block copolymer made *in situ* by anionic living polymerization [52].

The catalyst may even be insoluble in the medium, as in the case of Ziegler–Natta type reactions. For example, propylene can be polymerized in refined kerosene using a partially oxidized diethyl aluminum chloride/titanium trichloride catalyst. In one example the colloid stabilizer was a polypropylene-b-octene block copolymer made *in situ*. The particle size was 0.3 μm, and the polymer had a degree of crystallinity as high as 58% [53].

> In step-growth polymerization there is a very different particle nucleation condition from that in chain-growth.

In the case of 'condensation' polymers just about all of the known methods for their preparation have been adapted to form NADs. In step-growth polymerization, however, there is a very different particle nucleation condition from that in chain-growth: the oligomers are built up gradually throughout the reaction rather than by individual chains being initiated and growing rapidly in the presence of much monomer. In step-growth polymerization, for example, the average degree of polymerization, j, is just 2 at 50% conversion, 3 at 75% conversion, etc. Thus it may not be until 95% or higher conversion before j_{cr} is reached! Then particle nucleation may take place in a very short time interval. For example a solution of dimethyl terephthalate in refined gasoline was reacted with a emulsion of an aqueous solution of hexamethylene diamine in gasoline. Upon azeotropic distillation of water and methanol, a dispersion of polyhexamethylene terephthalamide was obtained [54].

An alternative route to condensation polymers involves a chain-growth mechanism of ring-opening reaction. As an example, ε-caprolactam has been polymerized with dibutyl zinc as catalyst in heptane with polylauryl methacrylate or poly(vinyl chloride-co-lauryl methacrylate) as stabilizer. Uniform particles of about 1 μm diameter were obtained [54].

4.2.5 Methods of preparation of NADs

4.2.5.1 PMMA latex in aliphatic hydrocarbon

The apparatus used for this preparation is shown in Fig. 30, Section 3.1.1. The stabilizer *precursor* is a 'macromer' which is a polylauryl methacrylate copolymer with a few pendant methacrylate double bonds. It is made in two stages: (1) a 97/3 lauryl methacrylate/glycidyl methacrylate copolymer is formed first by free radical (AIBN initiated) solution polymerization in aliphatic hydrocarbon. (2) This is followed by a ring-opening reaction of the pendant epoxy groups with methacrylic acid in the presence of N,N-dimethyl lauryl amine catalyst. The latex is then made in two stages: (i) The seed stage involves the copolymerization of MMA/MAA (98/2 w/w) with 22% of the macromer stabilizer precursor over a period of 45 min. (ii) This is followed a monomer feed of more 98/2 MMA/MAA with a small amount of octyl mercaptan chain transfer agent over 3 h. The final latex has a solids content of 52–55% and a modal particle size of 200 nm.

4.2.5.2 Polyethylene terephthalate (PET) latex in aliphatic hydrocarbon

Bis(2-hydroxyethyl) terephthalate is melted and emulsified into aliphatic hydrocarbon with a boiling temperature of 250°C using a steric stabilizer formed from poly(12-hydroxystearic acid)/MMA/GMA (glycidyl methacrylate) 100/90/10. The emulsion is then heated to distill out ethylene glycol and to bring about the polymerization. A final reflux temperature of 250°C is reached in about 2.5 h to achieve a dispersion of polymer of about 30% solids and with particles in the range of 2–20 μm diameter. (NB: strictly speaking this is a *dispersion*, or *suspension*, not 'emulsion,' polymerization because the reaction does not involve nucleation of particles, since they are all pre-formed in the initial emulsification. It is analogous to miniemulsion polymerization (Section 2.1.2.2), but the analogy is not perfect.)

4.3 Inverse emulsion polymerization

4.3.1 Polymerizing water-soluble monomers in dispersion

> It is desirable to bring the advantages of emulsion polymerization to water-soluble monomers.

There often arise occasions when it is desirable to bring the advantages of emulsion polymerization to water-soluble monomers. Many of these are crystalline solids, e.g. acrylamide, methacrylic acid, trimethyl-2-aminoethyl methacrylate quaternary ammonium chloride and sodium styrene sulfonate, which may not disperse easily in organic media to provide a supply of monomer to the latex particles. What can be done is to employ a concentrated aqueous solution of the monomer as the monomer reservoir, and in fact the final latex particles themselves exist as aqueous solutions of polymer dispersed in an organic liquid medium. The emulsifier used should be a water-in-oil type, such as sorbitan mono-oleate or -stearate. A schematic diagram of the various species existing during an inverse emulsion polymerization is presented in Fig. 40. The surfactant is ordinarily above the critical micelle concentration, CMC, so that micelles swollen with the aqueous monomer solution will be present along with droplets of the same solution. The latex particles during the reaction contain both monomer (diffusing in from the droplets) and polymer. Free radicals are generated in the continuous, organic phase, grow by polymerization, and form new particles (by heterogeneous nucleation) or enter pre-existing ones.

The first systematic study of these systems was published in 1962 by Vanderhoff and coworkers of the Dow Chemical Co. [56]. They showed that the basic Smith–Ewart recursion formula (Eq. 41) could be applied to describe the kinetics of the polymerization. The particles are on the order of 50 nm in diameter, and their internal viscosity is relatively low, allowing for rapid mutual termination of the free radicals. Therefore the average number of radicals per particle, $\bar{n}$, should be relatively small. Table 4.1 shows some of the results obtained by Vanderhoff *et al.* for sodium styrene sulfonate.

These data clearly show that the kinetics for the three lower temperatures are all in the Smith–Ewart Case 2 domain. As the temperature rises to 70°C and the initiation rate increases, α' becomes large enough so that the kinetics shift to the situation shown schematically in Fig. 21, Section 2.2.6 (see also Section 2.2.7 for α'). Polyacrylamide is made commercially in much the same way for use as a flocculant. The resulting latex appears like any other, i.e. it is milky-white and low to moderately high in viscosity, even though it may

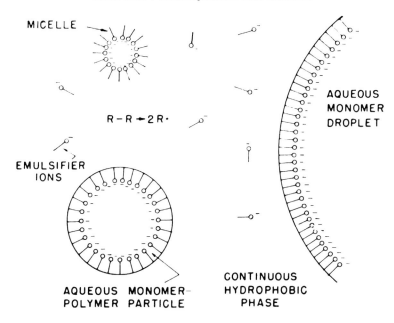

Fig. 40. Schematic representation of inverse emulsion polymerization [56].

contain polymer of a molecular weight of over one million daltons. The same polymer dissolved in water can produce a gel at 1% concentration!

4.3.2 Inverse *micro*emulsion polymerization

> Microemulsions are extremely dynamic, in that their structures are constantly fluctuating.

Microemulsions are different from ordinary emulsions in several ways. They are:

- thermodynamically stable, so that they can be formed by any order of mixing of their ingredients;
- inordinately small in particle size;

TABLE 4.1
Inverse emulsion polymerization kinetics of sodium styrene sulfonate [56]

Temp. (°C)	$\bar{n}$	k_p (10^3 M^{-1} s^{-1})	k_t (10^6 M^{-1} s^{-1})
40	0.50	2.77	1.10
50	0.50	3.54	2.05
60	0.53	5.04	3.53
70	0.61	7.47	6.37

- transparent;
- considered to be systems of swollen micelles; and
- extremely dynamic, in that their structures are constantly fluctuating. When we draw a picture of nicely spherical particles, it must be understood that these represent a time-average condition.

Inverse microemulsions are simply water-in-oil systems, and for our purposes the micelles contain an aqueous solution of a monomer. Since they are transparent, photoinitiation can be employed. This has the advantage that initiation can be stopped at any time, by turning off the light source, to observe relaxation kinetics. Alternatively, rotating sector or strobe techniques may be used to obtain radical lifetimes and absolute values of the rate constants [57]. To obtain such systems, relatively high amounts of emulsifier must be employed, usually along with a co-surfactant which in some cases is the monomer itself. The combination of the two can bring the aqueous/organic interfacial tension to negative values, leading to spontaneous emulsification.

> A j-mer may diffuse in and out of 10 to 1000 micelles before it adds a single monomer unit.

Particle formation in inverse microemulsion polymerization is by the heterogeneous nucleation of the swollen micelles because they are present in astronomical numbers, i.e. on the order of 10^{21} to 10^{22} particles per liter, so that even though they are very small, the value of $N_s r_s$ (number concentration and radius of the 'seed' micelles) is large enough so that the probability of oligomeric radical capture is essentially one (Section 2.1.2.1). It should be kept in mind that the dynamic nature of the micelles combined with the *relatively* slow rate of addition of monomer units means that a j-mer may diffuse in and out of 10 to 1000 micelles before it adds a single repeat monomer unit, depending upon the value of k_p! Thus there is plenty of time for the system to rearrange as polymerization progresses.

Thus once a micelle contains a 'permanent resident' propagating radical, it is regarded as a primary particle, and it grows by imbibing monomer and water from its surroundings by either of two mechanisms:

- diffusion through the organic medium; or
- collision with other swollen micelles.

By the end of the polymerization, there are still many micelles, but they are smaller because they have had their contents of monomer and water depleted. But the peculiar aspect of these systems is that even at the end of the reaction, the probability of particle nucleation for every new radical generated is still unity. This means that essentially each particle contains only one polymer molecule and that particle nucleation occurs continuously throughout the reaction, in complete contrast to the steady state picture of Smith and Ewart. These mechanisms are shown schematically in Fig. 41.

The evidence for this comes from the work of Candau, Leong and Fitch ('CLF'), who investigated acrylamide inverse microemulsion polymerization

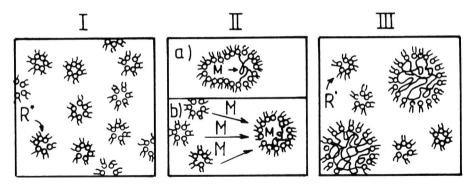

Fig. 41. Particle nucleation and growth in inverse microemulsion polymerization. (I) Initial condition: micelles (diameter ~6 nm). (II) Particle growth by (a) micellar collisions; (b) diffusion. (III) Final state: polymer (solution) in particles and empty micelles [59].

and analyzed the particle sizes of initial and final micelles and of latex particles, as well as the molecular weight of the polymer. From these they calculated how many particles and polymer molecules there were at the end of the reaction. This gave them approximately one polymer chain per latex particle, as shown in Table 4.2 [58]. The results raise the question of how the growing chains are terminated. Carver and coworkers found that chains were terminated by chain transfer to the medium when it had sufficient activity to do so, e.g. toluene, whereas with hexane or heptane fewer chains were formed in this way; with benzene as the medium, no chains were formed by transfer [59]. The propagating chain radical within the particle frequently will come into contact with the interfacial zone, rich in organic medium, as a result of its thermal motion. The ultimate fate of the free radicals was not determined.

> In some microemulsion polymerizations only one polymer chain exists per particle.

TABLE 4.2
Microemulsion polymerization of acrylamide [59]

Exp. No.	R_1 (Å)[a]	R_2 (Å)[a]	N_0 ($\times 10^{-21}$ L^{-1})	N ($\times 10^{-18}$ L^{-1})	N_m ($\times 10^{-21}$ L^{-1})	N_p	$\bar{n}$[b]
1	37	250	1.03	2.0	4.8	1.5	0.12
2	20	140	10.30	6.8	15	0.80	0.03
3	50	198	0.55	6.6	4.4	0.65	0.04
4	43	237	0.80	3.4	5.1	1.15	0.05

[a] 1 Å = 10^{-10} m.
[b] Calculated assuming constant N.
R_1, initial particle radius; R_2, final particle radius; N_0, initial number of micelles; N, final number of polymer particles; N_m, final number of micelles; N_p, number of polymer molecules per particle.

TABLE 4.3
Compared characteristics of polymerization in inverse emulsions and microemulsions

	Emulsion	Microemulsion
Mechanism		
Reaction time	Few hours	Few minutes
Particle nucleation	Early stage?	Continuous
Latex		
[E]	2–8%	>10%
Stability	Poor	High (thermodynamic?)
Aspect	Turbid	Transparent
Particle size	>100 nm	<100 nm
IP	> 2	~1.15–1.20
N_p	Few thousands	Very low (~1)
(Co)Polymer		
$\bar{M}_w$	High	High
Monomer sequence distribution	First-order Markov statistics	Bernouillian statistics

The values of $\bar{n}$, the time-average number of radicals per particle, calculated from the kinetics of the reaction (total number of radicals at a given time divided by the number of particles), has no physical significance under these circumstances. While it is polymerizing, each particle contains only one radical. Two consequences of the great number of particles are:

- the rate of reaction is very fast. For example, acrylamide can achieve 100% conversion in 10–12 min; and
- the molecular weight is extremely high, on the order of 1–10 million dalton.

Thus there are a number of distinctions between regular emulsion and microemulsion polymerizations. These are summarized in Table 4.3, in which the emulsifier concentration is represented by [E], weight-average molecular weight by $\bar{M}_w$, the number of polymer molecules per particle by N_p, and the induction period (in min) by IP (F. Candau, personal communication).

4.4 Emulsion polymerization in supercritical fluids

Supercritical fluids are substances which have been heated above the critical temperature under compression, so that they have the same density as the liquid, but have no meniscus. Their solvent qualities can be excellent, and these, in turn, depend upon the pressure and temperature (and therefore the density), so that for instance the Flory–Huggins χ-parameter can be fine-tuned by simply regulating the pressure. Carbon dioxide, in particular, has been found to be an excellent supercritical medium because it is cheap, it is a broad spectrum solvent, and it can be used under relatively mild conditions, e.g. $T < 100°C$ and $P < 350\,\text{bar}$ (1 bar = 10^5 Pa). Furthermore, CO_2 can be released into the atmosphere with relatively little environmental harm (it is, of course, not entirely benign since it is believed to be a greenhouse gas). Under these conditions, for example, MMA monomer is soluble but its polymer is not. Fluoropolymers and silicone polymers are soluble in supercritical CO_2.

> With an appropriate steric stabilizer NADs can be produced in supercritical CO_2.

Thus with an appropriate steric stabilizer NADs can be produced in this medium. For example, De Simone and coworkers found that PMMA particles of about $2\,\mu$m could be prepared using AIBN (azobis-iso-butyronitrile) as initiator and poly(1,1-dihydro perfluoro-octyl)acrylate as stabilizer [60, 61]. The stabilizer in this case acts like a graft copolymer in which the lyophobic backbone adsorbs onto the PMMA particle interface as it forms, while the perfluoro-octyl chains extend into the dispersing medium. The particles are larger than normal for emulsion polymerizations, so that the kinetics are undoubtedly in Smith–Ewart Case 3, although this was not investigated. It is likely that the large particle size arises from a large value for R_f in Eq. 12, and that the size can be reduced with a more efficient stabilizer, i.e. one with longer lyophilic chains and with perhaps more PMMA backbone repeat units per side chain (for a more complete discussion of steric stabilization, see Section 7.3). The final product, after extracting out unreacted AIBN and unbound stabilizer with CO_2, is obtained as a dry powder upon venting off the medium to the atmosphere or by capturing it for recycling.

There remains some question as to the advantages of supercritical media for these purposes as compared with, for instance, low-boiling aliphatic hydrocarbons which can be employed at atmospheric pressure and relatively low temperatures.

4.5 Applications

4.5.1 Automotive finishes

The automobile industry uses huge amounts of organic, polymeric finishes annually. These have to protect against corrosion and weathering as well as to have an attractive appearance in many colors. Polymethyl methacrylate and its copolymers have been found to supply these performance properties very well. Originally these were applied to the primed metal of automotive parts as lacquers, i.e. solutions of the polymer(s) dissolved in good, organic solvents. Because of the viscosity/molecular weight relationships, the solutions were at a concentration of 12–20% solids, and the $\bar{M}_w$ was fairly low. Thus 80–88% of the system was being evaporated into the atmosphere, creating serious pollution problems. By the addition of low viscosity–high $\bar{M}_w$ NADs of PMMA, the solids content could be greatly increased, and – as it turned out – some of the application and appearance properties were enhanced. NADs alone have poor coalescence of the particles upon evaporation of the medium, leaving a rough surface which is not glossy and considerable interstitial void volume which means the film is full of holes, and is therefore not providing good protection. The combination of dispersed particles and solution polymer eliminates these defects: the soluble polymer fills the interstices among particles throughout the film and by so doing also provides a smooth, glossy surface.

> The combination of dispersed particles and solution polymer eliminates film defects.

Incidentally, there is pressure being applied by governmental regulatory agencies to get rid of *all* volatile organic materials, with the result that increasing emphasis is being placed upon aqueous dispersions and powder coatings for automotive topcoats. Water-borne prime coats, made from alkyd emulsions which are electrodeposited, have been used for many years.

> Pressure is being applied by governmental regulatory agencies to get rid of *all* volatile organic materials.

4.5.2 Industrial coatings

The advantages of water, its low cost and environmentally desirable properties are offset by many disadvantages when compared with organic media:

- high latent heat of evaporation ($580 \, \text{cal} \, \text{g}^{-1}$*) compared with that of many organic liquids (generally less than $100 \, \text{cal} \, \text{g}^{-1}$);
- evaporation rate which depends upon relative humidity;
- high freezing temperature and its expansion upon freezing;
- a single boiling temperature compared with an almost infinite array available among organics; and
- emulsifiers and stabilizers used in aqueous systems impart water-sensitivity to paint films.

Thus a number of coatings applications continue to demand non-aqueous dispersions, but these are most likely to be used in industrial settings where the volatile organics can be condensed and recycled. Some coatings systems are combinations of soluble and dispersed polymers. The soluble polymer may carry crosslinkable functional groups, such that upon baking the film is thermoset.

> The soluble polymer may be crosslinkable, so that upon baking the film is thermoset.

---

* $1 \, \text{cal} = 4.184 \, \text{J}$; $1 \, \text{cal} \, \text{g}^{-1} = 4184 \, \text{J} \, \text{kg}^{-1}$.

Acrylic powders of PMMA and its copolymers can be prepared by nonaqueous emulsion polymerization using a continuous reactor combined with an evaporator/recycler [62]. Polystyrene powders have also been made by similar means. These, in turn, can be employed in powder coatings because the melt properties can be regulated by means of chain-transfer agents to give good film formation upon heating the polymer considerably above its T_g.

Other applications of NADs in coatings include adhesives, textile fiber coatings or 'sizes,' and as encapsulants for pigment particles.

4.5.3 Flocculants

> PAM in low concentrations causes 'bridging' among particles to form aggregates with relatively high sedimentation velocities.

The ultrahigh molar mass, water-soluble polymers such as polyacrylamide (PAM) are useful as flocculants. For example in sewage treatment, an extremely dilute dispersion of solids of fine particle size is to be purified by sedimentation of the solids. But without aggregation the particles are so small that they take too long to settle out. The addition of the PAM in low concentrations causes 'bridging' among particles to form aggregates with relatively high sedimentation velocities. The PAM molecules, because of their great length in solution, attach themselves to several sewage particles simultaneously. Elastic forces within the polymer chains tend to bring its various segments together, thus producing flocs, as shown in Fig. 42. If an attempt is

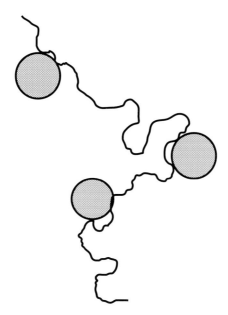

Fig. 42. Bridging flocculation of particles by high molar mass soluble polymer.

made to dissolve bulk polymer of, say, ten million dalton molar mass, it would take an extremely long time, even with heating and stirring. This is because as solvent penetrates the polymer, it forms a stiff gel with many internal chain entanglements. To form a solution all the chains must disentangle from each other, with very long reptation times. An inverse emulsion of PAM, on the other hand, can be destabilized by adding an appropriate surfactant, such that upon putting the NAD into water, it inverts, and the PAM dissolves rapidly. Because very low concentrations are required, the amount of oil phase is ordinarily negligible, for example in sewage treatment applications. The PAM dissolves easily in this case because (1) it is already in solution within the particles, although concentrated, and (2) the surface area offered by the inverse emulsion particles is enormous. Inverse *micro*emulsion polymers are even better for these purposes since they have few or only one polymer molecule per particle. They have a cost disadvantage because of their higher emulsifier content.

Other applications of flocculants include paper making and mineral processing. In the former case, cellulose fibers and clay particles are brought together by cationic polyelectrolytes, such polyvinyl benzyl trimethyl ammonium chloride and polydiallyl dimethyl ammonium chloride, both of which can be formed by nonaqueous emulsion polymerization.

References

49. Barrett, K.E.J. (ed) (1975). *Dispersion Polymerization in Organic Media*, John Wiley, London.
50. Fitch, R.M. and Kamath, Y.K. (1975). *J. Colloid Interface Sci.* **54**, 6.
51. Barrett, K.E.J. and Thomas, H.R. (1969). *J. Polym. Sci.*, A1, 7, 2630.
52. Firestone Tire and Rubber Co. (1965). British Patent 1 007 476.
53. Hercules Inc. (1969). British Patent 1 165 840.
54. Barrett, K.E.J. and Thompson, M.W. (1975). In *Dispersion Polymerization in Organic Media* (K.E.J. Barrett ed), p. 223, John Wiley, London.
55. Union Carbide Corp. (1972). U.S. Patent 3 632 669.
56. Vanderhoff, J.W., Tarkowski, H.L., Shaffer, J.B., Bradford, E.B. and Wiley, R.M. (1962). *Adv. Chem. Ser.* 34, 32. American Chemical Society, Washington, DC.
57. Carver, M.T., Dreyer, U., Knoesel, R., Candau, F. and Fitch, R.M. (1989). *J. Polym. Sci., Polym. Chem. Ed.* **27**, 2161.
58. Candau, F., Leong, Y.S. and Fitch, R.M. (1985). *J. Polym. Sci.* **23**, 193.
59. Carver, M.T., Hirsch, E., Wittmann, J.C., Fitch, R.M. and Candau, F. (1989). *J. Phys. Chem.* **93**, 4867.
60. DeSimone, J.M., Maury, E.E., Menceloglu, Y.Z., McLain, J.B., Romack, T.J. and Combes, J.R. (1994). *Science* **265** (5170), 326.
61. DeSimone, J.M., Maury, E.E., Combes, J.R. and Menceloglu, Y.Z. (1994). U.S. Patent 5 312 882.
62. I.C.I. (1971). British Patent 1 234 395.

Chapter 5

Characterization of Polymer Colloids

5.1 Light scattering

Colloids generally appear optically clear, hazy or translucent, or turbid depending upon the particle size and concentration, with larger sizes exhibiting greater opacity. Even apparently clear colloids will show a definite 'beam' when a collimated ray of light passes through them in a dark room – even when the colloids are very dilute. This is the so-called 'Tyndall Effect,' and is diagnostic for all colloids. Polymer colloids are no exception. The refractive index of the disperse phase must be different from that of the medium in order for these optical effects to be seen; otherwise the colloids are truly transparent regardless of particle size and concentration.

5.1.1 Rayleigh theory

As the light ray enters the colloid its oscillating electrical field induces dipoles in the molecules of the particles and causes them to oscillate. An oscillating dipole radiates electromagnetic waves in all directions except in the direction of its axis, i.e. one cannot 'see' that a dipole is oscillating when looking at it end on. The intensity of this secondary radiation, the scattering, is a strong function of the incoming frequency, with the result that if white light enters the colloid, blue wavelengths tend to be scattered and red ones tend to be transmitted. Lord Rayleigh was the first to develop a quantitative theory for light scattering of dilute dispersions of particles that are small relative to the wavelength of the incident light (diameter $c. < \lambda/10$, where λ is the wavelength). The particles must have no color (nonabsorbing) and must not be electrically conducting. The theory explained why the sky is blue and why the sun is yellow in the sky and red on the horizon. This scattering of light is also the reason why smoke and haze and small particle size polymer colloids all have a bluish cast.

A brief synopsis of Rayleigh's theory [63] is as follows:

Let i be the intensity of the scattered radiation. Then i is proportional to the kinetic energy of the moving electronic charge in the oscillating dipole:

$$i \propto \tfrac{1}{2}mv^2$$

where m is the equivalent mass and v is the velocity, so that the intensity is proportional to the square of the velocity of the moving electronic charge. The electric field, E, is equal to the amplitude of oscillation, A, and if A is doubled at constant wavelength, v is doubled.

$$\therefore A \propto v, \quad \text{and} \quad i \propto v^2 \propto A^2 \propto E^2$$

This applies to a moving electrical field. For an oscillating field:

$$i \propto E^2 = E_0^2 \cos 2\pi(\nu t - x/\lambda) \tag{59}$$

where ν is the frequency of the radiation, x/λ is the distance in multiples of the wavelength and E_0 is the electrical field at time, $t = 0$. Equation 59 translates into Eq. 60 for a so-called 'forced oscillator,' one in which the oscillation is induced by an external field:

$$i_s = \left(\frac{16\pi^4 a^2}{\lambda^4 r^2}\right) \cos^2 2\pi(\nu t - r/\lambda) \sin^2 \theta_z \tag{60}$$

where a is the particle radius, r is the distance from the particle and θ is the angle of observation relative to the transmitted beam. Equation 60 shows the inverse fourth power dependence of the scattered intensity upon the wavelength of the incident light and the dependence on the inverse square of the distance, because of radiation in all directions, in ever widening circles.

Now we take a large number, N (dm^{-3}), of independent scatterers (dilute enough so that the scattered light from one particle is not re-scattered by another, so-called multiple scattering) with a refractive index n suspended in a medium of refractive index n_0, and observe the time-averaged intensity of scattered light at various angles, θ. What we observe is shown in Fig. 43, where the plane of polarization of the incident beam is indicated by the double arrow on the left, and the induced oscillation by the double arrow in the center (in the same plane). The scattering intensity is given by the distance from the center by the solid radius at the angle θ. The figure gives the results of Eq. 60 averaged over time. For example, if the incident radiation is unpolarized (a & b combined in Fig. 43), the intensity of scattered radiation is the same at all angles of observation. However if the incident beam is polarized, then the scattering intensity is uniform in the plane perpendicular to the plane of polarization, while in the plane of polarization it has zero intensity at $\theta = 90°$, and maximum intensity at $0°$ and $180°$.

If one takes a highly diluted, small particle size latex in a cylindrical cell (a beaker will do) into a darkened room and illuminates it with a collimated light beam, and either observes through a polarizing filter (a pair of polarizing sunglasses can serve well) or polarizes the incident beam, one can easily observe these effects. It is especially interesting that with unpolarized incident light and isotropic, spherical particles randomly dispersed, one obtains polarized scattered light! The same is true of the blue sky: the light scattered from it is polarized, especially at a $90°$ angle from the sun.

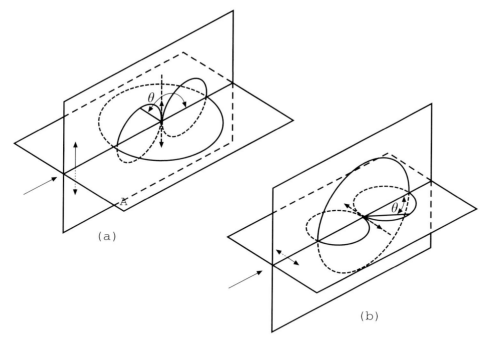

Fig. 43. Rayleigh scattering: angular dependence of the intensity of scattered light as a function of angle θ (see text for details).

> The Rayleigh scattering intensity has a sixth power dependency on the particle size.

For quantitative work what is commonly done is to take the ratio of intensities at 90° and 0°, the so-called Rayleigh ratio, R_{90}. This eliminates certain geometrical factors related to the scattering volume, the size of the incident beam and the cell path length. One then obtains for the system described above:

$$R_{90,\text{uv}} = 9\pi^2 \left(\frac{n_0}{\lambda_0}\right)^4 \left(\frac{m^2 - 1}{m^2 + 2}\right) \sum_i N_i v_i^2 \qquad (61)$$

$$R_{90,\text{uv}} \propto a^6$$

where the subscript uv refers to unpolarized incident light and vertically polarized scattered light; $m = n/n_0$, the ratio of polymer and solvent refractive indexes; λ_0 is the vacuum wavelength of the incident light; and N is the number concentration of particles of volume v. Note that the Rayleigh scattering intensity has a very strong, sixth power dependency on the particle size. One practical consequence of this is that dust particles strongly interfere with light scattering measurements, and great care is needed to filter out dust.

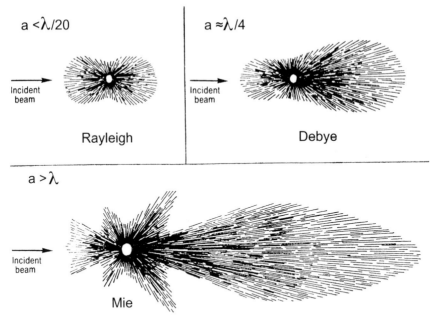

Fig. 44. Angular dependence of light scattering intensity from a single particle for various particle size domains [64].

As the particles become somewhat larger than $\lambda/20$, the scattering intensity begins to depend upon the angle of observation. This is shown in Fig. 44 for various particle size domains, where the incident radiation is plane polarized. As the particle diameter approaches the wavelength of light, there is more scattering in the forward direction. This is often referred to as the Rayleigh–Debye (or sometimes as the Debye–Gans) region and the theory is due to an extension of the Rayleigh theory by Peter Debye.

5.1.2 Rayleigh–Debye theory [65]

> At larger particle sizes there appear maxima and minima in the intensity at various angles.

In the pure Rayleigh theory the particles are so small that all the molecules within them act together as a single induced oscillating dipole. In larger particles there is sufficient distance between dipoles so that an the incident wave crest, for example, will reach one before the other with the result that the two dipoles will oscillate out of synchrony. Thus their respective scattered waves will be out of phase and interfere with each other, as shown in Fig. 45. The degree to which they are out of phase will depend upon the angle θ. For particles in the Rayleigh–Debye range it turns out that reinforcement of the scattered waves occurs in the forward directions, while there is diminution at all other angles. At still larger sizes forward scattering still has the highest

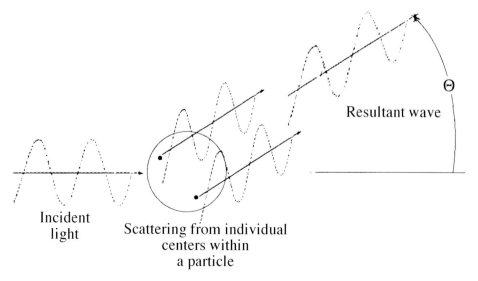

Fig. 45. Angular dependence of scattered light intensity results from interference of light waves from different parts of the same particle.

intensity, but now there appear maxima and minima in the intensity as one observes at various angles. This is usually referred to as the Mie region (to be discussed below) after the man who developed the comprehensive theory for particles of all sizes, both absorbing and nonabsorbing.

To obtain the angular dependence of the scattering intensity, one defines the 'scattering vector,' Q, as

$$Q \equiv \frac{4\pi n_0}{\lambda_0} \sin\frac{\theta}{2}$$

where λ_0 is the vacuum wavelength of the light and n_0 is the refractive index of the medium. The ratio $\lambda_0/n_0 = \lambda$ is the wavelength in the medium. The scattering intensity for a monodisperse colloid ($\sum_i N_i v_i^2 = Nv^2$), from Eq. 61, becomes

$$I(Q) = KNv^2 \times P(Q)$$

$$K \equiv 9\pi^2 \left(\frac{n_0}{\lambda_0}\right)^4 \left(\frac{m^2-1}{m^2+2}\right) \qquad (62)$$

where $P(Q)$ is the so-called 'form factor' which gives the dependence on particle size and angle of observation as follows:

$$P(Q) \equiv \left[\frac{3(\sin Qa - Qa\cos Qa)}{(Qa)^3}\right]^2$$

where a is the particle radius. A plot of the experimental equivalent of Eq. 62 for a monodisperse polystyrene latex with a particle radius $a = 200$ nm is

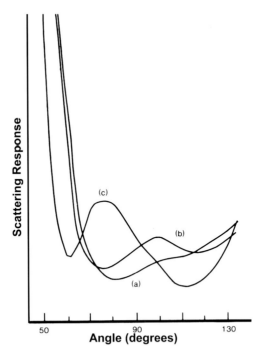

Fig. 46. Angular dependence of light scattering intensity for polystyrene latex ($a = 200$ nm). (a) Pure PS latex; (b) swollen with THF for 10 min; (c) swollen with THF for 167 min. [66].

shown in Fig. 46, curve (a) in which there is a minimum at 80° and the hint of a maximum at 100°.

Upon adding tetrahydrofuran (THF), the particles swell, the peaks shift and both maxima and minima become more pronounced as a result of two factors: (1) the size of the particles increases (to 233 and 280 nm in (b) and (c), respectively) and (2) the refractive index changes. These curves were taken from a study of the rate of swelling of polystyrene by various solvents, which showed light scattering to be an effective method for obtaining data on the diffusion of small molecules across the polymer/water interface and into polymers [66]. The method is nondestructive and instantaneous, which for kinetic studies is essential.

5.1.3 Mie theory

The general theory of light scattering which applies to all colloids of both light-absorbing and nonabsorbing particles of any size and shape is referred to as the Lorenz–Mie theory. It states that for vertically polarized incident and scattered radiation (V_v scattering):

$$I_v = \left[\frac{\lambda^2}{4\pi^2 r^2}\right] \left\{\sum_{n=1}^{\infty} \frac{2n+1}{n(n+1)} [a_n \pi_n + b_n \tau_n]\right\}^2 \tag{63}$$

where the coefficients a_n and b_n involve Ricatti–Bessel functions and their derivatives, while π_n and τ_n are derived from Legendre polynomials and their derivatives. Although these represent a series of complicated equations, computer programs are available which allow for easy computation of any colloidal situation [67].

Because of the richness of data obtainable from angular dependence measurements it is possible in this size range to obtain both size and refractive index of the particles. The latter gives information about the composition of the polymer, if it is not known.

An example of the scattering patterns which are exhibited by monodisperse colloids is shown in Fig. 47, in which the log of the intensity, I (vertical axis) is plotted against scattering angle, θ, and the size parameter $\alpha = 2\pi a/\lambda$. It can be seen that the smaller sizes show no peaks and valleys, but definite forward scattering. As the size increases, so does the complexity in the angular dependence of I, and the forward scattering increases by almost four orders of magnitude! This can be visualized by rotating Fig. 47 through 180° to look at its 'backside,' as seen in Fig. 48. The inversion of scattering data in these regions to obtain particle size and refractive index is not trivial, but computer programs are available to accomplish the task [67].

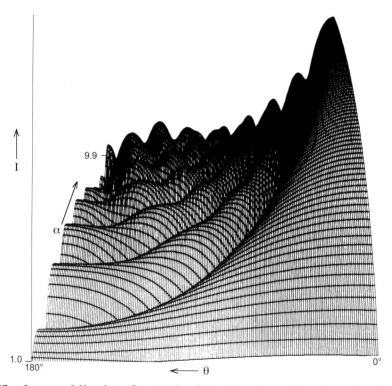

Fig. 47. Lorenz–Mie plot of scattering intensity, I, as a function of angle θ and size parameter α for $m = 1.200$ (M.L. Pruitt, R.N. Rowell, J.R. Ford and R.L. Rowell, unpublished work).

80 POLYMER COLLOIDS: A COMPREHENSIVE INTRODUCTION

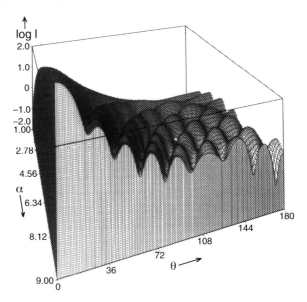

Fig. 48. Diagram of Fig. 47 rotated through 180° (M.L. Pruitt, R.N. Rowell, J.R. Ford and R.L. Rowell, unpublished work).

5.1.4 Higher order Tyndall spectra (HOTS)

> HOTS can be used as a good approximate method for determining particle size.

The angular dependence of light scattering is a function of the refractive index of the polymer, n, in the particles, as given by Eqs 61, 62 and 63. The refractive index in turn is usually different for different wavelengths of light. In the Mie region, then, a peak position for red is at a slightly different angle θ from that for green, etc. The result is that incoming white light is scattered so as to produce beautiful colors at various angles, which are readily seen in very dilute, monodisperse colloids in a darkened room with a collimated beam of white light. Thus HOTS can be used as an approximate method for determining particle size, almost at a glance. For most common polymers the rule is that the particle size is 220 × (number of red bands) nm in diameter.

Experimentally, one adds a fraction of a drop of latex to c. 200 ml of *clear*, distilled water in a *clean* beaker and, starting at $\theta = 0°$, one walks around to higher angles, counting the number of red bands in the scattering plane. With experience it is possible to estimate fractional numbers of red bands from their angular position. As particle size increases, the bands shift to higher θ. Thus a latex with 2.5 red bands must be quite uniform in size and have a diameter of $2.5 \times 220 = 550$ nm. One caveat is that HOTS can be observed sometimes even when the colloid is not monodisperse, for example either with a bimodal system in which each size is narrowly distributed, or where a population of Rayleigh-region particles is present along with monodisperse larger particles.

5.1.4.1 Polydispersity effects

So far we have been discussing the behavior of monodisperse colloids, i.e. those with very narrow particle size distributions. Polydispersity has no effect on the shape of the intensity distribution in the Rayleigh region, as long as all particles are within the appropriate size range. In the Debye region the general features will remain but the angular dependence of the scattering intensity will reflect an average of all the sizes with the largest ones strongly dominating because of the sixth power dependence. In the Mie region there will be a smoothing out of the peaks.

In these cases an average particle size is obtained which is the mass-average, $\bar{a}_w$, in the Rayleigh region and the number average, $\bar{a}_n$, in the Mie region:

$$\bar{a}_w = \left(\frac{3 \sum_i N_i v_i^2}{4\pi \sum_i N_i v_i} \right)^{1/3} \quad \text{and} \quad \bar{a}_n = \left(\frac{3 \sum_i N_i v_i}{4\pi \sum_i N_i} \right)^{1/3} \quad (64)$$

For a general reference on light scattering, see the book by Kerker [68].

5.1.5 Experimental

An example of an apparatus for obtaining time-average, angular dependence of light scattering intensity is shown in Fig. 49. It is comprised of a black box – literally – with a port to admit light from the source. If the source is not a laser, then a frequency filter or diffraction grating is used to select a narrow wavelength

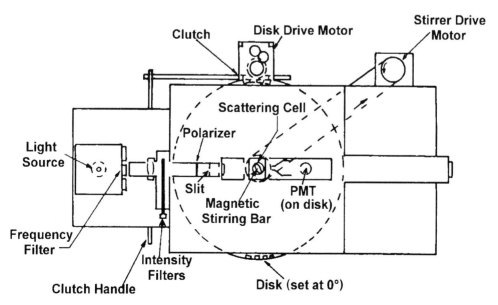

Fig. 49. Apparatus for obtaining time-average, angular dependence of light scattering.

of incident light. The photomultiplier (PMT) detector is mounted on a disk which can be rotated to any angle from 0° to 180°. There is an automatic cutoff at very low angles to prevent the transmitted beam from damaging the detector.

In this particular apparatus provision has been made for stirring the contents of the cell by means of a drive magnet situated below the box. This is so that reagents can be mixed, for instance, in determining the kinetics of swelling of polymer particles by solvents or the kinetics of coagulation following the addition of salt. A drive motor also has been attached to the edge of the disk so that the scattering angle can be scanned automatically, with the output from the PMT going onto a strip-chart recorder, digital oscilloscope or into a computer for data analysis.

The incident beam intensity is regulated by the insertion of neutral filters. It can also be polarized by the insertion of an appropriately oriented polarizing filter. The beam passes through a slit to define a scattering volume within the cell in conjunction with the solid angle subtended by the detector. This is required when making measurements of the absolute intensity of the scattered light.

5.1.6 Dynamic light scattering

Whereas 'static' light scattering, discussed in Sections 5.1.1–5.1.5 above, is time-independent, there are effects which can be studied which vary with time. These have been called 'quasi-elastic light scattering,' QELS, 'photon correlation spectroscopy,' PCS, and more simply, dynamic light scattering, DLS.

If one uses a monochromatic incident light beam and analyzes the *spectrum* of the scattered radiation, it is seen to be line-broadened. This comes about because of the thermal motions of the scattering particles which cause Doppler shifting of the light to higher or lower frequencies, depending upon whether the particle is moving towards or away from the detector at the moment of observation. Thermal fluctuations of the solvent molecules also lead to line-broadening, but these effects are easily separable from those due to the particles because of the large difference in velocities between molecules and colloidal particles.

The line shape is Lorentzian and can be expressed as

$$I(Q,\omega) = \frac{ADQ^2}{\omega^2 + D^2Q^4} \tag{65}$$

where

$$Q \equiv (4\pi n_0/\lambda_0)\sin(\theta/2)$$

and where Q is the same as that in Eq. 62, ω is the frequency shift, A is a constant and D is the Fickian diffusion coefficient of the particle. Theoretically one can obtain the value of D directly from the half width at half height of the scattered beam:

$$\omega_{1/2} = DQ^2 \tag{66}$$

The hydrodynamic radius, assuming spherical symmetry – otherwise the *equivalent* hydrodynamic radius – is obtained directly from the diffusion coefficient by means of the Stokes–Einstein equation:

$$a_h = \frac{k_B T}{6\pi\eta D} \tag{67}$$

The diffusion constant D is affected by any solvent the particle carries along with it, especially in the case where there may be a hairy layer of steric stabilizer or a solvent-swollen shell on the particles. This can be a valuable tool to measure the approximate thickness of such layers by comparison of the a_h values for bare particles and those covered by a lyophilic layer of polymer. Caution must be exercised in interpreting the results, however, since the exact position of the shear plane in both cases is not known; it does not necessarily correspond to δ_h. Figure 50 represents schematically the difference in surface layer thickness, δ_h, between two particles, one with surface ions and the other with surface polymer, both carrying some solvent molecules inside of the shear plane. The core particles are the same size, but they have different hydrodynamic radii.

Experimentally one ordinarily uses photon counting and correlates the average intensities at various time intervals, τ, with a pulse amplifier–discriminator. Equation 65 gives the intensity of the scattered light as a function of its frequency, and is thus often referred to as the power spectrum. From the power spectrum one can obtain the real time-average of the product $I(t) \times I(t+\tau)$.

The derived correlation function, $g(\tau)$, is directly related to the diffusion coefficient of the particles:

$$g(\tau) = 1 + \exp(-2DQ^2\tau) \tag{68}$$

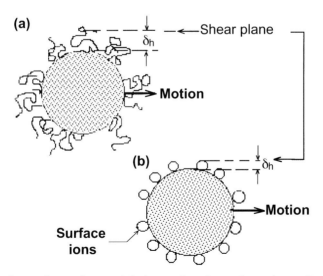

Fig. 50. The shear plane of a particle in motion depends on its surface composition: (a) surface polymer layer, large δ_h; (b) surface ionic layer, small δ_h.

where

$$Q = (4\pi n_0/\lambda_0) \sin(\theta/2)$$

It is possible to purchase apparatus commercially which not only gathers dynamic light scattering data in this fashion, but also calculates the autocorrelation function, the diffusion coefficient, and then from Eq. 67 obtains the hydrodynamic particle size. It must be noted again that this discussion relates to monodisperse systems. Calculation of the breadth of the particle size distribution can be made from the slope of the $g(\tau)$ curve. It is claimed that one can even obtain bimodal size distributions with a fair degree of accuracy from the changes in the slope as a function of $t + \tau$.

> A cloud of counterions, along with their solvation spheres, moves with the particle as it diffuses about.

Polymer particles carrying an electrical surface charge, ordinarily due to ionic end groups on the polymer chains, usually hundreds to thousands in number, must maintain electroneutrality by being surrounded by a diffuse cloud of an equal number of counterions (see Section 7.2.2 below for a more complete discussion of the electrical double layer). This cloud of counterions, along with their solvation spheres, moves with the particle as it diffuses about in its Brownian motion. Most of this baggage will thus be within the shear plane, so that the hydrodynamic radius will be much larger than that of the core particle.

Adding an inert salt to the colloid screens the counterions from the surface charges, greatly compressing the diffuse part of the electrical double layer and thus largely eliminating this effect. However adding salt decreases the stability of the latex such that aggregates may begin to form (see Section 7.5). It is possible to check for this by monitoring the *time-average intensity* of the scattered light, recalling the sixth power dependence on particle radius (Eq. 61).

> The diffusion coefficient is dependent upon the concentration of the colloid.

The diffusion coefficient is also dependent upon the concentration of the colloid, since the motion of one particle can affect that of its neighbors. The effect may originate from electrodynamic or hydrodynamic interactions or from both. The easiest way to obviate this concern is to determine the a_h (Eq. 67) as a function of concentration and extrapolate to zero concentration. In other cases it may be that the cooperative diffusion coefficient is the one desired, e.g. to correlate with rheological behavior.

So there are several caveats to be aware of in performing light scattering experiments, and especially in interpreting the data. Commercial dynamic light scattering instruments often give a single value 'particle size' which may or may not be the dimension desired. On the other hand, intelligent use of data, especially the combination of both time-average and dynamic light scattering data, can provide a rich source of information not only about the core particle, but also about the nature of the particle surface and its associated polymeric or electrical double layer. For a general reference on dynamic light scattering, see the book by Berne and Pecora [69].

5.2 Neutron scattering

5.2.1 General considerations

A beam of neutrons exhibits wave-like behavior and thus can be diffracted and scattered, but without electrical interactions. The particles do have a magnetic moment as a result of a nuclear spin of 1/2, so that they interact with the nuclei of atoms. The scattering behavior is otherwise very similar to that of light, and the same types of equations can be used to describe it. Because of the high energies of nuclear particles, the wavelengths of neutrons are two to three orders of magnitude smaller than those of light, from 1 to 5 Å (0.1 to 0.5 nm) for thermal, and 5 to 20 Å (0.5 to 2.0 nm) for cold neutrons, so that they may be used to observe structures on a much smaller scale than that for light.

There can be both elastic and inelastic scattering. In the latter case the scattered neutrons have lost momentum relative to those of the incident beam. The scattering vector, $\mathbf{Q}$, in this case measures the loss in momentum:

$$\mathbf{Q} \equiv \frac{1}{\hbar}(m\mathbf{v}_0 - m\mathbf{v}) \tag{69}$$

where $\hbar$ is Planck's constant divided by 2π, m is the neutron mass and $\mathbf{v}$, its velocity. For elastic scattering there is no loss in momentum ($\mathbf{v} = \mathbf{v}_0$), but scattering occurs at an angle θ, and the magnitude of $\mathbf{Q}$ becomes:

$$Q \equiv \frac{4\pi}{\lambda} \sin\frac{\theta}{2} \tag{70}$$

in direct analogy to the scattering vector for light given in Eq. 65. The wavelength, λ, is related to the momentum through the de Broglie relationship:

$$\lambda = \frac{2\pi\hbar}{mv} \tag{71}$$

5.2.2 Coherent scattering length

The equivalent of the refractive index is the so-called coherent scattering length, b. It is related to the scattering cross-section, σ_{coh}, of the nuclei which are interacting with the incoming neutron waves:

$$\sigma_{coh} \equiv 4\pi b^2 \tag{72}$$

TABLE 5.1
Coherent scattering lengths for selected isotopes

Isotope	Symbol	b (10^{-12} cm)
Hydrogen	^{1}H	−0.374
Deuterium	^{2}H	0.667
Carbon	^{12}C	0.665
Oxygen	^{16}O	0.580
Chlorine	^{35}Cl	1.18
Sulfur	^{32}S	0.280

The incoherent scattering appears as background noise, and must be subtracted from the total scattering intensity. Tables of b-values have been published for many isotopes. Some values of interest are given in Table 5.1.

5.2.3 Coherent neutron scattering length density

To calculate the scattered intensity, it is the coherent neutron scattering length *density*, ρ, of a molecule which is used:

$$\rho \equiv \sum_i \frac{b_i}{V} \qquad (73)$$

where V is the molecular volume. So, for example the value of ρ for water would be calculated as:

$$\rho = \frac{[2 \times (-0.374) + 1 \times 0.580] \times 10^{-12}\,\text{cm}}{18\,\text{g\,mol}^{-1}/(1\,\text{g\,cm}^{-3} \times 6.02 \times 10^{23}\,\text{mol}^{-1})} = -5.6 \times 10^9\,\text{cm}^{-2}$$

Values for this and other molecules are given in Table 5.2.

TABLE 5.2
Coherent scattering length densities for selected molecules

Molecule	Molecular formula	ρ (cm^{-2})
Water	H_2O	−0.56
'Heavy' Water	D_2O	6.40
Dodecane	$C_{12}H_{26}$	−0.46
Deuterododecane	$C_{12}D_{26}$	6.71
Polystyrene	$[C_8H_8]_n$	1.42
Deuteropolystyrene	$[C_8D_8]_n$	6.47
Polymethylmethacrylate	$[C_5H_8O_2]_n$	1.07

5.2.4 Angular dependence of neutron scattering intensity

The angular dependence of neutron scattering intensity can now be written down in direct analogy, again, to light scattering (Eq. 62). The condition here, in Eq. 74, is that the system is dilute enough to avoid interparticle interactions:

$$I(Q) = A(\rho_p - \rho_m)^2 N v^2 \times P(Q) \tag{74}$$

where

$$P(Q) \equiv \left[\frac{3(\sin QR - QR\cos QR)}{(QR)^3}\right]^2$$

The subscripts p and m on the coherent scattering length densities correspond to the particles and medium, respectively; N and v are the number concentration and volume of the particles; A is an instrumental constant. The quantity $A(\rho_p - \rho_m)^2$ corresponds to the refractive index term, K, in Eq. 62.

5.2.5 Neutron scattering experiments

A plot of $I(Q)$ as a function of Q shows the richness of data obtainable, and the great sensitivity to particle size, as seen in Fig. 51. It is interesting to note that the range of Q (0–0.35) corresponds to angles of $\theta = 0°$ to $c.\,3.5°$, i.e. these data were taken at very low angles. The technique is thus often referred to as small angle neutron scattering, or SANS.

The experimental conditions typically employed are $\lambda = 0.3$–1.0 nm; distance between the sample cell and the detector is 5–38 m; and the detector is a two-dimensional array of sensors. The signal is radially averaged, and then the

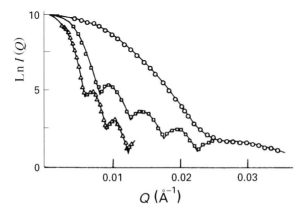

Fig. 51. Angular dependence of neutron scattering intensity for monodisperse polystyrene colloids. Diameters: ○, 34.6 nm; □, 132 nm; △, 202 nm [70].

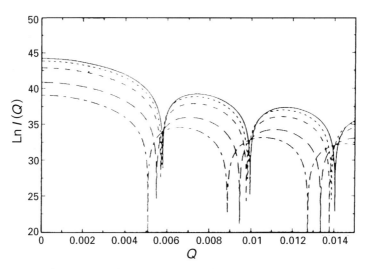

Fig. 52. Theoretical SANS intensity curves for: PS/deuteroPS core/shell latexes in D$_2$O/H$_2$O mixtures: ···, 20%; – –, 40%; — —, 60%; — – —, 65% D$_2$O; and —, equivalent PS latex [71].

background scattering from the pure medium is subtracted from that of the colloid.

The experimental curves in Fig. 51 may be compared with one calculated from Eq. 74. This is shown as the solid curve in Fig. 52. Although the coordinates of the two figures are not the same, it is clear that the minima in the experimental curves are not as sharp and deep as in the theoretical one.

The reasons for this are either or any of the following: (a) polydispersity of the particle size; (b) some spread in the wavelength of the source; (c) finite size of the detector elements and (d) the angular divergence of the incident beam. It is important in interpreting experimental data to determine the relative contributions of these factors. It is also possible by computer modelling and curve fitting to determine the particle size distribution of a colloid from the degree of spreading of the minima.

5.2.5.1 Core/shell particle morphologies

The other curves in Fig. 52 are calculated for core/shell particles with a radius of 80 nm in which the core is a 67 nm radius polystyrene latex and the shell is deuteropolystyrene. From Table 5.2 it can be seen that the core and shell polymers have very different coherent scattering length densities, ρ_p. If the ρ_m value of the medium matches that of the shell ('contrast matching'), as in Fig. 53b, then only the core can be 'seen' in a neutron scattering experiment. On the other hand, if the ρ_m of the medium matches that of the core, as in Fig. 53c, then only the hollow shell can be observed. The scattering spectra in these two cases will be very different because of the difference in the form factors, $P(Q)$. A homogeneous particle, on the other hand, displays only a

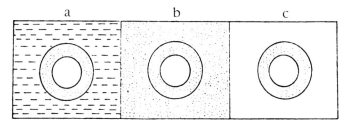

Fig. 53. A core/shell particle under various contrast-matching conditions: (a) no match; (b) medium and shell matched; (c) medium and core matched [70].

change in *intensity* as the 'contrast' of the medium is varied; the *shape* of the $I(Q)$ vs. Q curve does not change.

5.2.5.2 Contrast matching technique

A powerful experimental technique, then, is to systematically vary the contrast of the medium – in this case, a mixture of H_2O and D_2O. Subsequently, fitting of the experimental curves, taking into account the spreading of the theoretical curves as discussed above, provides detailed information about the internal morphology of the particles [70]. Two important results have been obtained from the work of Mills *et al.* [71]:

(1) Equilibrium swelling of latex seed particles gives essentially radially uniform concentration of monomer or solvent. There is no observable 'repulsive wall effect' near the particle surface that would lead to a higher concentration of monomer near the surface as proposed by previous workers, and

(2) Even with uniform monomer swelling of seed particles, one can obtain a core/shell morphology because of 'surface anchoring' of polymer chain ends. This arises from the fact that hydrophilic (usually ionic) free radicals enter from the continuous, aqueous phase, causing most Stage II polymerization to occur near the surface. If the polymer chain dimensions so formed are substantially less than the particle radius, then they will tend to be concentrated near the surface forming a shell.

5.2.6 The structure factor in concentrated polymer colloids

As colloidal particles come closer together as a result of higher concentrations, they begin to interact because of the overlap of their electrical double layers or their steric stabilizer layers, or a combination of both. For example, in a monodisperse polymer colloid of negatively charged particles, all the particles repel each other to the same degree so that they will move away from each other to the same extent, which leads to a uniform, ordered array in which the particles find themselves, perhaps, in a body-centered cubic (bcc) arrangement. The system is still quite fluid, and the particles are not touching each other. The distance between rows of particles in such an array is on the order of the wavelength of visible light, so that they exhibit Bragg diffraction, such as that shown in Plate 2. These are true liquid crystals.

For neutron scattering the spatial correlations can be expressed in terms of a 'structure factor,' $S(Q)$:

$$S(Q) = 1 + \frac{4\pi N}{Q} \int_0^\infty [g(r) - 1] r \sin Qr \, dr \tag{75}$$

in which $g(r)$ is the particle pair correlation function, and r is the center–center interparticle distance. The correlation function gives the probability of finding another particle at the distance r from a given particle:

$$g(r) = \frac{N(r)}{N} \tag{76}$$

where N is the overall average particle number concentration and $N(r)$ is the concentration as a function of radial distance in the immediate vicinity of the given particle. The angular dependence of the neutron scattering intensity now becomes:

$$I(Q) = A(\rho_p - \rho_m)^2 N v^2 \times P(Q) \times S(Q) \tag{77}$$

in comparison to Eq. 74. Fourier transformation of Eq. 75 allows one to calculate the radial distribution function from experimentally determined values of $S(Q)$, as given in Eq. 78.

$$g(r) = 1 + \frac{1}{2\pi^2 rN} \int_0^\infty [S(Q) - 1] Q \sin Qr \, dQ \tag{78}$$

In Fig. 54 are shown results for a polystyrene colloid at three different concentrations. The medium was 1×10^{-4} M aqueous NaCl. It can be seen that as the concentration was increased, the system went from an essentially disordered state (no peaks) to one in which there was a high degree of order (sharper peaks). At 1% there are no nearest neighbors closer than 500 Å, and random distribution ($g(r) = 1$) at interparticle distances greater than 1000 Å. At 13% there is a high concentration of nearest neighbors at 500 Å and a second 'shell' of neighboring particles at 1000 Å. From such data it is possible to calculate free energies of interaction between particles (pair potentials) [72].

5.2.7 Conclusion

Thus neutron scattering is seen as a powerful technique for investigating the composition (by measurement of ρ_p) and morphology of latex particles. It can be used to determine the dimensions of a surface, lyophilic polymer layer serving as a steric stabilizer (Section 7.3), the properties of an adsorbed

CHAPTER 5 CHARACTERIZATION OF POLYMER COLLOIDS

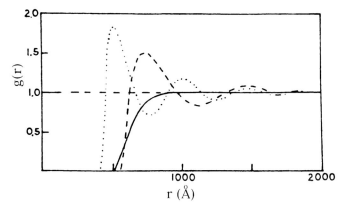

Fig. 54. Radial distribution function for a PS colloid at 0.01 (—), 0.04 (– –) and 0.13 (···) by volume [70].

surfactant layer, and even the electrical double layer properties. Furthermore, neutron scattering can be employed to investigate the degree and dimensions of ordering in more concentrated polymer colloids by means of the structure factor, $S(Q)$. A slight drawback is that an atomic reactor is required – not exactly a benchtop technique! But major national and international centers are available to researchers both in the United States and in Europe. A general reference to the fundamentals of neutron scattering, but somewhat out-of-date for recent developments, is the book by Bacon [73].

5.3 Microscopy

The particles of polymer colloids are so small (between about 10 nm and 1000 nm) that special instrumentation is required to visualize them.

> The phenomenon of thermal motion is fundamental to everything that happens in chemistry and chemical physics.

5.3.1 Optical microscopy

The limit of resolution of optical microscopes using white light and glass refracting lenses is about 200 nm. This is good enough to see the thermal, or Brownian motion and ordering of particles, even if it is not good enough to resolve their dimensions accurately, nor especially to see any internal morphology. Everyone should have the opportunity to observe the phenomenon of thermal motion because it is so fundamental to everything that happens in chemistry and chemical physics. A very dilute polymer colloid of c. 300–600 nm diameter particles on a glass slide, under a cover slip and under an objective lens with about 2000× magnification can be used.

5.3.1.1 The ultramicroscope

It is possible to 'see' latex particles, even when they are not resolved, by the spot of light which they reflect under appropriate illumination. The phenomenon is analogous to that of viewing an artificial satellite passing in the night sky: the naked eye cannot resolve the object itself, but it is nevertheless easily seen because of the reflection of the sun's light. For example, 2.0 nm gold colloidal particles can be observed readily under 'dark field' illumination. The Brownian motion of such particles, incidentally, is extremely rapid and spectacular. The optical set-up classically has been referred to as the 'ultramicroscope.' These kinds of optics are used in microelectrophoresis (Section 8.2). A modern and more sophisticated version of the ultramicroscope is the laser scanning microscope.

5.3.1.2 Confocal laser scanning microscopy

As the name implies, a laser beam is focused at some depth within the fluid colloidal system, and scanned at that depth to produce a cross-sectional view. A series of such views at different depths of a polystyrene colloid [74] is shown in Fig. 55. The medium has the same density as that of the PS by mixing in

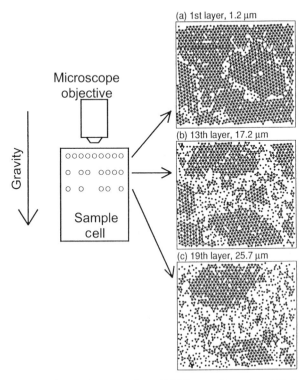

Fig. 55. Laser scanning micrographs of 300 nm diameter PS latex, 0.7% conc. in density-matched H_2O/D_2O mixture [74].

D$_2$O, so that there can be no effects due to gravity. One observes in this experiment a high concentration of particles at the *top* of the cell along with almost perfect ordering of the particles. There is decrease in the concentration and a decrease in the degree of order at 25.7 μm depth, equivalent to 19 layers if perfect ordering existed.

The conclusion from these experiments is that, under the given experimental conditions, the latex particles are attracted over large distances to the glass surface of the top of the cell even though both are electrically negatively charged. Ise and coworkers propose that this is due to electrostatic attractions of the counterions between the particles and the glass surface (more on this in Section 7.2.3.3).

Another example of these long-range electrostatic attractions is manifested in the formation of large voids in disordered PS latexes [75] as shown in Fig. 56. In this case the particles are large, 960 nm in diameter, with a very high negative surface charge density of 12.4 μC cm^{-2}. The latex was allowed to stand undisturbed for 4 h. The voids are huge compared to the size of the particles – on the order of 25 μm in diameter – indicating, once again, long-range attractions, in this case among the particles even though they are all charged alike. Because these effects occurred far from the glass interface, this phenomenon does not appear to be an artifact caused by the glass.

Experimental details for these experiments may be of value:

- the confocal laser scanning microscope was made by Carl Zeiss of Oberkochen, Germany;
- the optics included a 100 × oil immersion objective lens;
- the light source was a 5 mW argon laser;
- the D$_2$O/H$_2$O mixture was density-matched for polystyrene; and
- the latex was extensively purified by membrane dialysis against 'Milli-Q' reagent-grade water, and subsequently kept in contact with highly purified mixed-bed ion exchange resin beads to remove any further traces of extraneous ions and polyelectrolytes introduced during polymerization.

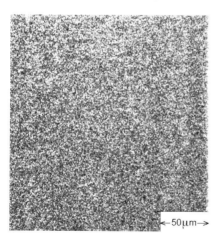
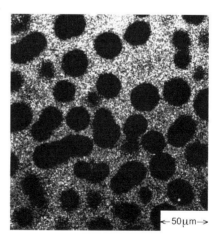

Fig. 56. Void structures in PS colloid observed with laser scanning microscopy at 0 mm (left) and 10 mm (right) from top of cell [75].

5.3.2 Electron microscopy

5.3.2.1 Transmission electron microscopy (TEM)

Because of the limit of resolution of the light microscope – visible light has wavelengths from about 400 to 800 nm – larger than, or about the same dimensions as, colloidal particles – researchers have resorted to the use of electron beams. These have several advantages and some disadvantages when compared with light. Advantages of electron microscopy are:

- Much shorter wavelengths allow for resolution of details down to a few nanometers.
- Electron beams penetrate polymers, so that the internal structure of latex particles often may be seen via transmission electron microscopy.
- 'Staining' techniques can be employed in which electron-dense materials, e.g. osmium tetroxide, OsO_4, or ruthenium tetroxide, RuO_4, react with certain portions of particles to enhance their contrast.

Disadvantages are:

- Electron beams can be destructive of polymers, causing depolymerization and evaporation under the high vacuum conditions of the instrument. Thus, for example, measurements of particle dimensions must be done with great care that shrinkage has not occurred.
- Organic polymers are not very electron-dense, with the result that thin samples – and small diameter polymer particles – are difficult to 'see.' Staining or shadowing (see below) is required.

> 'Staining' techniques in which electron-dense materials react with portions of particles can be used to enhance contrast.

Microtomy and staining techniques. Notwithstanding the disadvantages, TEM is perhaps the most commonly used method for investigating particle size and morphology. Examples of TEM micrographs are shown in Figs 28 and 29 (Section 2.3). Both of these represent the use of microtome sectioning and heavy metal staining to show the internal morphologies of multiphase polymer colloidal particles. A microtome is a highly refined 'meat slicer' which can cut sections of appropriate thickness, e.g. 20 nm, so that transmission micrographs can be obtained. The latex particles must be embedded in an incompatible plastic matrix such as an epoxy resin. This means that the particles must be transferred from the aqueous medium to that of the embedding resin without disturbing their morphology (by swelling, for example).

Positive staining involves the selective reaction of a heavy metal compound with the polymer or with a selected polymer domain in the particles. For instance, in the PMMA/PS core/shell particles of Fig. 29 the microtomed sections were exposed to the vapors of RuO_4 above a solution of that compound [76]. It reacts with polystyrene by oxidation and probably subsequent ester formation to form ruthenium derivatives of the polymer. It does not react with PMMA. Excellent contrast is thus obtained between the shell and the core. It would be almost impossible to discern the morphology in an unstained sample.

> Negative staining involves an electron-dense material that does not react with the polymer.

Negative staining involves an electron-dense material that does not react with the polymer, but rather is introduced to the aqueous phase and forms a residue among the particles upon evaporation of the water. Thus by contrast the particles stand out light against a dark background. This can be used to measure the size distributions of very small particles which are otherwise difficult to resolve in the TEM. An example of a polystyrene latex, negatively stained with an aqueous solution of potassium polyphosphotungstic acid, KPT, is shown in Fig. 57. The particle size of this colloid was determined, by measuring 500 particles, to be 68 nm with a standard deviation of $\sigma = 9.2$ nm [77]. The technique, which was pioneered in biological research laboratories, reportedly will resolve structures less than 1 nm in size.

Other negative stains have been used, such as uranium nitrate or acetate and ammonium molybdate. Because these and KPT are all strong electrolytes, they tend to coagulate the colloid. It is therefore preferred that the stain be added *after* the particles have been deposited upon the TEM substrate film. An older method for enhancing detailed structures (but not internal morphology), and still used a great deal, is called shadowing.

Shadowing. After the latex particles have been deposited upon the TEM base film, of e.g. nitrocellulose or Formvar®, and dried, they are placed in a high vacuum and a solid material is evaporated onto them at a low angle in such a manner as to produce what appear to be shadows. The source material is usually carbon, or a carbon/platinum alloy in the form of a small – *c.* 1 mm – pellet which is heated white hot. The vacuum must be sufficiently good

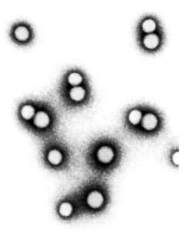

Fig. 57. Polystyrene latex negatively stained with KPT, 49 000× [77].

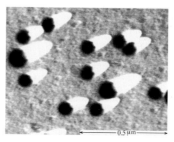

Fig. 58. PMMA NAD particles with PT/C shadowing, 66 500×.

($<5 \times 10^{-6}$ Torr*) so that the mean free path of the vapor molecules is greater than the distance between source and sample, i.e. so that they travel in straight lines and form sharp shadows. The angle of incidence is usually about 15–20°. For high resolution the PT/C alloy is preferred because the deposit has a finer grain size than that of C alone. An example of the results is shown in Fig. 58.

> Since the Pt/C alloy is heated to incandescence, its vapor can cause thermal degradation of the polymer particles.

Since the Pt/C alloy is heated to incandescence, it can cause thermal degradation of the polymer particles as it is deposited, resulting in shrinkage or distortion which may aggravate that caused by the electron beam. So when measurements are made of particle diameters, they should be made on the width of the *shadows*, not on those of the original particles.

Instrumentation and sample preparation. Transmission electron microscopes can be purchased commercially essentially 'off the shelf,' but they are large, expensive, complex and delicate instruments. They require vibration-free mounting and they must maintain high vacuum and scrupulous cleanliness to avoid artifacts. The source of electrons is usually a hot wire coated with a material of low ionization potential, such as rare earth oxides. The electrons are accelerated towards a positive plate with a potential of a few tens of eV to a million or more eV. Most TEMs operate at from 50 to 300 eV. The electron beam is focused by passage through a series of solenoids, and it then passes through the sample and onto a plate coated with a phosphor to produce a visible image. Photographs can be obtained by placing photographic film in the plane of focus, electrons having the same effect on silver halide as visible light. Image processing software may also be used to enhance resolution and contrast, as well as to determine particle size distributions, areas of domains between prescribed grayness levels, etc. The latter allows one to determine, for example, the degree of phase separation within individual particles.

Sample preparation involves first forming an extremely thin film of nitrocellulose or Formvar® by evaporation of a solution ($c.\,0.3\%$ concentration) spread onto a microscope slide. This film is cut around the edges with a razor blade and then the slide is carefully introduced at an angle under the surface

---

* 1 Torr ≈ 133.322 Pa.

CHAPTER 5 CHARACTERIZATION OF POLYMER COLLOIDS

of very clean water in such a manner that the plastic film separates and floats on the liquid surface. An electron microscope 'grid,' a thin, round screen of about 1 cm diameter with a mesh size of, e.g. 300, is placed on the floating film. The film-cum-grid is then lifted off by contacting from above with a piece of paraffin film. As it is lifted it breaks away from the rest of the film which remains floating and available to make more grids. Subsequently the grid-supported film is coated with carbon or carbon/platinum in an operation essentially identical to shadowing described above. The purpose of this is to strengthen the film and to prevent distortion under the electron beam.

The surface of the film is now hydrophobic, however, which means that a droplet of aqueous colloid placed upon it will not wet out. This is obviated by exposing the film to intense ultraviolet radiation (e.g. a high pressure mercury arc lamp) in the air for 20–30 min to oxidize the surface. A drop of very dilute ($c.\,0.01\%$ solids) latex is placed on the newly hydrophilic film surface from a glass capillary, and after a minute or two the drop is contacted with the edge of a piece of absorbent paper to leave only a thin wet layer on the support. It is then allowed to air-dry, which takes only a few minutes. Thereafter staining may be undertaken, or the grid may be used immediately for examination.

> The area under observation may be only a few nanometers square, but it needs to be representative of the entire macroscopic sample.

The area under observation may be only a few nanometers square, but it needs to be representative of the entire macroscopic sample. Thus sample preparation techniques are extremely important, especially when the latex particles are not uniform in size or morphology. A common error occurs when a drop of diluted colloid is placed on the base film and the sample is allowed to evaporate to dryness. As the medium evaporates the residual droplet contracts under surface tension such that its edge moves inward. The moving edge, in turn, can sweep with it larger particles towards the center of the sample, leaving behind smaller ones near the edges of the original droplet. Thus it is important to scan over the entire preparation to search for artifacts of this type. When doing particle size distributions, one must ensure good statistical sampling by using several preparations and counting a sufficiently large number of particles, depending upon the degree of dispersity and the accuracy required.

> The drop's moving edge can sweep with it larger particles towards the center of the sample, leaving behind smaller ones.

5.3.2.2 Scanning electron microscopy (SEM)

The electron beam can be scanned over a surface in the same way as that in a video monitor, with the exception that the surface is the sample investigated.

98   POLYMER COLLOIDS: A COMPREHENSIVE INTRODUCTION

The secondary electrons from the sample surface are then focused onto a detector which, in turn, transmits the signal to a video monitor! The major advantages of this technique are:

- there is essentially no destruction of the sample material;
- surface contours are readily seen, often without the need for shadowing;
- a stereoscopic, 3-D image can be made by detecting from two different positions, rendering corresponding photographic images, and observing these with a twin lens viewer; and
- chemical composition as a function of position on the micrograph can be obtained by simultaneous analysis of the X-rays which are emitted when any matter is bombarded with electrons (the Moseley effect).

There are some disadvantages of the SEM as well:

- only the surface of the sample is observed, so that no internal morphology can be discerned;
- nonconducting materials can accumulate a static charge which distorts the electron beam. Some method must be employed to either swamp out this effect by flooding the area with extraneous electrons or by coating the surface with a metallic conductor such as gold (by a shadowing-like method).

Some examples of SEM micrographs are shown in Figs 1b, c & d in Section 1.2. Often a combination of TEM and SEM is employed, for example to observe the progress of coalescence of particles during film formation by SEM, and changes in particle morphology at the same time by TEM. In fact, it is possible to obtain an instrument commercially which has both capabilities – a scanning transmission electron microscope (STEM).

There are several good texts on electron microscopy to which the reader is referred for further information [78, 79].

5.3.3 Atomic force microscopy (AFM)

The atomic force microscope uses an extremely sensitive probe to 'feel' the force field which varies with the contours of the material above a solid surface. It can thus produce a profile of the surface. A schematic diagram of the apparatus is shown in Fig. 59. The sample is mounted on a substrate which can be moved a few nanometers in any direction by means of an attached piezoelectric XYZ scanner. The sample is brought up to the flexible cantilever probe to a position where – without touching – a force of interaction is detected. The sample is scanned by moving it in raster mode, and the force between the probe and sample surface is measured. This can be done by keeping the probe tip in the same position while moving the sample up or down. The displacement distance is then a measure of the force, according to Hooke's Law. The flexible cantilever probe position is accurately measured by means of the reflected laser beam image on the split photodiode, as shown in Fig. 59.

CHAPTER 5 CHARACTERIZATION OF POLYMER COLLOIDS

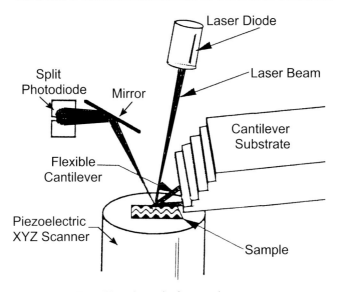

Fig. 59. Atomic force microscope.

The output of the AFM can be given as a 'photograph' of the surface as viewed from above, as well as a profile with a resolution of 1–2 Å. This is depicted in Figs 60 and 61 in which a dried film derived from a polybutyl methacrylate latex is observed [80]. By means of AFM Juhué and Lang were able to determine:

- that the particles are distorted into dodecahedrons immediately after the film has dried;

Fig. 60. AFM micrograph of PBMA dried latex film. Particle diameter = 140 nm [80]. Micrograph courtesy of the Institute Charles Sadron, Strasbourg.

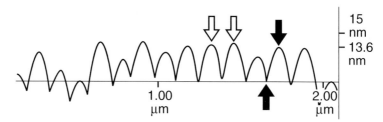

Fig. 61. Depth profile of freshly dried PBMA film by atomic force microscopy [80].

- that the particles are about the same size even after annealing for 30 min at a temperature far above the T_g for PBMA (open arrows in Fig. 61); and
- that the 'peak-to-valley' distance (solid arrows in Fig. 61) is greatly reduced upon annealing.

Note that the 'peak-to-valley' distance in Fig. 61 is 13.6 nm (the vertical axis is exaggerated).

Other studies have examined the degree of perfection of particle packing as a function of surfactant concentration (surface coverage of the particles) and particle size distributions.

Thus it can be seen that atomic force microscopy may be extremely useful in determining the effects of a great variety of experimental variables upon film formation from polymer colloids. It no doubt also holds great promise in the examination of film surfaces after exposure to use, either to weathering or to physical damage.

5.4 Hydrodynamic exclusion chromatography (HEC)

5.4.1 Capillary hydrodynamic fractionation (CHDF)

When a liquid flows through a capillary tube at low velocities, the drag against the tube walls creates a shear gradient such that the flow rate is slower near the wall than at the center of the tube. The profile of the velocity gradient is parabolic, and this type of behavior is known as laminar flow or Poiseuille flow after the man who described it. This is represented schematically in Fig. 62, and is to be contrasted to turbulent flow which occurs at higher velocities and in larger tubes. The transition between laminar and turbulent flow occurs when the Reynold's number, Eq. 79, exceeds approximately 2000.

$$R = \frac{ru\rho}{\eta} \qquad (79)$$

where r is the tube radius, u is the overall fluid velocity, ρ is the fluid density and η is the viscosity (cf. Section 3.1.4.1).

If Brownian particles are introduced into this laminar flow system, they will diffuse in a random manner. The radial component of that motion brings the

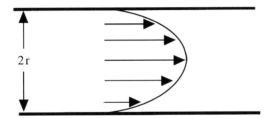

Fig. 62. Parabolic velocity gradient in laminar flow.

particles to different axial velocities. Smaller particles can diffuse closer to the walls than larger ones can (to a minimum distance of r_p, the particle radius), so that on a time average smaller particles will sample slower velocities, and therefore will elute at longer times from the end of the capillary tube. This is shown in Fig. 63. Thus physical segregation according to particle size can be achieved. In order for effective separation to occur, however, the tube must be only a few particle diameters, i.e. a few tens to hundreds of nanometers, in radius – a very fine tube indeed! In order to obtain sufficient sample for detection, a bundle of parallel tubes is used.

Detection is by some optical means. If the particles absorb light, then measurement of transmitted intensity at a given wavelength may be used. For example polystyrene absorbs in the ultraviolet. A more powerful technique is to measure the intensity of scattered light at several angles using a fiber-optic array. Application of the Mie equations then provides concentration, size and refractive index of the particles. The refractive index, in turn, gives information about the chemical composition of the polymer.

In small capillaries there is a large surface area/volume ratio, so that interactions of the particles with the capillary walls may be important. There are two kinds of interactions: (1) electrostatic and (2) van der Waals attraction. In the former, there is a repulsion of like-charged species at distances of a few, to a few tens of nanometers due to the overlap of the diffuse parts of their electrical double layers according to the DLVO theory (Section 7.2.3). There may also be an electrostatic attraction due to the counterions between particles and wall, according to Sogami and Ise [81]. Obviously if the particles and the wall carry opposite charges, they would strongly attract and the particles would adsorb onto the wall, rendering any elution virtually impossible. The van der Waals attraction is felt at very short distances, on the order of a few nanometers, and can ordinarily be overcome by the addition of

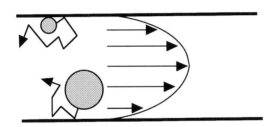

Fig. 63. Particle radial diffusion in laminar flow.

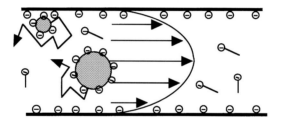

Fig. 64. Surfactant added to overcome wall effects.

appropriate concentrations of a surfactant which coats the capillary walls as well as the particles (Fig. 64). A concentration of $c.\,1\,\text{mM}$ sodium dodecyl sulfate and a colloid concentration of $c.\,0.1\%$ work well. Because the repulsions must not be too long-range (otherwise all the particles could be forced to the center of the tube), a careful balance of these effects is required.

Particle separations can be excellent using this technique, as shown in Fig. 65, in which a 50:50 mixture of two monodisperse PS latexes of 86 nm and 238 nm diameters were separated completely [82].

The curves in Fig. 65 can be translated to a particle size distribution by means of appropriate calibration through application of the Mie scattering equations to the detector response. Qualitatively one can observe that the peak area and height at 600 half-seconds is greater than that at 490, because of the larger particle size involved. The result is a very gratifying, accurate determination of the sizes and concentrations [82], as shown in Fig. 66. Here the peak heights and areas are equal because the weight percentages of each

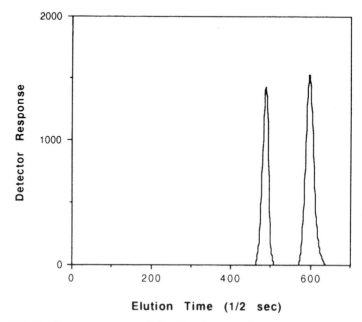

Fig. 65. CHDF chromatogram of 50:50 mixture of monodisperse PS colloids with diameters of 86 and 238 nm [82].

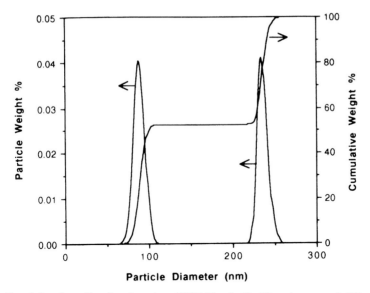

Fig. 66. Particle size distribution by CHDF of 50:50 mixture of PS colloids of diameters 86 and 238 nm [82].

kind of latex were equal. It must be kept in mind that CHDF provides *hydrodynamic radii* which may differ from those obtained by other techniques, e.g. transmission electron microscopy, TEM.

A comparison of CHDF and TEM results taken on a core/shell PS/PEG-600 latex (PEG-600 is polyethylene glycol of 600 Da molecular mass) shows a small difference between the two [83] (Fig. 67). Surprisingly the difference is in

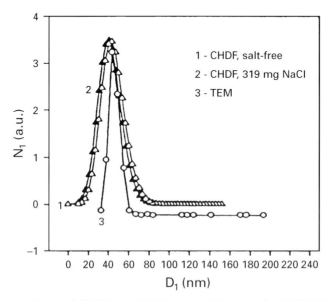

Fig. 67. Comparison of CHDF and TEM particle sizes for PS/PEG-600 core/shell latex at two salt concentrations [83].

the opposite direction from what might be expected: PEG is water-soluble and therefore in an expanded state in the aqueous medium, especially under conditions where it is not competing for water of hydration with added salt (curve 1, Fig. 67). The TEM micrograph data (curve 3) were taken on dried samples in which the PEG polymer would have been in a fully collapsed state. Thus one could expect the particle diameter, D, to have been smaller for the TEM than for the CHDF, especially considering that there is the possibility for electron-beam damage with consequent diminution in the observed size (Section 5.3.2.1). But in any case, the differences are small, and in the absence of an analysis of errors, one may conclude that the two methods agree quite well.

5.5 Sedimentation

> It is only on the moon or in a vacuum that a feather will fall as quickly as a rock.

Galileo found when he threw two cannon balls of different sizes from the Tower of Pisa that they both landed at the same time. This was because his measurements were not precise enough to detect the difference in 'wind resistance' between the two. If he had chosen a plastic foam ball for one of them, history might have been different! Although it is true that the acceleration due to gravity is the same for all objects, it is only on the moon or in a vacuum that a feather will fall as quickly as a rock. The viscous drag of the fluid medium acts upon all objects moving through it, including diffusing colloidal particles. This drag is proportional to the velocity, v, of the particles, as anyone who has put his hand out the car window while under way knows:

$$F_{res} = \phi v \tag{80}$$

where ϕ is known as the viscous drag coefficient, which for a sphere of radius r was given by Stokes as:

$$\phi = 6\pi\eta r \tag{81}$$

The driving force for sedimentation, F_{sed}, comes from the particle mass, m, and the acceleration, g, which may be due to gravity or to a centrifugal force:

$$F_{sed} = ma = \tfrac{4}{3}\pi r^3 \rho_{rel} g$$
$$F_{sed} = \tfrac{4}{3}\pi r^3 (\rho_p - \rho_0) g \tag{82}$$

where ρ_{rel} is the relative density of the particle in the medium, usually called the 'buoyant density,' and where ρ_p and ρ_0 are the densities of the particle and medium, respectively. It turns out that colloidal particles quickly (within milliseconds) reach a terminal velocity, v_t, in which the two forces are balanced, i.e. $F_{sed} = F_{res}$. Solving combined Eqs 80–82 with this equality for the sedimentation velocity gives:

$$\tfrac{4}{3}\pi r^3 (\rho_p - \rho_0)g = 6\pi \eta r v_t \quad \text{and} \quad \therefore v_t = \frac{2}{9}\frac{(\rho_p - \rho_0)r^2 g}{\eta} \tag{83}$$

So the sedimentation velocity is quite sensitive to the hydrodynamic radius of the particle. Equation 83 is built upon the assumption that the colloid is dilute, and that there are no interparticle interactions.

5.5.1 Ultracentrifugation

> If we wish to obtain measurements of terminal velocities, it is necessary to use a centrifuge.

Sedimentation velocities for colloidal particles in the Earth's gravitational field are so slow that their Brownian motion, combined with convection currents due to thermal fluctuations in a room, are enough to keep the particles suspended indefinitely. If we wish to obtain measurements of terminal velocities, it is necessary to use a centrifuge in which the acceleration is proportional to the square of the angular velocity, ω^2, and to the radial distance of the particles from the center of rotation, R. Since the field can vary depending upon the instrument used and the rotational speed chosen, it is convenient to use the sedimentation coefficient, S, which is simply the terminal sedimentation velocity per unit accelerating field:

In a centrifuge,

$$g = \omega^2 R \quad \text{and} \quad v_t = \frac{dR}{dt},$$

$$\therefore S \equiv \frac{v_t}{g} = \frac{d \ln R}{\omega^2 dt} = \frac{2}{9}\frac{\rho_{rel} r^2}{\eta} \tag{84}$$

This technique was applied by Bassett and Hoy to the investigation of acrylic latexes which, upon the addition of alkali, increased greatly in viscosity. It was thought that this effect was due to the expansion of the particles as surface carboxyl groups (introduced by copolymerization of acrylic acid in the second stage of emulsion semi-batch polymerization) are converted to ionic polyelectrolyte [84]. They used an adaptation of Eq. 84 to express the sedimentation coefficient as a function of the core density, ρ_c, the shell thickness, x, and the shell density, ρ_e:

$$\frac{S}{S_0} = \frac{r^3(\rho_c - \rho_0) + (3r^2 x + 3rx^2 + x^3)(\rho_e - \rho_0)}{r^2(\rho_c - \rho_0)(r+x)} \tag{85}$$

in which S_0 denotes the sedimentation coefficient for the unswollen particle, obtainable from data at low pH or from TEM micrographs. If the shell is highly swollen, then $\rho_e \approx \rho_0$, and Eq. 85 simplifies greatly to:

$$\frac{S_0}{S} \approx \frac{r+x}{r} \qquad (86)$$

As a particle swells upon increasing the pH the viscous drag increases, so that the particle slows down. Thus the inverted quantity S_0/S is used so that a plot shows an increase in *relative* (inverted) sedimentation coefficient with expansion of the particles upon swelling, as shown in Fig. 68 for three polymer colloids with varying acrylic acid content from 1% to 3% in the shell region.

The swelling is due to the expansion and solvation of high acid content polymer as it is converted to the salt. Part of the forces involved are the electrostatic repulsions among charged carboxylate anions. These are partially screened out as neutral salt is added, such that we would anticipate a reduction in the degree of swelling at high ionic strengths. This is in fact what Bassett and Hoy observed, as shown in Fig. 69.

In another study Liang and Fitch looked at the thickness of a layer of water-soluble, cellulosic polymer adsorbed onto the surface of polystyrene colloidal particles [85]. From earlier studies on diffusion of small molecules through the adsorbed layer, and on dynamic light scattering measurements of

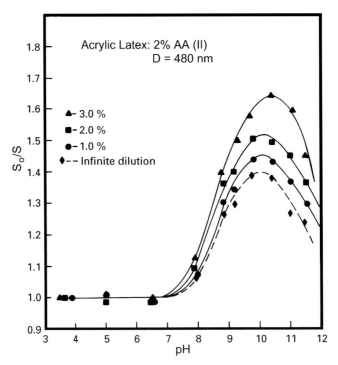

Fig. 68. Relative sedimentation coefficients for acrylic core/shell colloids with varying acrylic acid contents [84].

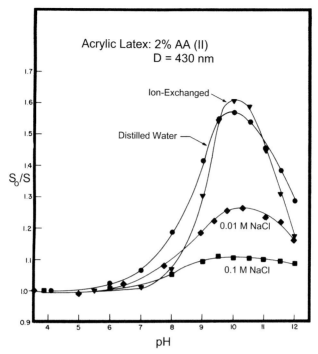

Fig. 69. Relative sedimentation coefficients for acrylic core/shell colloids with varying salt concentrations [84].

particle sizes, they were getting data which indicated that the adsorbed polymer layer was getting thinner upon remaining in contact with the cellulosic polymer solution for several days. To confirm these results they also determined sedimentation velocity on their samples. Sure enough, the aged samples actually sedimented more quickly than the fresh ones. Their data are given in Table 5.3, and were obtained using a Beckman Spinco Model E ultracentrifuge, with a double sector cell with sapphire windows and an ultraviolet detector. The quantity $S'f/S$ in Table 5.3 is practically the same as S/S_0 in

TABLE 5.3
Time-dependence of sedimentation velocities for PS colloids coated with cellulosic polymer [85]

	PS-B2 latex (uncoated)	HMHEC-coated PS-B2 latex	
		1 day after prep.	7 days after prep.
Temperature (°C)	22.0	20.9	19.2
Viscosity (cP)[a]	0.9548	1.0561	1.109
Speed (rpm)	3000	3000	3000
$d \ln r/dt$ (s^{-1})	6.13×10^{-5}	3.98×10^{-5}	4.47×10^{-5}
$S'f/S$	1	0.718	0.847

[a] 1 cP = 10^{-3} Pa s.

Eq. 85, except that S_0 refers to the uncoated particle, and with the added correction for any change in solution viscosity: $f = \eta/\eta_0$.

It is believed that this phenomenon occurs as a result of the rearrangement of the 'HMHEC' – hydrophobically modified hydroxyethyl cellulose – molecules after initial adsorption onto the PS particles. The HMHEC probably exists as giant micelles, and these are the species which first adsorb. Over time – in this case, days – they rearrange so that only a polymer monolayer remains, and the rest of the molecules in the original micelle desorb. It is interesting to note that the viscosity of the solution rises slightly after 7 days (Table 5.3), which supports this view.

5.5.2 Disk centrifugation

A disk centrifuge photosedimentometer, DCP, is comprised essentially of two concentric, parallel glass disks with a small space between them containing a liquid, the 'spin fluid,' through which the particles move. The disks are rotated around their center at speeds from 500 to 15 000 rpm. Provision is made for the introduction of a sample colloid at the center of rotation, as shown schematically in Fig. 70.

Particles are separated radially by their sedimentation velocities, according to their size and relative densities. A light source on one side and a photodetector on the other view the passage of particles at a fixed point, thereby producing a curve of turbidity as a function of time at a given radial distance, from which may be calculated the particle size distribution from the Lorenz–Mie equations (Eq. 63) and the sedimentation rate equation (Eq. 84). The equation for the particle diameter, d (m), assuming spherical geometry, is given below:

$$d = \left(\frac{18\eta \ln(R_d/R_i)}{\omega^2 \rho_{rel} t}\right)^{1/2}$$

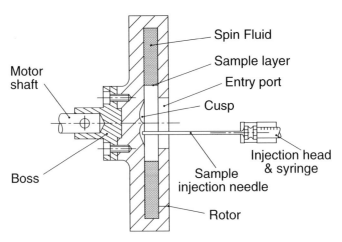

Fig. 70. Cross-section of a disk centrifuge photosedimentometer.

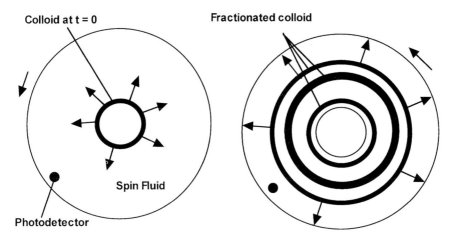

Fig. 71. Schematic representation of particle separation in a DCP using the LIST technique.

where t is sedimentation time (s), η is spin-fluid viscosity (kg m^{-1} s^{-1}), R_i and R_d are radial distances (m) initially and at the detector, respectively; ω is the rotational speed (s^{-1}), and ρ_{rel} is the buoyant density (kg m^{-3}) of the particles. A schematic diagram of the particle fractionation process is shown in Fig. 71. The spin fluid is introduced first and then a small amount of colloid. The colloid initially forms a thin line on the miniscus of the spin fluid, after which the particles migrate through the medium. This is known as the line start (LIST) technique [86].

If the density of some or all of the particles is less than that of the spin fluid, then the colloid is initially mixed with the spin fluid – the homogeneous start (HOST) method. Particle sizes in the range of approximately 10 nm to 60 μm can be measured in a matter of 10–30 min. When low density particles are centrifuged, turbulence may arise, in which case it has been found that employment of a density gradient in the spin fluid can eliminate the problem [87]. Hansen has shown that corrections should be applied, using the Mie equations, for the amount of forward scattering reaching the photodetector because of its finite size. Furthermore, if the light source has a broad spectrum, as in an incandescent bulb, integration across the wavelength distribution of the light, the sensitivity distribution of the photodiode, and the wavelength dependence of the scattered light must be made for greatest accuracy [87].

The method has the advantage that it is fast, and can be done on a relatively small, desk-top apparatus. A major strength is that a full particle size distribution can be obtained in a single measurement, unless the distribution is too broad, in which case different speeds, fluid viscosities, or buoyant densities may have to be employed. As with the other sedimentation methods, it is the Stokes equivalent spherical hydrodynamic diameter which is obtained, so that it is strictly applicable only to hard spheres. Nevertheless the information obtained can be very valuable [88].

Usually it is assumed that all particles have the same density. However, if this is not the case – for example in batch copolymerizations where compositional drift may occur – deconvoluting the data becomes a challenge. If the particle size distribution is very narrow (e.g. by TEM or HOTS), then a broader distribution on the disk centrifuge signals a density variation which can be translated into a compositional variation.

5.5.3 Sedimentation field flow fractionation

5.5.3.1 General principles

A powerful and versatile technique which combines the strengths of both hydrodynamic chromatography and sedimentation methods is called sedimentation field flow fractionation (SdFFF or SdF^3). The accelerating field in the SdFFF apparatus is applied normal to the axis of a capillary channel through which the colloid sample is flowing. The channel thickness is much larger than the tube radius in CHDF – usually between 75 and 260 μm – so that hydrodynamic exclusion is in most cases relatively unimportant. The result is a two-dimensional separation: one due to sedimentation radially and the other due to capillary flow, such that the larger particles are forced closer to the capillary walls where the Poiseuille flow is slowest. This is represented schematically in Fig. 72. The particles exhibit Brownian motion (Fig. 63), which in this case tends to spread them out. Because of the forces and dimensions of the SdFFF apparatus, the particles rapidly achieve sedimentation equilibrium, i.e. because of back-diffusion against the sedimentation force, the particles do not simply deposit on the channel wall but have a logarithmic concentration distribution near the wall, as shown schematically in Fig. 73. In this example the 'heavier' particles, Y, have an average distance from the lower wall of l_Y, while the 'lighter' ones, X, have a greater distance, l_X, because of their larger diffusion coefficient, D_X. Thus the X particles sample higher velocity regions than do the Y particles and move more rapidly down the channel. From these considerations the retention time, t_r, is approximately given by

$$t_r = t^0 \frac{w}{6l} = \frac{|F|w}{6kT} \tag{87}$$

where t^0 is the minimum elution time (for carrier fluid molecules), w is the channel thickness, F is the force acting upon a particle, and kT is the thermal energy.

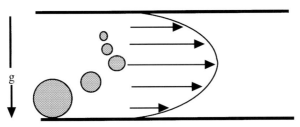

Fig. 72. Particle separation in sedimentation field flow fractionation.

CHAPTER 5 CHARACTERIZATION OF POLYMER COLLOIDS

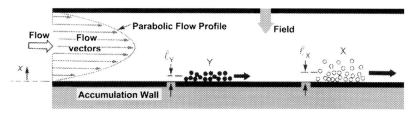

Fig. 73. Particle separation by SdFFF [89].

So as the particles flow down the tube they are separated axially as well as radially, and thus can be collected as preparative fractions. Because the separation is by sedimentation, it relies upon both the viscous resistance (a function of the radius) and the relative densities of the particles. Thus two particles of the same size but of different composition can be separated.

5.5.3.2 SdFFF apparatus

The apparatus, schematically represented in Fig. 74, was developed by Giddings at the University of Utah and has been commercialized. The channel is placed in a steel cylinder and is fitted with inlet and outlet tubes as shown. In Fig. 74 three colloids are shown, one of which, Z, is less dense than the fluid medium. The Z particles are thus 'floaters' and migrate to the upper wall of the channel, but of course are still eluted. An example of the high degree of resolution is given in Fig. 75 in which three peaks are observed for the SdFFF output for a PMMA latex. The principal peak, a, corresponds to singlets, whereas peaks b and c are doublets differing in size by approximately 10%, and peak d represents triplet aggregates. These assignments are borne out by

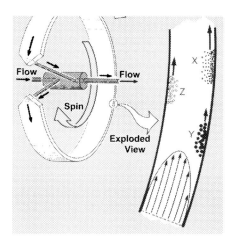

Fig. 74. SdFFF apparatus [89].

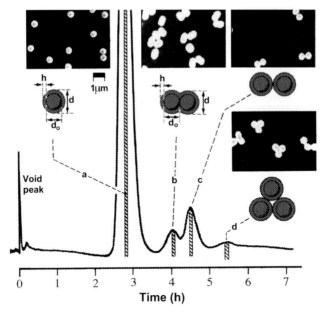

Fig. 75. SdFFF separation of PMMA latex and corresponding TEM micrographs from the eluates [89].

the corresponding TEM micrographs in Fig. 75 taken from the eluted samples [90]. Barman and Giddings speculate that the two kinds of doublets arise because some were formed in an early stage of emulsion polymerization, with diameter of d_0, and subsequently accumulated more polymer (added layer of thickness h). The other doublets were formed later, and aggregated after having grown to size d.

5.5.3.3 Other field flow fractionation methods

Several other modes of field flow fractionation have been developed, including:

- flow FFF
- thermal FFF
- electrical FFF
- steric FFF
- lift-hyperlayer FFF.

Flow, thermal, and some modes of steric FFF do not require a centrifugal field. Flow FFF, which can resolve particles as small as 1 nm, uses a cross-flow through semi-permeable membranes on either side of the channel, as the driving force instead of a centrifugual or gravity force. Steric FFF relies on a mechanism almost identical to that in capillary hydrodynamic chromatography in which the size of the particles, forced close to the lower channel wall by gravity or centrifugation, determines the different velocity layers in which

CHAPTER 5 CHARACTERIZATION OF POLYMER COLLOIDS

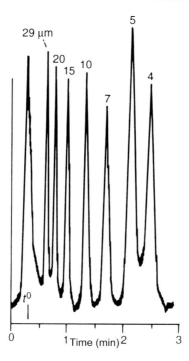

Fig. 76. High speed steric FFF separation of seven monodisperse PS colloids [89].

they will move. Steric FFF can separate micron-sized particles in a matter of a few minutes, as shown in Fig. 76.

5.5.3.4 Adsorption of polymers by SdFFF

A final example will show the sensitivity and power of sedimentation field flow fractionation. If a water-soluble polymer is introduced to a monodisperse PS latex, some of the polymer will adsorb at the PS/water interface and thereby affect the retention time, t_r. With a couple of SdFFF measurements it is then possible to determine both the mass and thickness of the adsorbed layer. Beckett, Ho, Jiang and Giddings have done this with several biopolymers, including human γ-globulin and a PS latex with a diameter of 0.198 μm. They used a Model S101 Fractionater (FFF Fractionation, Salt Lake City, Utah, USA) with a channel thickness of 127 μm and 86.5 cm length. They employed a field strength of 390 g at 1500 rpm and a flow rate of 0.73 mL min^{-1}. Sample fractograms are shown in Fig. 77 in which the bare latex and the same latex in the presence of two concentrations of the biopolymer were examined. The increase in retention times correspond to added mass of 146×10^{-18} g and 336×10^{-18} g (attograms!) of adsorbed γ-globulin and surface thicknesses, h, of 8.5 and 19.9 Å, respectively [91].

For a general introduction to field flow fractionation, the review article by Giddings in *Science* is recommended [89].

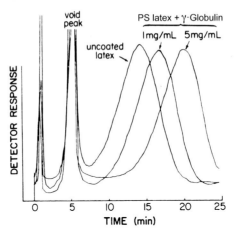

Fig. 77. SdFFF fractogram of PS latex and the same latex in the presence of two concentrations of γ-globulin [91].

5.5.4 Conclusion

Sedimentation can be a powerful tool in examining the hydrodynamic properties of polymer colloids. The apparatus involved may in some cases require a significant investment, and the technique can be demanding. But it can be an asset to the colloid scientist's armamentarium. The lower limit of applicability is where the particles are so small that their diffusion rates are on the order of the sedimentation rate. Even this effect can be used to advantage in a sedimentation *equilibrium* experiment in which, after relatively long sedimentation times in an ultracentrifuge, the two rates have equalized. The concentration gradient so formed provides absolute values of the molecular mass. Most general texts in colloid science have a section on sedimentation (see for example, [92]).

5.6 Nuclear magnetic resonance spectroscopy (NMR)

5.6.1 Motional averaging and peak width

NMR spectroscopy has been used for the study of the overall composition as well as the sequence distributions of polymers, the latter by looking at the fine structure of chemical shifts due to the influence of neighboring repeat units along a polymer chain. In the solid state any individual atomic spin that is being looked at is surrounded by nearest neighbor magnetic dipoles whose orientations are random and locked in place because of the high viscosity of the solid phase. This results in a broad distribution of chemical shifts so that no NMR peaks are observed. Only when the polymer is in solution or well above its glass transition temperature, T_g, is there sufficient motional averaging of neighboring dipoles to eliminate this effect, whereupon sharp peaks can be obtained. An example of thermal averaging is shown in Fig. 78,

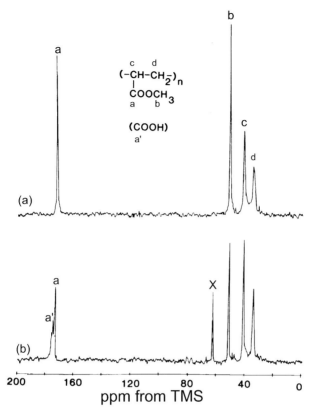

Fig. 78. ^{13}C NMR spectra of aqueous colloids of (a) PMA and (b) PMA/AA at 87°C [93].

in which proton-decoupled ^{13}C NMR spectra were taken on a latex of polymethyl acrylate, PMA (spectrum a), and a partially hydrolyzed PMA latex, and therefore a copolymer of PMA and polyacrylic acid, PAA (spectrum b) [93]. The spectra were obtained at 87°C, sufficiently above the T_g that the peaks are very sharp. From the areas under the peaks one can calculate polymer composition – and therefore the degree of hydrolysis in this case – by means of conventional, computerized peak resolution and area determination techniques. A general reference on determining compositions and sequence distributions is the book by Becker [94].

5.6.2 NMR spectra of aqueous polymer colloids

5.6.2.1 Morphology of hydrolyzed acrylate latex particles by NMR

We are interested here, however, in the application of NMR techniques to whole polymer colloids. A case in point is that mentioned above in which a PMA latex has been partially hydrolyzed, converting some of the methyl ester repeat groups to carboxyls. PMA is quite hydrophilic, and therefore the latex particles are swollen rather appreciably with water. The peculiar aspect of these colloids is that the hydrolysis is catalyzed by sulfonic acid groups on the surface of the particles, which incidentally gives interesting zero-order kinetics (discussed later in Section 6.2.4.2). The question in this case was whether the resulting particles had a core/shell morphology (as a result of hydrolysis progressing from the surface inward) or had a uniform, random distribution of acid groups as a result of uniform swelling by water and reaction with it.

The temperature-dependence of the NMR spectra, as shown in Fig. 79, indicated that at room temperature there was essentially no spectrum for the PMA latex, whereas for the partially hydrolyzed (26%) derivative there were peaks. These were in both the carbonyl (c. 175 ppm) and C–H (c. 30–60 ppm) regions (Fig. 79b). This indicated that the PMA/AA derivative had some groups which were mobilized, presumably by water-swelling even at room temperature.

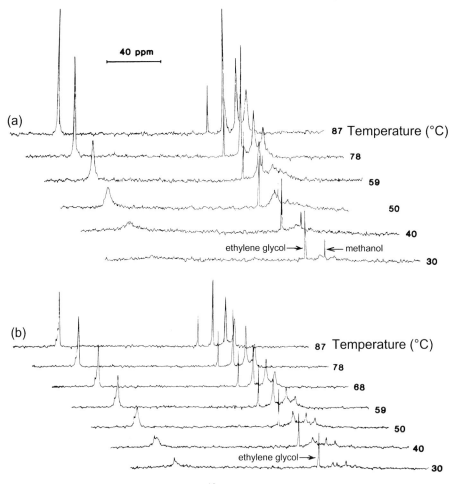

Fig. 79. Temperature-dependence of ^{13}C NMR spectra of (a) PMA and (b) 26% hydrolyzed PMA latexes. Curves are averages of 128 scans [93].

But how many PAA groups were involved, all of them, or just some of them, and where were they located? To answer these questions the integrated areas under all of the peaks were compared with the peak area of an internal standard, ethylene glycol, introduced into the sample in a glass capillary tube so as not to enter the polymer and plasticize it. The temperature-dependence of these relative integrated areas is shown in Fig. 80. The curves for 0% and 26% hydrolysis (−×− and −O− curves, respectively) correspond to sets (a) and (b) shown in Fig. 79.

It is clear from the data at 30° in Fig. 80 that segmental mobility greatly increases with the degree of hydrolysis. Since only a small portion of the polymer is involved, it must exist in a separate domain, presumably a shell around each particle. The carbonyl spectra, on the other hand, indicate that this mobility enhancement is associated with a *co*polymer of PMA/AA, rather than of pure PAA. So, although the exact morphology cannot be determined by these experiments alone, ^{13}C NMR does provide valuable information difficult to obtain otherwise.

5.6.2.2 NMR peak assignments

Incidentally, nuclear spin-lattice relaxation times, T_1, and nuclear overhauser effect, NOE, for the carbon–hydrogen couples can help to make peak assignments by calculating corresponding rotational correlation times, τ_r. Assuming isotropic reorientation and dipolar relaxation, the T_1 and NOE values are given by:

$$\frac{1}{T_1} = 0.10\hbar^2 \gamma_H^2 \gamma_C^2 r^{-6} x \qquad (88)$$

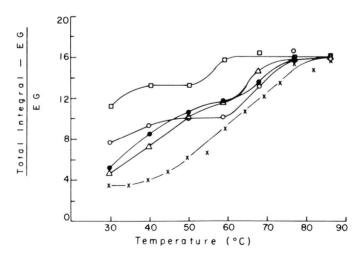

Fig. 80. Normalized integrals of ^{13}C NMR spectra for various degrees of PMA latex hydrolysis: ×, 0%; ●, 13%; △, 20%; ○, 26%; □, 36% [93].

where

$$x \equiv \frac{\tau_r}{1+(\omega_H-\omega_C)^2\tau_r^2} + \frac{3\tau_r}{1+\omega_C^2\tau_r^2} + \frac{6\tau_r}{1+(\omega_H+\omega_C)^2\tau_r^2}$$

and

$$\text{NOE} = 1 + \frac{\gamma_H}{\gamma_C x}\left(\frac{6\tau_r}{1+(\omega_H+\omega_C)^2\tau_r^2} - \frac{\tau_r}{1+(\omega_H-\omega_C)^2\tau_r^2}\right) \quad (89)$$

where γ_H and γ_C are the gyromagnetic ratios and ω_H and ω_C are the resonance frequencies in radians per second for hydrogen and carbon atoms. Calculation of the correlation time for methine carbon comes out to be about 0.58 and 1.51 ns from the T_1 and NOE data at 87°C, respectively. These are adequately in agreement, and support the assignment of the methine carbon peak to 40 ppm, as given in Fig. 78. The average value of about 1×10^{-9} s then can be taken as the correlation time for local motions in the polymer chains at 87°C.

5.6.3 Films derived from two-stage emulsion polymerizations

In some cases NMR spectroscopy may be applied to dried preparations from polymer colloids to investigate particle morphologies. 'Core/shell' particles of PB/MMA were prepared by two-stage, seeded emulsion polymerizations, either by (1) semi-continuous or (2) batch addition of MMA to a polybutadiene (PB) seed latex by Riess and coworkers [95]. Derived films from these compositions exhibit more or less good high-impact resistance depending upon the polymerization process and the overall PB/PMMA ratio. The synthetic techniques, in turn, affect the particle morphology, as we have seen, and the morphological details may have a very large influence upon dynamic mechanical properties (more on this in Section 2.4).

> Morphological details may have a very large influence upon dynamic mechanical properties.

Whether the particles have truly core/shell morphology, as in Fig. 29, or microdomain structure, as in Fig. 28, was not investigated. The question was, what is the nature of the polymer at the interface between the two domains (and presumably, how does that affect dynamic mechanical properties of the derived plastic, although this was not investigated either)?

Riess et al. employed relaxation measurements to get some idea of the mobility of PMMA chain segments within the particles. This was similar to the technique of Tarcha et al. [93] above, and is schematically represented in Fig. 81. An initial magnetic pulse in the x direction, H1x, of $\pi/2$ radians duration is given to the protons. Then another pulse of variable 'contact time,' t_c, in the y direction, H1y, is given during which cross polarization with the

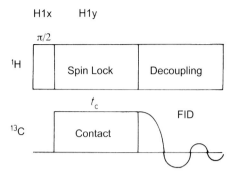

Fig. 81. NMR cross-polarization experiment scheme [95].

^{13}C carbons occurs. This is followed by a free induction decay, FID, of the carbon atom nuclei, registered as a decrease over time in the NMR peak intensity, as shown in Fig. 83. During all of this the sample is being spun at the magic angle. From 200 to 400 scans (2.4 s each) per point were required. The characteristic relaxation time, $T_{1\rho}(H)$, is proportional to the negative of the inverse of the slope of those curves (-1/slope).

When Riess *et al.* examined the dependence of $T_{1\rho}(H)$ of the methoxy carbon – which was representative of all the PMMA carbon atoms – they found a linear increase in $T_{1\rho}(H)$ with the amount of PMMA in the two-stage latexes, as shown in Fig. 82. They concluded that this could be due to either or both of two effects:

(1) the PMMA protons interact with neighboring polybutadiene nuclei magnetic dipoles; or
(2) the mobility of the PMMA chains is enhanced in the interphase regions because of the very low T_g of PB.

An additional observation, from Fig. 83, is that the method of preparation affects the $T_{1\rho}(H)$ of the PMMA chain segments. Recalling the negative inverse slope relationship, it can be seen that the order of increasing relaxation time is: semi-continuous < batch < blend. All had the same overall composition of PB/PMMA = 43/57. In fact the relaxation time for the blend, which was made

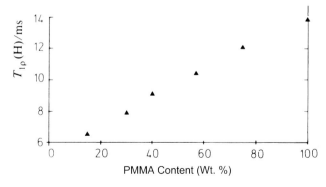

Fig. 82. ^{13}C NMR relaxation time for the methoxy carbon in PB/PMMA 'core/shell' particles [95].

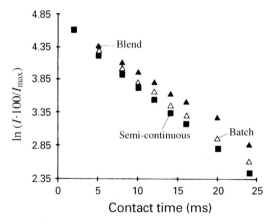

Fig. 83. ^{13}C NMR peak intensities vs contact time [95].

by co-coagulation of two homopolymer latexes, was the same as that for pure PMMA. Thus it appears that there is more PMMA in the interfacial regions in the colloid prepared by the semi-continuous process.

Because these preparations were all made by coagulation of the polymer colloids, filtered, washed, dried and then compacted into sample tubes, there is the possibility that distortion of the particle morphologies could produce artifacts. Nevertheless, it is clear that NMR can be a powerful tool for examining detailed structures within latex particles, both in the colloidal and dried states. Furthermore, even though the amount of polymer within the interfacial zone is relatively very small, on the order of 6–7 nm thick, NMR appears to be a sensitive technique for its detection and analysis [96].

Finally, it should be noted that during emulsion polymerization, because copolymer sequence distributions are affected by the techniques employed, whether two-stage, batch, semi-continuous, continuous, or power feed, ^{13}C NMR can be routinely used to monitor and analyze for sequence distributions by classical techniques [97].

5.7 Dielectric spectroscopy

5.7.1 Electrical double layer relaxations

When an oscillating electrical field of varying frequency is applied to an aqueous colloid of charged particles, the (complex) impedance will be comprised of two parts, one in phase and one out of phase. These are mathematically designated the real and imaginary components. The processes involved will be: (a) charge transfer at the electrodes; (b) transference of ions; (c) distortions of the electrical double layers at the various interfaces in the system (Fig. 84); and (d) orientations of the permanent dipoles present.

As the frequency is scanned, different characteristic relaxations will be observed corresponding to these processes. Generally the data are analyzed in terms of equivalent electrical circuits comprised of resistors and capacitors

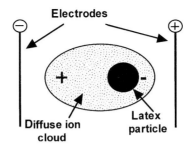

Fig. 84. Distortion of the diffuse part of the electrical double layer in an applied field.

connected in series and parallel. Most of the research on charged colloidal particles has been focused upon the counterions which are either adsorbed on the surface or constitute a diffuse ion cloud surrounding the particle, the so-called electrical double layer, EDL (Section 7.2.2). The characteristics of this double layer are extremely important to colloidal stability, diffusion, rheology, order–disorder phenomena, etc. We shall concentrate on the low-frequency domain in which the relaxation involves the ensemble motions of the diffuse part of the EDL, as shown in Fig. 84.

5.7.2 Experimental

In experimental dielectric spectroscopy the electrochemical cell can be quite simple, with the *caveat* that great care must be taken to ensure purity of the sample. It is best to use a Teflon® cell to avoid contamination by ions dissolved from glass, and an inert gas blanket to avoid dissolving carbon dioxide from the atmosphere. The electronics and data acquisition are controlled from a computer. A sinusoidally oscillating electrical field, on the order of tens of millivolts per centimeter, not much above the thermal energy of the system, kT, is applied across the electrodes. An electrochemical interface ensures that no polarization of the electrodes occurs.

> Great care must be taken to ensure purity of the sample.

The data obtained are in the form of the frequency-dependent total impedance, $Z_T(\omega)$, which can then be represented by equivalent electrical circuits of the Randel type:

each of which is theoretically expressed by Eq. 89:

$$Z_T(\omega) = R_0 + \frac{R_\infty}{1 + (i\omega C R_\infty)^{1-\alpha}} \qquad (89)$$

where $i \equiv \sqrt{-1}$, α is the Cole–Cole distribution constant, R_0 is the solution resistance, R_∞ is the interfacial resistance and C, the interfacial capacitance of the Randel circuit. Equation 89 is a complex function, and must be resolved into its real, Z', and imaginary, Z'', components, representing the elastic and viscous – or loss – parts of the impedance:

$$Z'(\omega) = R_0 + \frac{R_\infty + \omega^\beta C^\beta R_\infty^{1+\beta} \cos(\beta\pi/2)}{1 + (\omega C R_\infty)^{2\beta} + 2(\omega C R_\infty)^\beta \cos(\beta\pi/2)}$$

$$Z'' = \frac{\omega^\beta C^\beta R_\infty^{1+\beta} \sin(\beta\pi/2)}{1 + (\omega C R_\infty)^{2\beta} + 2(\omega C R_\infty)^\beta \cos(\beta\pi/2)} \qquad (90)$$

where $\beta = 1 - \alpha$. A plot of Z' vs Z'' gives a perfect semicircle on a 'complex plane' with its center lying on the real axis at a distance of $(R_0 + R_\infty/2)$ from the origin when $\alpha = 0$. It turns out that the mathematical analysis converges better when the reciprocal of the impedance, i.e. the complex admittance, $Y_T(\omega)$, is used. A perfect semicircle is also obtained when $Y''(\omega)$ is plotted as a function of $Y'(\omega)$, and $\alpha = 1$, as shown in Fig. 85, with the center now at $(Y_0 + Y_\infty/2)$. When a series of monodisperse polystyrene colloids with the same particle size but varying surface charge density (via sulfonate anions as polymer end groups at the particle surface) were investigated, the results

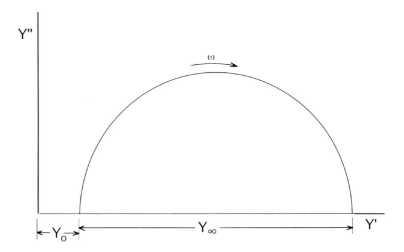

Fig. 85. Complex admittance diagram for a single Randel circuit.

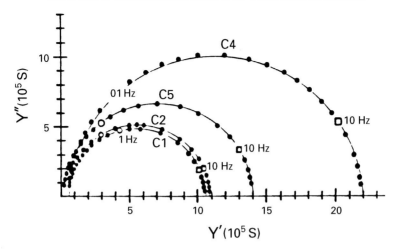

Fig. 86. Experimental complex admittance diagrams for model polystyrene colloids (Table 5.3) [98].

shown in Fig. 86 were obtained [98]. The characteristics of the colloids are given in Table 5.4.

The result was a series of perfect semicircles whose radii are a direct function of the surface charge density on the particles, as predicted by the theory. The dielectric measurement takes about 2 min to scan the frequency range involved. It can also be seen that the maximum in the dielectric loss, Y'', occurs at 1–3 Hz, indicating the likelihood that this relaxation corresponds to the motions represented in Fig. 84. Yoshino has calculated the fundamental frequency of the standing wave motions of the diffuse part of the double layer to be on the order of 1 Hz [99], in agreement with these measurements. These colloids were highly purified and ion-exchanged so that no extraneous electrolyte was present and all the

TABLE 5.4
Characteristics of model polystyrene colloids

	C1	C2	C5	C4
$\bar{D}_n$ (nm)	208	206	198	208
P	0.0335	0.0050	0.0070	0.0070
SCD (μC cm^{-2})	6.6	12.5	16.5	21.1

$\bar{D}_n$ is number-average particle diameter determined by dynamic light scattering.
P is the polydispersity parameter:

$$P \equiv \frac{\langle R^6 \rangle \langle R^4 \rangle}{\langle R^5 \rangle^2} - 1.$$

SCD is surface charge density.

counterions were H^+. Thus they can be regarded as being in the 'sulfonic acid form.'

5.7.3 Ionic strength

It turns out that when simple salt solutions are placed in the cell, with no colloid present, the product $\theta \log \omega^*$ is a measure of the ionic strength of the solution, regardless of the nature of the salt. Here θ is the phase angle between the real and imaginary responses, and ω^* is the characteristic relaxation frequency for the salt solution.

Curve A in Fig. 87 shows this relationship, where the quantity $\theta \log \omega^*$ has been given the name of 'apparent ionic strength.' This remarkable relationship applies over at least four orders of magnitude in salt concentration and holds for many salts, e.g. NaCl, NaOH, MgCl$_2$, and HCl. When the apparent ionic strength of the PS colloids is determined, they fall on curve C in Fig. 87, which is parallel to curve B. The latter is a theoretical curve generated by simply taking 1/2 the slope of curve A.

The four colloids of Table 5.3 as well as two others from the laboratory of Professor Ise in Japan all lie on curve C. Apparently the *low frequency* dielectric experiment, responding to *counterions* only, 'sees' only half of the total ions present. The fixed ions on the surface of the particle do not respond to this frequency range and to these low fields. This provides a rapid, quantitative measure of both the ionic strength of a salt solution and the surface charge on colloidal particles.

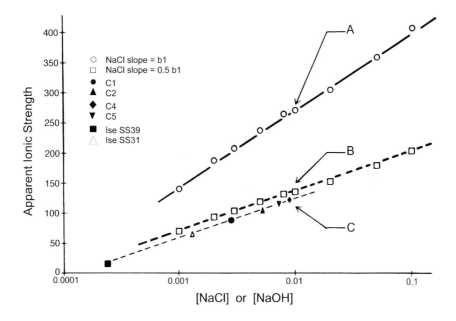

Fig. 87. Apparent ionic strengths, $\theta \log \omega^*$, of salt solutions and PS colloids vs log concentration [98].

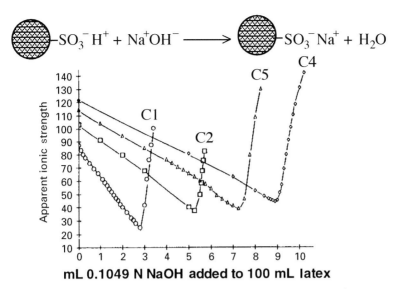

Fig. 88. Titration of sulfonic acid form of PS colloids. $\theta \log \omega^*$ plotted against volume (mL) of 0.1049 N NaOH added to 100 mL latex [98].

5.7.4 Titration, ion-exchange and counterion condensation

When these acid-form latexes are titrated with a base, there is a steady decrease in $\theta \log \omega^*$ until the equivalence point is reached, after which it rises rapidly, as shown in Fig. 88. The rise is due to the added ions of Na^+OH^- but why does the apparent ionic strength of the medium decrease? It appears that some of the new sodium counterions, which have replaced H^+ during the titration, have adsorbed onto the surface of the PS particles. This is a well-known phenomenon, often referred to as 'counterion condensation.'

> Sodium counterions have adsorbed onto the surface of the PS particles.

Another way to look at this is to consider the degree of ionization, f, of a particle as being the fraction of counterions *not* adsorbed. Values for the same series of PS colloids are presented in Table 5.5. The results are in good

TABLE 5.5
Degree of ionization of Na^+ form of PS model colloids

	C1	C2	C5	C4
Surface charge ($\mu C\,cm^{-2}$)	6.6	12.5	16.5	21.1
Degree of ionization, f	0.14	0.12	0.086	0.085
Effective surface potential, Ψ_{eff} (mV)	13.1	21.3	20.2	25.5

agreement with those obtained by other methods, and indicate that about 85–90% of all sodium counterions are adsorbed. This layer of counterions is often referred to as the Stern layer after the man who proposed the idea in order to explain why the effective electrical potential, Ψ_{eff}, of charged colloidal particles was often so much less than that calculated from the titrated charge.

The values of Ψ_{eff} calculated taking into account f, given in Table 5.5, are in agreement with those obtained by independent measurements, e.g. by electrophoretic mobility or kinetics of coagulation.

5.7.5 Specific ion effects

If sodium ions adsorb so strongly to the polystyrene surface bearing sulfonate anions, what about other alkali metal cations? Titration with LiOH and CsOH reveal in their dynamic dielectrical response unexpected results as shown in Fig. 89. The smallest drop in apparent ionic strength comes from titration with Cs^+ ion which, because of its large ion size and higher polarizability, might have been expected to adsorb more strongly. Although this phenomenon may be attributable to *solvated* ion sizes, it is not completely understood. Nevertheless the results do show that dielectric spectroscopy is a powerful tool with which to study specific ion effects [100].

5.7.6 Shear-dependence

There are many other applications of dielectric spectroscopy of colloids. In 1987 a NATO Workshop group stated that, 'The study of concentrated

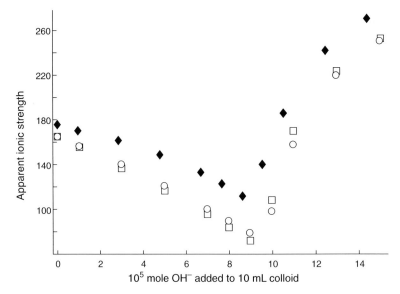

Fig. 89. Titration of sulfonic acid form of PS colloids. $\theta \log \omega^*$ vs volume of base added: ○, NaOH; □, LiOH; ◆, CsOH [98].

polymer colloids subjected to shear fields should give information on the relaxation times of the different processes associated with changes in the structure of the dispersion' [101]. A qualitative example of this is shown in the shear-dependent, complex permittivity of bentonite clay dispersions in which the clay particles were ion-exchanged with a homologous series of alkyl amine hydrochlorides to produce 'organoclays' with surface alkyl chains from 6 to 16 carbon atoms in length [102]. These were dispersed in an aliphatic hydrocarbon fluid. Typical results for a C-10 modified clay at 12% solids are shown in Fig. 90. The solid circle points are experimental and the open circles are calculated from theory, using essentially Eqs 89 and 90, with adjustment of the parameter α.

The explanation for the dramatic changes under relatively low shear in the complex permittivity curves in Fig. 90 were ascribed to the break-up of association structures among the clay particles. The frequency range scanned in this case was from 20 Hz to 100 kHz, and it is not known exactly what dipolar processes were responsible for the effects observed.

Dielectric spectroscopy also has been used to determine surface charge density [98], particle size [103], double layer effects in weak electrolyte polymer colloids [100], order–disorder phenomena, the 'hairiness' of the particle surface [104], and the compression of the double layer in the presence of inert salts. Incidentally, other relaxations are found at higher frequencies, i.e. in the kilohertz and megahertz regions. The kilohertz relaxations are associated also with the surface layers on the particles, whereas the megahertz responses may have more to do with permanent dipoles within the polymer itself. The

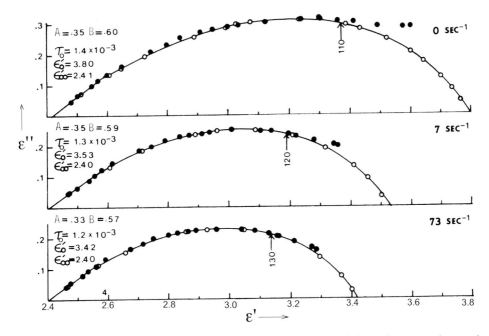

Fig. 90. Complex plane plot of shear-dependent permittivity of organo-bentonite dispersions in polybutene [102].

technique is nondestructive, fast (generally only a few seconds or minutes per measurement); it can be used on concentrated colloids; and it uses cells of simple construction. The basic theory for the application of dielectric spectroscopy to colloids in general can be found in the treatise by Dukhin and Shilov [105].

5.8 Surface chemistry

5.8.1 Overview

In any good scientific investigation it is important that the experimental system being studied be well defined. Because polymer colloids can be synthesized with very narrow particle size distributions and their surface chemistry can be predetermined, as we shall see in the next chapter, they have become highly desired as model colloids for a host of different kinds of studies. But before any serious analysis and characterization is done, the latexes must be purified. Because of the nature of free radical emulsion polymerization, there are always a variety of by-products which have to be removed. These include initiator fragments, added salts and buffers, oligomeric species which are statistically always present in radical polymerizations, and polyelectrolyte formed by the polymerization of water-soluble monomers (when present) in the aqueous phase.

> The Fourth Law: the perversity of inanimate objects in the universe tends towards a maximum.

Colloids are characterized by high specific interfacial areas which are easily subject to contamination, and polymer colloids are no exception other than that they have relatively low interfacial free energies compared with, say, silver halide, metal or metal oxide sols. They are contaminated particularly by adsorption of ions, surfactants, polymers, gases and small molecules such as monomers and solvents. It is perhaps a result of the fourth law of thermodynamics (the perversity of inanimate objects in the universe tends towards a maximum!) that attempts at purification frequently lead to further contamination and unforeseen artifacts. Thus the reagents and devices employed for analysis and characterization of polymer colloids must themselves have been rigorously purified.

5.8.2 Cleaning of polymer colloids

5.8.2.1 Ion-exchange

Most polymer colloids, but by no means all, carry negative charges at the surface generated by ionic functional groups introduced during the emulsion polymerization. Purification by mixed bed ion-exchange resins does two things: (1) it supposedly removes all extraneous electrolyte (polymeric and small ions), and (2) it exchanges all the counterions for H^+, in the case of anionic particles and for OH^- in the case of cationics. I say 'supposedly' because ion exchange resins themselves are *always* contaminated with polyelectrolyte as a result of the severe conditions employed in their manufacture (e.g. fuming sulfuric acid action on PS beads produces sulfonation and oxidation products). In their landmark work Vanderhoff, van den Hul, Tausk and Overbeek demonstrated this contamination and showed how ion-exchange resins could be purified to obviate the problem [106]. The separate cation- and anion-exchange resin beads are washed with distilled water and then successively with 3 N NaOH, hot water (always distilled to very high purity), methanol, cold water, 3 N HCl, hot water, methanol, and cold water. This cycle usually has to be repeated five times before the conductivity of the wash water is equal to that of the distilled water. The mixed bed then produces wash water with a conductivity even less than that of the distilled water.

> Ion exchange resins are always contaminated with polyelectrolyte.

The mixed bed ion-exchange process removes all inert, extraneous salts and converts M^+ counterions to H^+, or X^- to OH^- as observed above. But the PS-based ion-exchange beads also have hydrophobic domains which can be quite effective in removal of *nonionic* surfactants and other contaminants. The purified colloids are then in a form in which they can be titrated. Since there is now only one titratable counterion for each surface fixed group, the titration equivalence point provides an accurate count of the number of fixed groups on the particle surface, from which it is easy to calculate the surface charge density. The latter is normally expressed in terms of microcoulombs per square centimeter ($\mu C\, cm^{-2}$) (cf. Table 5.3).

> Titration provides a count of the number of fixed groups on the particle surface.

An additional caution must be given: hydrophobic latex particles (e.g. polystyrene) with surface sulfonate groups are essentially ion-exchange resin 'beads' themselves. When placed in a mixed bed they can compete for the adsorption of contaminants, especially polyelectrolytes. Furthermore, as polyvalent 'macro-ions,' they may adsorb onto the much larger resin beads of opposite charge. Also, the removal of ionic surfactant from the latex during ion-exchange, may induce coagulation if the residual surface charge density is low and the particle size is small [107]. These two effects can: (1) reduce the solids content which, if not measured after ion-exchange, will lead to gross errors in calculation of surface group concentrations, and (2) lead to skewed particle size distributions for systems which are not monodisperse.

5.8.2.2 Dialysis and serum replacement

> Dialysis has the advantage of simplicity, but it is also fraught with hazards.

Dialysis through a semi-permeable membrane has been the classical method of choice for the purification of colloids. The membrane is permeable to small molecules, ions and oligomers but not to particles of colloidal dimensions. It has the advantage of simplicity, but it is also fraught with hazards. Typically, a section of 'sausage tubing' regenerated cellulose is tied at one end to make a bag, the colloid is placed therein, and this is placed in a beaker with stirred, distilled water. The water is replaced periodically, and the dialysate is tested for conductivity, light transmittance or some other property. When there is no trace of further permeation, usually a matter of several weeks, the colloid is deemed clean.

The caveats in this instance are:

(1) water-soluble polymer may be leached from the cellulose membrane and adsorb onto the latex particles;
(2) some surfactants are sufficiently well anchored to the particle surface that they do not desorb under these mild conditions;
(3) *ab*sorbed molecules such as residual monomer, oxidation products (e.g. persulfate initiator may oxidize styrene to benzaldehyde) and initiator fragments diffuse out of the particles at extremely low rates; and
(4) fungal contamination may occur over the long dialysis times [108].

The first problem can be overcome by careful cleaning of the membrane prior to use; the second, by ion-exchange; the third, by 'steam stripping' the latex prior to dialysis; and the fourth, by scrupulous cleanliness and/or the addition of formaldehyde when permitted. Dialysis remains a useful method for removal of most of the contaminants, and ion-exchange, for further purification.

Continuous hollow fiber dialysis. The process of dialysis is greatly speeded up by increasing the specific surface area of the semi-permeable membrane. This is accomplished with hollow fibers which are about the diameter of human hair and fabricated from cellulose. The hollow fibers are thoroughly cleaned prior to use by running water through them in contact with deionized latex. A bundle of these fibers, attached at both ends to tubing, is placed in a special beaker, which can be securely capped, which contains a magnetic stirrer, and which contains the latex. Distilled water is run through the interior of the fibers continuously, and in a matter of *hours* the conductivity of the dialysate is equal to that of the water. Typically, 100 mL of latex will be dialyzed against 30 L of water in 12 h. It should be noted that even 'conductivity' water contains traces of metal ions which can be efficiently picked up by the latex particles [109]. So it is important to ion-exchange the latex – as discussed above – after dialysis.

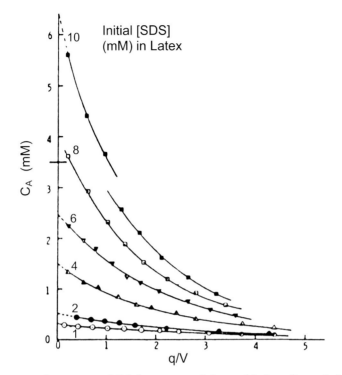

Fig. 91. Serum replacement of PS latex containing added sodium dodecyl sulfate. Concentration of SDS in effluent, C_A, as function of effluent volume, q/V [110].

> Using hollow fibers, in a matter of *hours* the conductivity of the dialysate is equal to that of the water.

Serum replacement. In serum replacement pure water is introduced directly into the colloid to dilute the serum, which is continuously being forced out through a membrane. For example 300 mL of the latex is confined with a 'Nucleopore' membrane in a stirred, flow-through cell, and the water is pumped through at a rate of 50–70 mL h^{-1}. A demonstration of the efficiency of this technique is shown in Fig. 91 in which the concentration in the effluent, C_A, of sodium dodecyl sulfate, SDS, previously added to a cleaned PS latex, is plotted against the volume of effluent [110]. Here q is the volume of effluent and V is that of the latex.

> Serum replacement is a convenient method for obtaining adsorption isotherms.

The rate of efflux is relatively slow, so that the system can be considered at adsorption equilibrium at all times. The net result is that this is a convenient

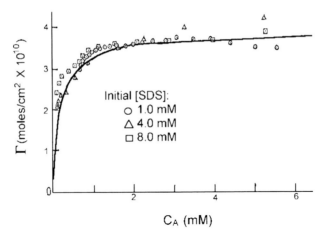

Fig. 92. Adsorption isotherm of SDS on PS latex by serum replacement [110].

method for obtaining adsorption isotherms. Actually, measurement of the surfactant concentration may be done conductometrically, simplifying the experimental procedure. Calculation of the SDS remaining adsorbed from conductometric data is shown as a function of its serum concentration in Fig. 92, in which the surface excess, Γ, is plotted against C_A. The curves are Langmuirian, so that it is possible to calculate from them the area occupied per surfactant molecule, a_s. This comes out to be for SDS 43 Å^2 on PS and 57 Å^2 on PMMA [110].

5.8.2.3 Centrifugation and washing

It turns out that if there is some contamination of a polymer colloid by polyelectrolyte, even dialysis and ion-exchange may not be sufficient to remove it. The usual criterion for purity is a constant titer after repeated ion-exchanges. However, Chonde and Krieger showed, using radioactively labelled materials, that the titer may be reduced even further by centrifugation and washing [111]. This technique involves using a preparative ultracentrifuge in which all the latex particles are spun down to a pellet, the supernatant serum is decanted, and then the particles are redispersed with ultrasound into clean, distilled water. This is repeated until a constant titer is achieved.

> Centrifugation and washing can only be applied to colloids with high surface charge or steric stability.

Although this is the ultimate method of cleaning latexes, it can only be applied to systems which will not coagulate under the conditions in which all adsorbed stabilizers are removed, and which are subjected to rather severe mechanical forces of high acceleration and compression, followed by ultrasonic agitation. Thus it should be used on colloids with relatively high stability due to bound surface ionic groups or steric stabilizing groups.

5.8.3 Surface group composition and concentration

The nature of the surface groups can be determined by various methods, some of which, such as ISS/SIMS or XPS (see below), are sensitive to just the first few ångstroms of the particle surface, whereas others, such as transmission spectroscopy, sample the entire particle.

5.8.3.1 Titration

By far the most commonly used method is that pioneered in the Netherlands [112], namely titration with acid or base. Detection by means of conductance is most commonly employed, since the slopes of the curves obtained provide information about the nature of the surface groups and the purity of the system. In Fig. 93 the titration with standardized barium hydroxide solution of the acid form of a polystyrene latex containing only surface sulfonate groups is shown. The barium 2^+ cation provides a sharper end point than sodium or potassium. It is important, incidentally, since only *micro*equivalents of base are being used, that adsorbed and dissolved carbon dioxide be rigorously purged from the cell by purified inert gas.

The equivalence point is readily obtained by extrapolating the linear portions of the ascending and descending curves, as shown in the figure. The sharpness of the end point is characteristic of strong acids. It is *presumed* that with a hydrophobic polymer and a low concentration of titratable groups,

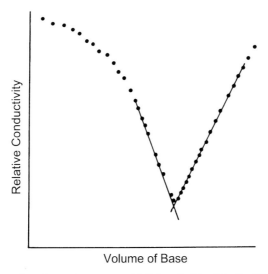

Fig. 93. Titration of strong acid PS colloid with $Ba(OH)_2$ solution.

that all of the groups are on the surface of the particles. Any buried groups would not have been ion-exchanged, and would therefore have no corresponding H^+ counterions.

When a latex containing weak acid groups is titrated, the curves obtained have very different slopes, just as in the titration of small molecule weak acids. Some experimental results by Stone-Masui and Stone [113] on polystyrene latexes containing both strong acid (sulfuric half-acid ester) and weak acid (carboxyl) surface groups are shown in Fig. 94. The first end point is due to strong acid. The slope then increases only slightly to the second end point corresponding to the weak acid. When only a small amount of weak acid is present the minimum in the curve may be just rounded, making it difficult to determine the equivalence point. This kind of information, combined with a knowledge of the chemistry of synthesis of the colloid and its subsequent treatment, often can provide a rather complete picture of the nature of the particle surface.

In this particular instance (latex L6 – highly purified – Fig. 94) no weak acid comonomer was involved; only potassium stearate was used as surfactant. The authors concluded – wrongly – that stearate became grafted onto the latex particle surface by free radical transfer reactions during the emulsion polymerization (cf. Section 5.8.3.3, SIMS, below). Others have found evidence of surfactant grafting, the extent of which undoubtedly depends upon the redox potentials of the initiator and polymer or monomer and the temperature of the reaction. Latex L6 was re-titrated two years later; after 10 months

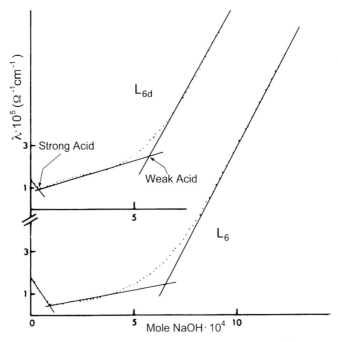

Fig. 94. Titration of PS colloids with both strong and weak acid surface groups [113].

dialysis (!) it was designated as L6d in Fig. 94. The amount of strong acid declined to less than half its original value, no doubt because of hydrolysis of the sulfate ester:

$$\underset{\substack{|\\ \bigcirc}}{\text{———CH}-\text{CH}_2\text{OSO}_3\text{H}} + \text{H}_2\text{O} \longrightarrow \underset{\substack{|\\ \bigcirc}}{\text{———CH}-\text{CH}_2\text{OH}} + \text{H}_2\text{SO}_4$$

The sulfuric acid so formed was removed by dialysis and ion-exchange. The weak acid concentration remained almost constant during that long period of purification [113].

Post-reactions of nontitratable groups may be used to determine their number and kind. For example, reaction of –OH groups with the anhydride of a dicarboxylic acid produces a half-acid ester which can then be titrated with base. A preferred alternative is to oxidize surface hydroxyl groups with persulfate in the presence of 10^{-5} M Ag^+ catalyst, followed by ion-exchange and titration [114].

Cationic colloids can likewise be titrated with acid after ion-exchange to the OH^- form. However, quaternary ammonium hydroxides tend to be unstable, losing amine. It is therefore important to conduct such titrations as soon after ion-exchange as possible, and to store cationic colloids in the halide or other inert salt form. A way to minimize this problem is to add an excess of acid to the latex in the OH^- form, and then back-titrate the excess acid with base.

The apparatus involved in conductometric titration is relatively simple. The cell must have provision for thermostatting, stirring and purging with inert gas. Platinum electrodes – which do not need blackening – are usually used, with a 60 Hz AC potential across them. A motor-driven micropipet may be used to deliver titrant at a constant rate, and the output can be sent to an appropriate recorder.

5.8.3.2 X-ray photoelectron spectroscopy (XPS)

This technique is also known as electron scattering for chemical analysis, or ESCA. If an X-ray photon falls upon a solid surface, electrons can be emitted. This is the photoelectric effect, and the electrons originate from various atoms which make up the solid. The ionization potential is determined by the nature of the atom and its bonding orbitals. The X-ray source is ordinarily Al $K\alpha_{12}$ or Mg $K\alpha_{12}$ which produces photons of 1486.6 and 1253.7 eV respectively. The sampling depth for a typical polymer such as polystyrene turns out to be such that 86% of the signal for C_{1s} atoms comes from within a layer 46 Å thick [115].

A typical photoelectron spectrometer is shown schematically in Fig. 95. The incident angle α and the observing angle θ can be varied by changing the source position and by rotating the sample, respectively. The electron energy spectrum is scanned by varying the potential across the energy analyzer. Since this is a high-vacuum technique, typically 10^{-9} to 10^{-10} Torr,

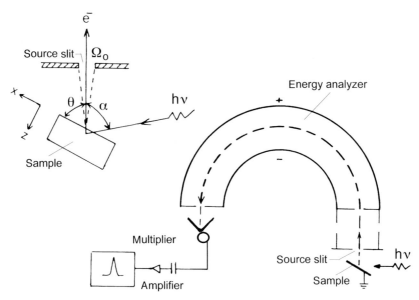

Fig. 95. Schematic diagram of an X-ray photoelectron spectrometer [115].

the samples must be in dry powder form. An XPS spectrum is shown in Fig. 96, taken on a PS latex made with persulfate initiator. The sulfur S_{2p} peak has about 1/1000 the intensity of the C_{1s} peak, but is still quantifiable. Comparison of the amount of sulfur in the surface layer with that in the particles as a whole indicates that nearly all of it is on the surface. Comparison of the S_{2p} peaks of a series of PS latexes, as seen in Fig. 97, shows measurable differences.

> It is possible to calculate surface concentrations from relative peak intensities.

It is possible to calculate surface concentrations from relative peak intensities as follows. The intensity of the experimental peak, I_X, relative to that of C_{1s}, designated I_R, is measured, and then the surface concentration, n atoms cm^{-2}, is calculated by Eq. 91 in which L is the luminosity of the analyzer, σ is the photoelectron cross-section, ϕ is its angular dependence, λ is the photoelectron mean free path in the polymer, and C_R is the volume concentration of the reference atoms, in this case, C_{1s}, calculated from the polymer density and its molecular composition

$$\frac{I_X^a}{I_R} = \frac{L_X}{L_R} \cdot \frac{\phi_X}{\phi_R} \cdot \frac{\sigma_X}{\sigma_R} \cdot \frac{n}{\lambda_R \sin\theta \cdot C_R} \tag{91}$$

For their latex L_h, Stone and Stone-Masui had found by titration the equivalent of 5.2×10^{13} S atoms cm^{-2} ($8.4\,\mu\text{C cm}^{-2}$). From their XPS measurements of

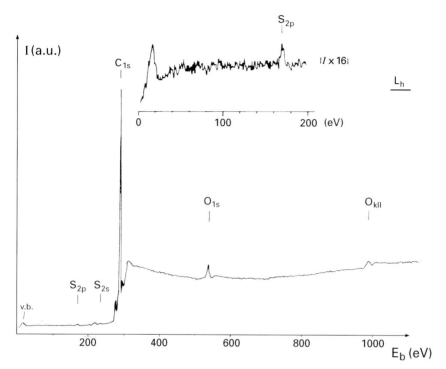

Fig. 96. XPS spectrum of polystyrene latex with surface sulfate groups [115].

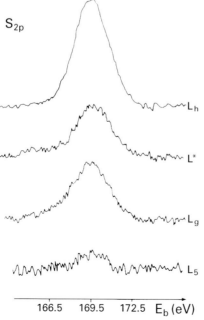

Fig. 97. XPS spectra of S_{2p} peaks in various PS latexes. L5 curve amplified 2× [115].

surface sulfur concentration, applying Eq. 91 to the spectrum in Fig. 96, they obtained $n = 4.0 \times 10^{13}$ S atoms cm^{-2} (6.5 μC cm^{-2}), in quite reasonable agreement considering the differences in techniques, and the fact that the sampling depth for XPS may be less than that for the titration [115]. A major source of error is probably surface roughness due to the fact that the XPS sample was made from freeze-dried latex powder. This produces an uncertainty in, or a distribution in, the angle of observation, θ.

Nevertheless, XPS has been shown to be a powerful technique for sampling monolayers with only a low coverage of atoms. For example 4.0×10^{13} S atoms cm^{-2} (from above) $= 2.5$ nm^2 per sulfate group, or about one-tenth of complete surface coverage. Pichot and coworkers have employed XPS methods to determine the surface concentration of aldehyde [116] and cellobiose [117] groups introduced to the surface by means of two-stage emulsion polymerization of monomers containing these functional groups with styrene. They were able to assign binding energies to C_{1s} peaks for $-C-H$, $-C-O$, acetal and carboxyl carbons, from which they could determine the relative concentrations of the functional groups at the particle surfaces [116, 117]. Pichot et al. also used N-atom analysis to obtain surface concentrations of a maltobionamide functional group [118]. A general reference for XPS is the book by Briggs and Seah [119].

5.8.3.3 Ion scattering spectroscopy/secondary ion mass spectrometry (ISS/SIMS)

In this technique an evacuated sample of dried latex is bombarded with a beam of ions of ^{3}He$^+$, ^{20}Ne$^+$ or argon. In ISS the incident ions exchange energy with surface atoms as they are reflected into an energy analyzer. In SIMS more energetic ions are used which are capable of displacing and ionizing atoms in the surface of the material. It is these secondary ions which are then analyzed. The same apparatus can be used for both kinds of analysis.

ISS. In ISS the energy spectrum of the ions scattered at a given angle, θ, is obtained by an electrostatic analyzer with an ion dectector. The energy, E, of the scattered ions of mass M_2 relative to those of the incident beam, E_0 and M_1, is given by Eq. 92, which is based upon the ballistic conservation of momentum. Various factors affecting scattered ion yield necessitate the use of calibration standards for quantitative work [120].

$$\frac{E}{E_0} = \frac{M_1^2}{(M_1 + M_2)^2} \left[\cos\theta + \left(\frac{M_2^2}{M_1^2} - \sin^2\theta \right)^{1/2} \right]^2 \quad (92)$$

ISS is a highly surface-selective method since ions are scattered from only the first one or two monolayers of atoms. A typical ^{3}He ISS spectrum is shown in Fig. 98, which is not of a latex, but of pure sodium dodecyl sulfonate [121]. ISS is insufficiently sensitive to sulfur to get strong peaks from surface sulfonate groups on latex particles. With a neon ion incident beam no sulfur

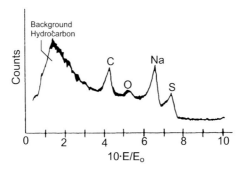

Fig. 98. ^{3}He ISS spectrum of crystalline sodium dodecyl sulfonate [121].

peak is observed at all because ^{20}Ne$^+$ is too massive. Nevertheless ISS can be used for the detection of a variety of surface elements.

SIMS. This method is also known as fast atom bombardment-mass spectrometry, FAB-MS. If an ion beam, which is more energetic than that used in ISS, impinges on a surface, atoms, radicals and molecules – both charged and uncharged – are ejected or 'sputtered' from the surface. The detection of these secondary ions by a mass spectrometer is the basis of SIMS. There is some evidence that the method may be quantitative.

As in ISS, however, the detection of sulfur is difficult and unreliable. Fifield *et al.* took an all-sulfonate PS latex with low surface charge density ($1.2\,\mu\mathrm{C\,cm}^{-2}$) and ion-exchanged the protons with Rb$^+$. This was done because SIMS is highly sensitive to Group I elements and because rubidium is not likely to be a contaminant. The spectrum is shown in Fig. 99, in which the two naturally occurring Rb isotopes are seen, and in the appropriate ratio of peak heights. The surface of the particles was first cleaned up by gentle irradiation with ^{3}He$^+$ for a period of 3 min to remove adsorbed gases and other incidental contaminants.

A comparison of this latex with another one having a surface charge density of $17\,\mu\mathrm{C\,cm}^{-2}$ gave the correct ratio of Rb peak heights, suggesting that this may be a good method for determining surface functional group concentrations.

It is also possible by SIMS to determine whether molecules are chemically bound or just adsorbed onto the surface, especially when they are not readily removed by normal cleaning procedures. A case in point is the adsorption of stearic acid on polystyrene latex particles, initially placed there as potassium stearate surfactant in the emulsion polymerization. The salt is converted upon

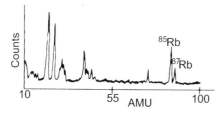

Fig. 99. ^{20}Ne SIMS spectrum of Rb$^+$ sulfonate form of PS latex [121].

ion-exchange to the highly insoluble, 18-carbon acid which is not removed by dialysis for as long as 10 months (cf. Section 5.8.3.1 above). When a dried sample of this latex was subjected to SIMS, the parent molecular ion of stearic acid was clearly seen, as shown in Fig. 100, and it disappeared only after heating the sample in a vacuum at 10^{-4} Torr for about 16 h near 100°C [122].

If the stearic acid were chemically bound, then there would not have been such a strong signal – or perhaps any, because of break-up upon ion bombardment – for the negative ion of mass 283, nor would it have disappeared upon heating.

5.8.3.4 Nuclear magnetic resonance spectroscopy (NMR)

Under certain circumstances NMR may be used to detect surface groups, as discussed in Section 5.6. For example, if the experimental temperature were below T_g the lines would be so broad as to be unobservable as seen in Fig. 79a. If, on the other hand, one observes a signal even below the presumed T_g, it must be because of surface plasticization by the medium, as seen in Fig. 79b. Then the NMR spectrum provides information about the surface groups, although the sampling depth is not known. Otherwise it would be much like absorption spectroscopy in which the entire latex particle is sampled, and compositional information can be helpful only if it is known that the groups in question are at the particle surface. Under such circumstances it is a good idea to have corroboration through some independent analytical method.

5.8.3.5 Surfactant adsorption

A cationic surfactant will be attracted to negatively charged groups on the surface of a latex particle. Hydrophobic cations will tend to adsorb strongly,

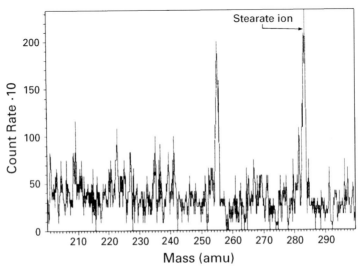

Fig. 100. SIMS spectrum of PS latex prepared with potassium stearate [122].

such that when an amount of cations equimolar to the surface anions has adsorbed, the surface charge will be neutralized. Thus it is possible to measure the concentration and polarity of the surface groups [123]. Any cationic surfactant adsorbed onto bare surface beyond the electroneutral point will bring about charge reversal. These changes can be followed by observing the direction and rate of travel of the latex particles in an externally applied electrical field (see Chapter 8).

The surfactant titration method is extremely sensitive, as can be seen from the ordinate axis in Fig. 101: the 'end point' occurs at almost 10^{-7} molar CTA. The degree of adsorption can be determined conductometrically, taking the difference between the amount of titrant surfactant in solution and that added as the amount adsorbed. If only the surface functional group concentration is required, a simpler method is to carefully titrate with surfactant of opposite charge until rapid coagulation is observed. This should represent the electroneutral state [124].

5.8.3.6 Dye partition

Palit and coworkers devised a method for determining the number of functional end groups per polymer molecule *in solution*. If all of the functional groups are at the particle surface then a calculation of the surface group concentration can be made. The method involves dissolving the polymer in a hydrophobic organic solvent and shaking this with an aqueous solution of a dye which is substantive to the end groups. An amount of dye equivalent to the number of end groups will thus transfer to the organic phase, where it can be measured by absorption spectroscopy [125]. For example, the dye

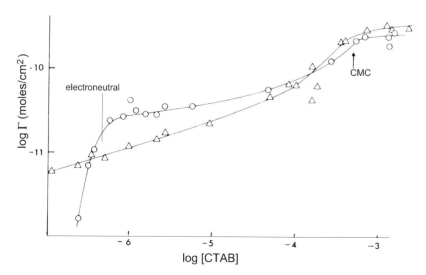

Fig. 101. Adsorption isotherm of cetyl trimethyl ammonium (CTA) onto PS bulk (–△–) and PS anionic latex particles (–○–) [123].

1,1'-diethyl carbocyanine ('Pinacyanol') will form a salt with carboxylate anion which has a strong absorbance at around 610–620 nm depending upon the organic solvent used.

There are problems with this technique, however, such that many workers have not been able to get agreement with independent measurements. Glass contains surface anionic groups which also may react with the dye, so that glass cells present a problem. Water can be extracted into the organic phase, and the resulting solution might dissolve traces of dye. Any residual surfactant may tend to solubilize the dye as well. There appear to be dependencies on the pH of the aqueous phase and upon the molecular weight of the polymer. There can, of course, be no anionic groups along the polymer chain (e.g. –COOH due to surface hydrolysis of acrylate) which would react with the dye. The method has also been developed for other end groups using other dyes and experimental conditions. Although the dye partition method has not been generally accepted as a quantitative tool, it may be valuable as a qualitative means of determining the nature of end groups present in polymers made by emulsion polymerization.

References

63. Lord Rayleigh (1871). *Phil. Mag.* **41**, 107, 274.
64. Brumberger, H., Stein, R.S. and Rowell, R.L. (1968). *Science and Technology*, Nov., p. 38.
65. P. Debye (1915). *Ann. Physik.* **46**, 809.
66. Liang, S.J., Fitch, R.M. and Ugelstad, J. (1984). *J. Colloid Interface Sci.* **97**, 336.
67. Ford, J.R., Rowell, R.N. and Rowell, R.L. (1984). *A Computer Program for the Inversion of Angular Light Scattering Data*, Department of Chemistry, University of Massachusetts, Amherst, MA.
68. Kerker, M. (1969). *The Scattering of Light and Other Electromagnetic Radiation*, Academic Press, New York.
69. Berne, B.J. and Pecora, R. (1975). *Dynamic Light Scattering*, John Wiley, New York.
70. Ottewill, R.H. (1987). In *Future Directions in Polymer Colloids* (M.S. El-Aasser and R.M. Fitch eds), NATO ASI Series E, No. 138, p. 253, Martinus Nijhoff, Dordrecht, Netherlands.
71. Mills, M.F., Gilbert, R.G., Napper, D.H., Rennie, A.R. and Ottewill, R.H. (1993). *Macromolecules* **26**, 3553.
72. Ottewill, R.H. (1985). *Ber. Bunsenges. Phys. Chem.* **89**, 517.
73. Bacon, G.E. (1977). *Neutron Scattering in Chemistry*, Butterworths, London.
74. Ise, N. (1994). *Int. Polym. Colloid Newsletter* **25** 1, 35.
75. Ito, K., Yoshida, H. and Ise, N. (1994). *Science* **263**, 66.
76. Lee, S. and Rudin, A. (1992). In *Polymer Latexes, Preparation, Characterization and Applications* (E.S. Daniels, E.D. Sudol and M.S. El-Aasser eds), ACS Symposium Series 492, p. 234, American Chemical Society, Washington, DC.
77. Scholsky, K.M. and Fitch, R.M. (1985). *J. Colloid Interface Sci.* **104** (2), 592.
78. Hayat, M.A. (1988). *Principles and Techniques of Electron Microscopy; Biological Applications*, 3rd edn, CRC Press, Boca Raton, FL.
79. Lyman, C.E., Newberry, D.E., Goldstein, J.I., Williams, D.W., Romig, A.D. Jr., Armstrong, J.T., Echlin, P., Fiori, C.E., Joy, D.C., Lifshin, E. and Peters, K.K. (1990). *Scanning Electron Microscopy, X-Ray Microanalysis, and Analytical Electron Microscopy: A Laboratory Workbook*, Plenum Press, New York.
80. Juhué, D. and Lang, J. (1994). *Colloids Surfaces A: Physicochem. Eng. Aspects* **87**, 177.

81. Sogami, I. and Ise, N. (1984). *J. Chem. Phys.* **81**, 6320.
82. Dos Ramos, J.G. and Silebi, C.A. (1993). *Polym. Int.* **30** (4), 445.
83. Tauer, K. (1994). *Proceedings of the 3rd International Symposium on Radical Copolymers in Dispersed Media*, p. 106, Groupe Français des Polymères, Lyon.
84. Bassett, D.R. and Hoy, K.L. (1980). In *Polymer Colloids II* (R.M. Fitch ed), p. 1, Plenum Press, New York.
85. Liang, S.J. and Fitch, R.M. (1982). *J. Colloid Interface Sci.* **90** (1), 51.
86. Coll, H. and Searles, C.G. (1986). *J. Colloid Interface Sci.* **65**, 110.
87. Hansen, F.K. (1991). *ACS Symposium Series*, No. 472, p. 168, American Chemical Society, Washington, DC.
88. Kusters, J.M.H. (1993). Ph.D. Thesis, Technische Universiteit Eindhoven, The Netherlands, p. 73.
89. Giddings, J.C. (1993). *Science* **260**, 1456.
90. Barman, B.N. and Giddings, J.C. (1992). *Langmuir* **8**, 51.
91. Beckett, R., Ho, J., Jiang, Y. and Giddings, J.C. (1991). *Langmuir* **7**, 2040.
92. Hiemenz, P.C. (1986). *Principles of Colloid and Surface Chemistry*, 2nd edn, Chapter 2, Marcel Dekker, New York.
93. Tarcha, P.J., Fitch, R.M., Dumais, J.J. and Jelinski, L.W. (1983). *J. Polym. Sci., Polym. Phys. Ed.* **21**, 2389.
94. Becker, E.D. (1980). *High Resolution NMR*, 2nd edn, Academic Press, New York.
95. Nzudie, D.T., Delmotte, L. and Riess, G. (1991). *Makromol. Chem., Rapid Commun.* **12**, 251.
96. Luaro, M.F., Pétiaud, R., Hidalgo, M., Guillot, J. and Pichot, C. (1994). *Proceedings of the 3rd International Symposium on Radical Copolymers in Dispersed Media*, p. 86, Groupe Français des Polymères, Lyon.
97. Ramirez, W. and Guillot, J. (1994). *Proceedings of the 3rd International Symposium on Radical Copolymers in Dispersed Media*, p. 179, Groupe Français des Polymères, Lyon.
98. Fitch, R.M., Su, L.S. and Tsaur, S.L. (1990). In *Scientific Methods for the Study of Polymer Colloids and Their Applications* (F. Candau and R.H. Ottewill eds), p. 373, Kluwer, Dordrecht, Netherlands.
99. Yoshino, S. (1993). *Polym. Int.* **30** (4), 541.
100. Su, L.S., Jayasuriya, S. and Fitch, R.M. (1993). *Polym. Int.* **30** (2), 221.
101. Krieger, I.M., Goodwin, J. *et al.* (1987). In *Future Directions in Polymer Colloids* (M.S. El Aasser and R.M. Fitch eds), NATO ASI Series E, No. 138, p. 110, Martinus Nijhoff, Dordrecht, Netherlands.
102. Kuo, R.J., Ruch, R.J. and Meyers, R.R. (1978). In *Emulsions, Latices and Dispersions* (P. Becher and M.N. Yudenfreund eds), p. 257, Marcel Dekker, New York.
103. Schwartz, G. (1962). *J. Phys. Chem.* **66**, 2636.
104. Rosen, L.A. and Saville, D.A. (1990). *J. Colloid Interface Sci.* **140**, 82.
105. Dukhin, S.S. and Shilov, V.N. (1974). *Dielectric Phenomena and the Double Layer in Disperse Systems and Polyelectrolytes*, John Wiley, New York.
106. Vanderhoff, J.W., van den Hul, H.J., Tausk, R.J.M. and Overbeek, J.Th.G. (1970). In *Clean Surfaces: Their Preparation and Characterization for Interfacial Studies* (G. Goldfinger ed), Marcel Dekker, New York.
107. McCann, G.D., Bradford, E.B., van den Hul, H.J. and Vanderhoff, J.W. (1971). In *Polymer Colloids* (R.M. Fitch ed), p. 29, Plenum Press, New York.
108. Hearn, J., Wilkinson, M.C., Goodall, A.R. and Cope, P. (1980). In *Polymer Colloids II* (R.M. Fitch ed), p. 379, Plenum Press, New York.
109. McCarvill, W.T. and Fitch, R.M. (1978). *J. Colloid Interface Sci.* **64** (3), 403.
110. Ahmed, S.M., El-Aasser, M.S., Micale, F.J., Poehlein, G.W. and Vanderhoff, J.W. (1979). In *Solution Chemistry of Surfactants* (K.L. Mittal ed), Vol. 2, p. 853, Plenum Press, New York; Ahmed, S.M., El-Aasser, M.S., Pauli, G.H., Poehlein, G.W. and Vanderhoff, J.W. (1980). *J. Colloid Interface Sci.* **73**, 388.
111. Chonde, Y. and Krieger, I.M. (1980). *J. Colloid Interface Sci.* **77**, 138.
112. Verwey, E.J.W. and Kruyt, H.R. (1933). *Z. Physik. Chem.* **A167**, 149.
113. Stone-Masui, J.H. and Stone, W.E.E. (1980). In *Polymer Colloids II* (R.M. Fitch ed), p. 331, Plenum Press, New York.

114. van den Hul, H.J. and Vanderhoff, J.W. (1970). *Br. Polym. J.* **2**, 121.
115. Stone, W.E.E. and Stone-Masui, J.H. (1983). In *Science and Technology of Polymer Colloids II* (G.W. Poehlein, R.H. Ottewill and J.W. Goodwin eds), NATO ASI Series E, No. 68, p. 480, Martinus Nijhoff, Dordrecht, Netherlands.
116. Charleux, B., Fanget, P. and Pichot, C. (1992). *Makromol. Chem.* **193**, 205.
117. Charreyre, M.T., Boulanger, P., Delair, Th., Mandrand, B. and Pichot, C. (1993). *Colloid Polym. Sci.* **271**, 668.
118. Revilla, J., Elaïssari, A., Pichot, C. and Gallot, B. (1995). *Polym. Adv. Tech.* **6**, 455.
119. Briggs, D. and Seah, M.P. (1983). *Practical Surface Analysis by Auger and X-Ray Photoelectron Spectroscopy*, John Wiley, Chichester.
120. Czanderna, A.W. (1975). *Methods of Surface Analysis*, p. 76, Elsevier, Amsterdam.
121. Fifield, C.C. and Fitch, R.M. (1981). *J. Dispersion Sci. Tech.* **2**, 267.
122. Stone-Masui, J.H. and Stone, W.E.E. (1993). *Polym. Int.* **30**, 169.
123. Connor, P. and Ottewill, R.H. (1971). *J. Colloid Interface Sci.* **37** (3), 642.
124. Greene, B.W. (1973). *J. Colloid Interface Sci.* **43** (2), 449, 462.
125. Banthia, A.K., Mandal, B.M. and Palit, S.R. (1977), *J. Polym. Sci., Polym. Chem. Ed.* **15**, 945.

Chapter 6

Chemistry at the Interface

6.1 Introduction

We have seen earlier in this book that polymer colloids carry various chemical groups at the polymer/medium interface. In this chapter the chemistry involved is examined from two major points of view: (1) controlled synthesis and (2) chemical reactions at the interface. The practical implications for doing so are many:

- Latex paints involve huge production volumes worldwide. The shelf stability of the paints, pigment dispersion, adhesion to surfaces by the derived films, and the barrier and dynamic mechanical properties of the films all are affected by the surface chemistry of the latex particles. Water-based printing inks have similar considerations.
- Water-borne adhesives interact with the bonded surfaces by means of their surface groups, and the moisture-sensitivity of the bonds is affected as well.
- Some synthetic fibers are made by emulsion polymerization followed by coagulation, washing, dissolution in a good solvent and spinning. Substantivity of dyes is affected by surface groups originating in the emulsion polymerization. The effluent water from coagulation and washing contains surfactants and other chemicals. Surfactants can be eliminated by chemically bonding the stabilizing groups to the particle surface.
- Molding resins are often made by emulsion or suspension polymerization and must also be coagulated and washed.
- Polymer colloids are used in important ways in clinical diagnostics, drug delivery and in chemotherapy.
- Physicists and chemists are using monodisperse polymer colloids for a host of experiments, for example in the rheology of complex fluids, measuring fundamental forces such as dispersion interactions, coagulation kinetics, steric interactions, and electrical double layer forces. A knowledge of surface group chemistry and surface charge densities is essential for such studies.

6.2 Origins of surface functional groups

A unique feature of emulsion polymerization is that the free radicals are generated in a phase different from that in which polymerization takes place.

The result is that all radicals derived from the initiator must enter the particles through the polymer/water interface. If the end group is sufficiently hydrophilic, it will be held at the interface like a cork on the end of a string, floating on water. An exception would be free radicals resulting from chain transfer with monomer or added solvent which would tend to be hydrophobic in most cases. It is possible, however, to add deliberately a surface-active chain transfer agent containing a functional group. Chain transfer to surfactant also can occur, and the resulting free radical will tend to stay at the particle surface. Of course, functional groups can be introduced via monomers containing them, although it is less likely that they will all be held at the interface, because monomers may have more hydrophobic moieties than initiator fragments. All of these potentialities are considered in detail below.

6.2.1 Initiator-derived functional groups

It was shown in Section 2.1.1 that an initiator fragment in the aqueous phase usually, if not always, reacts with one or more monomer units in the aqueous phase to form a surface-active free radical. It is this species which enters the particle through the interface, and it is its chemistry which determines to what extent it is retained at the interface, and what kind of functionality can be involved. This is schematically represented in Fig. 102, in which a sulfonate group is brought to the particle surface.

Some common initiators and the functional groups they produce are given in Table 6.1. The last item in the list does not produce reactive functional groups, but the PEG chains on the surface, when used, can greatly affect the adsorption of molecules and therefore of reactivity. The PEG moieties also are known to complex alkali metal cations, which may have implications for catalysis.

6.2.1.1 Surfactant-free PS colloids

Persulfate-initiated systems. It is possible with the appropriate initiator and emulsion polymerization conditions to make 'emulsifier-free' latexes. The

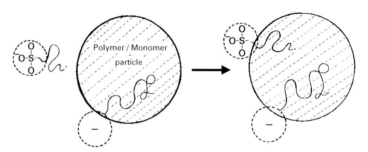

Fig. 102. Capture of surface-active oligoradical to introduce functional end group at the polymer/water interface.

oligomeric radical shown in Fig. 102 is surface-active after it has added just a few monomer units, as long as they are sufficiently hydrophobic (or lyophobic, to be more general). In the absence of surfactant, nucleation must be homogeneous (Section 2.1.1) followed by considerable, but limited, coagulation. Under these conditions, the rate of generation and entry of surface-active oligoradicals will determine the extent of coagulation and therefore of particle size: the more rapidly that the latex particles accumulate surface charge, the smaller they will be when they achieve sufficient charge to be colloidally stable. With electrostatic stabilization, the influence of ionic strength is important (Section 7.2.3), whereas if nonionic initiators, such as the PEGA, CerAlc or ACO's, are used, then the latexes are sterically stabilized (Section 7.3) and ionic strength will not be as important.

Some examples may help to illustrate these points: Goodwin et al. [127] found that monodisperse polystyrene colloids could be synthesized without using surfactant, with just persulfate and styrene in water, sometimes with added salt to control ionic strength. Exploration of the variables of temperature, initiator and salt concentrations led to an empirical equation which has practical value for anyone interested in making such colloids. The end groups at the particle surface were sulfate half-acid ester, alcohol and sometimes carboxyl, depending upon the pH and other conditions of the polymerization [127], as given in Table 6.1. Their equation for the particle diameter, D, is:

$$\log D = 0.238 \left[\log \frac{[\mu][M]^{1.723}}{[P]} + \frac{4929}{T} \right] - 0.827$$

where μ is the ionic strength of the aqueous phase, [M] is monomer concentration, [P] is persulfate concentration, and T is the absolute temperature. The master curve for this *empirical* equation, with the experimental data from which it was derived, is shown in Fig. 103. It appears to be valid over four orders of magnitude of experimental conditions [128], and has been used successfully by many groups around the world to make model polystyrene colloids of high uniformity and predetermined size.

Cationic colloids. Liu and Krieger, on the other hand, made cationic polystyrene latexes which they said were monodisperse. They used AIBA·2HCl (azoiso-butyramidinium; see Table 6.1) as the initiator without any surfactant [129]. The dependence of particle size on ionic strength is shown in Fig. 104, and on initiator concentration, in Fig. 105. The clear reduction in particle size with increasing initiator concentration and its increase with ionic strength both attest to the HUFT mechanism for particle formation, summarized by Eq. 12, in which coagulation plays a prominent role. Liu and Krieger found the surface charge density for these cationic colloids to be on the order of $1.6\,\mu\text{C}\,\text{cm}^{-2}$.

It should be noted that the amidinium group is formed by protonation of amidine, a fairly strong base. This means that the charge density is a function of pH. Furthermore, amidine is rather easily hydrolyzed to carboxyl, so that the stability of the surface chemistry of these latexes may be difficult to maintain.

TABLE 6.1
Selected initiators and derived surface groups [126]

Name	Abbreviation	Chemical formula	Derived surface groups
Persulfate	P	$S_2O_8^{2-}/M^{2+}$	$ROSO_3^-M^+$; ROH; RCOOH
Persulfate/bisulfite/iron	PBI	$S_2O_8^{2-}/HSO_3^-/Fe^{2+,3+}$	$ROSO_3^-M^+$; $RSO_3^-M^+$; ROH
Bisulfite/iron (III)	BI	HSO_3^-/Fe^{3+}	$RSO_3^-M^+$
Bisulfite/silver (I)	BAg	$HSO_3^-/Ag(NH_3)_2^+$	$RSO_3^-M^+$
Bisulfite/copper	BCu	$HSO_3^-/Cu(NH_3)_4^{2+}$	$RSO_3^-M^+$
Azoiso-butyramidinium	AIBA	$=\{N-\underset{CH_3}{\underset{|}{C}}-\underset{CH_3}{\underset{|}{C}}\overset{+}{N}H_2X^-}\}_2$ NH_2	$R-\underset{CH_3}{\underset{|}{C}}-\underset{CH_3}{\underset{|}{C}}\overset{+}{N}H_2X^-$ NH_2 R
4,4'-Azobis-4-cyano pentanoic acid	ABCPA	$=\{N-\underset{CN}{\underset{|}{\underset{|}{C}}}-CH_2-CH_2-COOH\}_2$ CH_3	COOH RCOOH
Hydrogen peroxide – iron (II)	(Fenton reagent)	H_2O_2/Fe^{2+}	ROH
Perphosphate	PPh	$H_2P_2O_8^{2-}M^{2+}$	$ROPO_3M^+H^+$ and ROH (?)
p-Sulfomethyl benzoyl peroxide	SMBP	$\left[\text{COO}-\!\!\!\!\bigcirc\!\!\!\!-CH_2-SO_3^-M^+\right]_2$	$RSO_3^-M^+$

Name	Abbr.	Structure	End group
Azobis-α-cyano-alkyl sulfonate	ACAS	=\{N−C(CH$_3$)−(CH$_2$)$_n$−SO$_3^-$Na$^+$\}$_2$ with CN	RSO$_3^-$M$^+$
Azobispolyethylene glycol i-butyrates (PEG = polyethylene glycol)	PEGA	=\{N−C(CH$_3$)(CH$_3$)−CH$_2$−CO−(O−CH$_2$CH$_2$)$_n$−OH\}$_2$	RCO(PEG)$_n$−OH
Cerium (IV)/alcohol	CerAlc	RR′CH−OH/Ce^{4+}	RO(PEG)$_n$−OH or RR′OH
Asymmetric azo-'Inisurf'	ACO−#	t-Bu−N=N−C(CH$_3$)(CN)−CH$_2$−CH$_2$−CO−(OCH$_2$CH$_2$)$_n$−O−φ−C$_9$H$_{19}$	End group: RCO(PEG)$_n$−O−φ−nonyl
Cu(II)/thiosulfate	CuThio	Cu^{2+}/S$_2$O$_3^{2-}$	RSSO$_3^-$M$^+$

150   POLYMER COLLOIDS: A COMPREHENSIVE INTRODUCTION

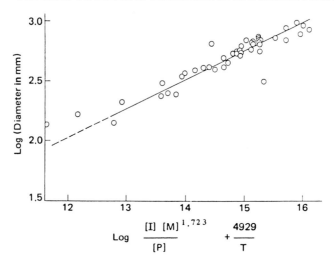

Fig. 103. Surfactant-free PS latex particle size as a function of a variety of experimental conditions [128].

Sulfonate surface groups. One of the most stable surface groups is sulfonate, which resists hydrolysis and oxidation. It is derived from the $\cdot SO_3^-$ radical, itself readily obtainable from the oxidation of sulfite ion:

$$SO_3^{2-} + Ox^{n+} \rightarrow Ox^{(n-1)+} + \cdot SO_3^- \qquad \cdot SO_3^- + M \rightarrow \cdot MSO_3^-, \qquad \text{etc.}$$

Oxidizing agents which have been used include persulfate, iron (III), silver diammine, and copper (II) tetrammine, as shown in Table 6.1. Ferric ion, and cupric to a lesser extent, have the problem that as multivalent cations they

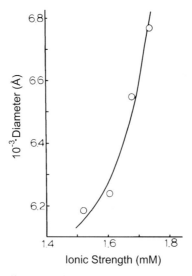

Fig. 104. PS latex diameter dependence upon NaCl concentration [129].

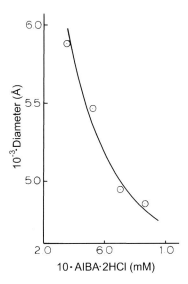

Fig. 105. PS latex diameter dependence upon AIBA · 2 HCl concentration [129].

tend to cause coagulation of anionic colloids. In the case of ferric, a surfactant must be used to obtain a stable colloid at all. Silver amine avoids this problem by being univalent, but it has the disadvantage of high cost, if this is to be a commercial process, along with the fact that the '$Ox^{(n-1)+}$' is silver metal in this instance which has to be removed. Sometimes silver deposits as a beautiful mirror on the walls of the flask, obviating the need for special techniques for its removal [130]. The resulting latexes have only surface sulfonate groups and are highly stable [131]. The water-soluble sulfonate derivatives of benzoyl peroxide (SMBP) and the azo compound (ACAS) work effectively, in a manner similar to that of AIBA, but they remain laboratory curiosities.

6.2.2 Comonomer-derived functional groups

6.2.2.1 Advantages and disadvantages of using comonomers

Comonomers bearing functional groups tend to be the easiest means for introducing such groups to the latex particle surface. There are some concerns, however:

- Small comonomer molecules tend to be water-soluble and therefore capable of forming soluble polyelectrolyte during emulsion polymerization. Acrylic acid and sodium styrene sulfonate are examples.
- Functional groups may not be held strongly at the interface, and consequently they can become 'buried' within the particles.
- With ordinary batch emulsion polymerization there is no way to control separately both particle size and surface group concentration.

A way to avoid most, if not all, of these problems is to use a strongly surface-active comonomer. Table 6.2 lists several common functional monomers.

POLYMER COLLOIDS: A COMPREHENSIVE INTRODUCTION

TABLE 6.2
Comonomers which give surface functional groups

Name	Structure	Abbreviation
Weakly surface-active		
Acrylic acid	$CH_2=CH \cdot COOH$	AA
Acrolein	$CH_2=CH \cdot CHO$	A
Methacrylic acid	$CH_2=C(CH_3)-COOH$	MAA
Sulfoethyl methacrylate	$CH_2=C(CH_3)-COOCH_2CH_2SO_3^- M^+$	SEM
Sulfomethyl styrene	$p\text{-}(CH_2=CH)\text{-}C_6H_4\text{-}CH_2SO_3^- M^+$	SMS
Acrylamido methyl propane sulfonate	$CH_2=CH-CONH-C(CH_3)_2-CH_2SO_3^- M^+$	AMPS
Styrene sulfonate	$p\text{-}(CH_2=CH)\text{-}C_6H_4\text{-}SO_3^- M^+$	SS
Methacroyloxy ethyl trimethyl ammonium	$CH_2=C(CH_3)-COOCH_2CH_2-N^+(CH_3)_3\ X^-$	META
p-Formyl styrene	$p\text{-}(CH_2=CH)\text{-}C_6H_4\text{-}CHO$	FS

CHAPTER 6 CHEMISTRY AT THE INTERFACE

TABLE 6.2 – continued

Name	Structure	Abbreviation
Strongly surface-active		
Acrylamido stearate	$CH_2=CH-CONH-C(H)(- (CH_2)_7 CH_3)(- (CH_2)_8 COO^- Na^+)$	AAmS
Styryl undecanoate	4-vinylphenyl–$CH(CH_3)-(CH_2)_8-COO^- M^+$	SU
Sulfodecyl styryl ether	2-vinylphenyl–$O-(CH_2)_{10}-SO_3^- M^+$	SDSE
Trimethyl ammonium decyl styryl ether	2-vinylphenyl–$O-(CH_2)_{10}-N^+(CH_3)_3\ X^-$	TADSE
Maleoyloxy ethyl dimethyl dodecylbenzyl ammonium	$H_{25}C_{12}$–C$_6$H$_4$–CH$_2$–$N^+(CH_3)_2$–CH$_2$CH$_2$–O–CO–CH=CH–COOH, X^-	MDDA
6-(2-Methylpropenoyl-oxy) hexyl b-D-cellobioside	cellobiosyl–O–(CH$_2$)$_6$–O–CO–C(CH$_3$)=CH$_2$	CHMA

6.2.2.2 Independent control of particle size and surface group concentration

By the use of highly surface-active, functional comonomers in two-stage emulsion polymerization it is possible to control independently both particle size and surface group (or charge) concentration. The idea is to make a seed latex in the first stage with a minimal amount of the comonomer to control particle size. In separate experiments the Gibbs surface tension/concentration curve, like that shown for sodium-SDSE (SSDSE) in Fig. 106, is obtained [132]. Subsequently the adsorption isotherm on the seed latex can be obtained by measuring the surface tension of seed latexes (thoroughly deionized) to which varying amounts of the comonomer have been added. The surface tension gives the concentration in solution from the Gibbs plot, and the difference between the amount in solution and the amount added gives the amount adsorbed.

The second stage of the emulsion polymerization is then conducted with the desired amount of added comonomer plus an additional amount of the principal monomer, e.g. styrene, to ensure bonding of the functional monomer to the particle surface. Choice of initiator is important in this operation since it can introduce unwanted functional groups. Furthermore, water-soluble initiators must be used to realize decent surface yields [132]. Even so, it is extremely important to clean the colloids after polymerization if water-soluble polyelectrolyte will interfere with subsequent use of the colloids, since there will always be some formed from the comonomer which is in solution [133]. It is thus probably a good idea to use a functional comonomer with as low a CMC as possible. For model colloids the following cleaning steps are recommended:

- steam stripping to remove unreacted monomers and other volatile by-products;
- filtration through glass wool to remove 'coagulum';

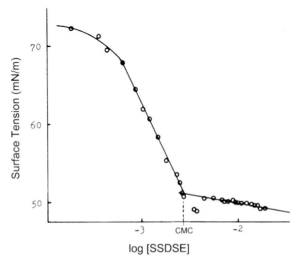

Fig. 106. Gibbs plot of SSDSE [132].

- ion exchange to remove ionic impurities and to convert counterions to the desired form;
- repeated centrifugation/decantation/redispersion to completely remove remaining impurities, especially polyelectrolyte; and
- filtration through a membrane filter (e.g. 0.6 μm) to remove coarse particles!

6.2.3 Chain transfer-derived surface groups

6.2.3.1 Coincidental transfer to surfactant

Chain transfer during emulsion polymerization can play an important role in the surface chemistry of latex particles. This is especially true of molecules which tend to concentrate at the interface. For example, emulsifiers with labile hydrogen atoms are particularly capable of becoming chemically bound through this process. These can be ordinary alkyl groups as in lauryl sulfate or sulfonate. It has been found that as many as half of the surface functional groups may be derived from SDS or SDSo (sodium dodecyl sulfonate) even when the emulsion polymerization is conducted at room temperature [134]. As we have seen earlier (Section 5.8.3.3, SIMS), strongly adsorbed surfactants, such as stearic acid, may be removed with such difficulty that they appear to be chemically bound. In many practical situations it may make no difference, but the scientist involved needs to understand that in the latter case, simply raising the pH changes the acid to its water-soluble salt, making it relatively easy to desorb.

6.2.3.2 Using 'transurfs'

One may deliberately wish to use chain transfer to a surface-active species in order to introduce functional groups to the particle/medium interface. For example, Fifield synthesized sodium 10-mercapto-1-decane sulfonate (SMDSo) and used it in a two-stage emulsion polymerization in a manner similar to that used with surface-active, functional comonomers described earlier: a polystyrene seed latex with low surface charge density was highly purified to remove all extraneous surface-active material. SMDSo and more styrene were added for the second stage polymerization. By this means the surface density went from 0.34 μC cm^{-2} for the seed to 5.3 μC cm^{-2}, a 16-fold increase [135]! Disadvantages of this technique are that the synthesis of the surface-active chain transfer agent may be costly, and – in the case of mercaptans – the compounds must be kept carefully away from air and other oxidants. The choice of initiator is important in all of these instances to ensure that it does not (a) overly react with the mercaptan (as peroxides would do), or (b) introduce undesirable surface groups. For example, Fifield used butanedione-2,3 ('biacetyl') and visible light to photogenerate nonionic, nonoxidizing free radicals at room temperature.

Nonionic transurfs also have been employed in batch polymerizations to form highly stable particles, some of which were found to be resistant to freezing and thawing as well as to high concentrations of polyvalent electrolyte. The chain-transfer surfactants used contained polyethylene oxide moieties of various chain lengths, for example:

$$HO-(CH_2-CH_2O)_n(CH_2)_{11}-SH$$

in which n was 17, 40 and 75. Latex particle sizes between 99 and 220 nm were obtained, which in some cases were monodisperse [136].

A useful review of the chemistry of 'surfmers,' 'inisurfs' and 'transurfs' has been prepared by Guyot and Tauer [137].

6.2.4 Chemical modification of surface groups

6.2.4.1 Reactions of persulfate

One may use conventional chemical reactions to modify the surface groups on polymer colloids. Sometimes this happens inadvertently. For example during polymerization with persulfate initiator several reactions occur: sulfate ion radicals produce organic sulfate ester groups which can be hydrolyzed at a rate which depends on the pH to form hydroxyl:

$$\cdot SO_4^- + nM \longrightarrow \cdot(M)_n-OSO_4^-$$

$$\cdot(M)_n-OSO_4^- + H_2O \longrightarrow \cdot(M)_n-OH + H_2SO_4$$

But in the presence of persulfate (more precisely, *peroxy*disulfate, $S_2O_8^{2-}$) the alcohol can be oxidized to carboxyl:

$$\cdot(M)_n-OH + 2\,S_2O_8^{2-} + H_2O \longrightarrow \cdot(M)_{(n-0.5)}-COOH + 4\,HSO_4^-$$

6.2.4.2 Hydrolysis of surface groups

Many common monomers contain groups which are hydrolyzable, so that their derived water-borne colloids may exhibit changes in surface chemistry over time. The products of hydrolysis may be released into the medium and change it as well. For example, latex paints based upon polyvinyl acetate are known to undergo pH drift as surface acetate ester groups are transformed into hydroxyl groups (forming 'polyvinyl alcohol') and acetic acid. Thus these systems must be properly buffered if malodor, corrosion in metal containers, and ultimate coagulation of the latex are to be avoided. Introduction of higher homolog vinyl esters can contribute considerably to retardation of this hydrolytic degradation. Other monomers susceptible to hydrolysis include the lower acrylate esters, acrylamide, acrylonitrile to a lesser extent, etc. The *meth*acrylates are much less easily hydrolyzed because of their chain stiffness which greatly impedes diffusion of small molecules and ions (H_2O and H_3O^+) into the polymer, and the α-methyl group which provides steric hindrance.

Autocatalyzed surface hydrolysis. When particles bearing strong electrolyte groups are ion-exchanged to the acid form the surface concentration of H^+ can become on the order of millimolar to molar, leading to the possibility that these protons can act to catalyze the hydrolysis of surface groups. For example, the methyl ester groups in polymethyl acrylate latex particles will be hydrolyzed to acrylic acid groups and methanol. It turns out that the kinetics are zero order, i.e. the rate is constant over long periods of time [138], as seen in Fig. 107. Incidentally, such systems may be useful for the controlled release of small, active molecules into the surrounding medium.

In this particular example, methanol is released. This is not a particularly exciting result, but other acrylate esters, such as benzyl, cyclohexyl, naphthyl and salicyl, also have been investigated. Although the salicylate did not show simple zero-order release kinetics, it nevertheless suggests that long-term pain relief may be achieved by the ingestion of such a functional polymer colloid [139].

The hydrolysis is undoubtedly catalyzed by surface acid groups because the rate becomes negligible upon exchanging H^+ counterions with Na^+. Furthermore, the (pseudo) zero-order initial rate of autocatalyzed hydrolysis has been shown to depend upon the surface acid concentration expressed as microequivalents of H^+ per square centimeter, as shown in Fig. 108. In this example polymethyl acrylate colloids with various surface concentrations of strong acid were titrated with base as a function of time to give initial rates. The log–log plot of these rates against surface acid concentration, σ'_0, gave a slope of 1.16, strongly suggesting first-order kinetics.

As mentioned earlier, the acid groups themselves are subject to hydrolysis if they are sulfate (half-acid esters of sulfuric acid):

$$ROSO_3^- H^+ + H_2O \rightarrow ROH + H_2SO_4$$

Sulfonate groups (sulfonic acids) generally do not undergo hydrolysis. So it is possible to synthesize latexes with both RSO_3^- and ROH groups in controlled ratios by judicious use of initiators such as PBI, followed by hydrolysis [134].

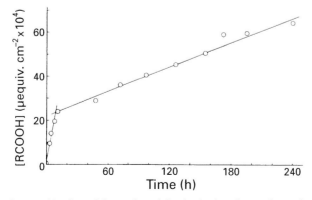

Fig. 107. Surface sulfonic acid-catalyzed hydrolysis of a polymethyl acrylate colloid at 90°C [138].

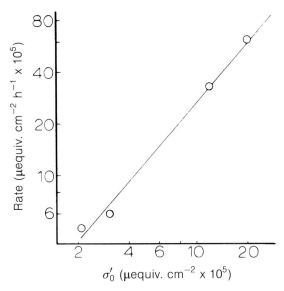

Fig. 108. Dependence of hydrolysis rate on surface acid concentration. Polymethyl acrylate at 90°C [138].

Externally catalyzed surface hydrolysis. The pH of the surrounding aqueous medium can have a pronounced effect upon the rates of hydrolysis of surface groups. For example, polyvinyl acetate latexes are notorious for 'pH drift' as the surface ester groups hydrolyze to form acetic acid. Release of the acid into the medium makes the process auto-acceleratory. The drop in pH tends to destabilize the latex, thereby shortening its 'shelf life.' To overcome this a buffer must be added. The 'aspirin' release rates of the polysalicyl acrylate colloids mentioned above are sensitive to slight changes in pH: the rate at pH 7.4 (blood pH) is measurably slower than that at pH 7.0 [139].

6.2.4.3 Derivatization of benzyl chloride groups

Surface amine and quaternary ammonium groups. Vinyl benzyl chloride (chloromethyl styrene) is a commercially available, highly reactive monomer. It readily copolymerizes with styrene to form aqueous latexes with little surface hydrolysis, if care is taken to maintain neutral pH and low temperatures. The monomer is not very surface-active, so that the chlormethyl groups are distributed more or less randomly throughout the latex particle. Introduction of ammonia or amines to the aqueous phase induces reaction with only the surface functional groups, e.g:

where the $-\phi-$ represents in this case the *p*-phenylene group. Similarly, quaternization can be achieved by reaction with a tertiary amine. It is a good idea, in order to optimize latex stability when the colloid is to carry a positive charge, to choose an initiator which introduces cations or at least uncharged groups to the particle surface. The amine-containing colloids can be used to bond with proteins and antibodies for medical diagnostics and targeted drug delivery systems, whereas the cationic colloids will have good compatibility with cationic pigment dispersions and may be substantive to certain kinds of negatively charged surfaces such as glass, wood and hair.

Mercaptan surface groups. The benzyl chloride group will react with thiourea to form a thiuronium salt. This in turn can be hydrolyzed in aqueous base to the mercaptan [140]:

$$\text{Particle}-\phi-CH_2Cl + H_2N-\underset{\underset{S}{\|}}{C}-NH_2 \longrightarrow \text{Particle}-\phi-CH_2S-C\underset{NH_2}{\overset{NH_2^+}{\diagup}} \quad Cl^-$$

$$\xrightarrow{NH_4OH} \text{Particle}-\phi-CH_2SH$$

Surface thiol groups have been used for binding biologically active molecules as well. For example, the cysteine thiol groups in a protein will form a disulfide bond with the latex $-SH$ group in the presence of an oxidizing agent such as ferricyanide. A flexible spacer group is often used to allow the surface thiol to have greater mobility and to allow a large protein molecule to approach the latex particle closely enough to react.

6.2.4.4 Redox reactions at the particle surface

Surface aromatic nitro groups can be reduced to amine by the action of ferrous ion. The latter becomes oxidized to insoluble hydrous iron oxides, which upon subsequent heating crystallize to magnetic γ-Fe_2O_3, maghemite. This reaction has been used to make magnetic latex particles or microspheres (larger than 1 μm diameter) by an ingenious process. Highly porous particles of PS are first made in which the internal as well as external surface is covered with the $-NO_2$ groups by nitration in mixed nitric and sulfuric acids. Subsequent treatment with aqueous Fe^{2+} and heating produces a deposit of maghemite within the pores of the particle, essentially filling them. The other product of the reaction – due to reduction of the nitro groups – is surface amine! There has been a host of applications of such particles in medical diagnostics, cell separation and chemotherapy [141]. An electron micrograph of a thin cross-section of a 4.5 μm diameter particle containing 20% iron ('M-450', Dynal A.S., Oslo, Norway) is shown in Fig. 109.

The remaining porosity is filled, and any additional surface functional groups are added in a final polymerization step. Practical applications of these functionalized, magnetic particles will be discussed in Section 6.4 below.

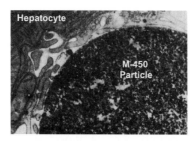

Fig. 109. TEM of ultrathin section of 'M-450' magnetic polystyrene particle with attached hepatocyte. From Danielsen, H. *et al.* (1986). *Scand. J. Immunol.* **24**, 179–187. Photo courtesy of the Radium Hospital, Oslo, Norway.

6.3 Heterogeneous catalysis

6.3.1 Acid-catalyzed hydrolyses of small molecules

With hydrophobic surfaces, very high specific surface areas, controlled particle size and the ability to vary not only the kind, but also the spacing of surface groups, polymer colloids are ideal candidates to serve as heterogeneous catalysts. The process of heterogeneous catalysis must involve five fundamental steps:

(1) diffusion of substrate molecules to the interface,
(2) adsorption and interaction with the catalyst at the surface,
(3) reaction within the adsorbed layer,
(4) desorption of the reaction products (to provide sites for subsequent events), and
(5) diffusion of products away from the surface.

The simplest and most obvious reactions are those catalyzed by surface acid groups. If there is 'bare' surface between the functional groups, this may serve as a site for adsorption and thus concentration of substrate in the vicinity of the catalytic sites. In some cases this concentration of substrate is the sole reason for acceleration in the overall rate of reaction. However, if adsorption is too strong, the product will not diffuse away, and it will inhibit further reaction.

An example of this kind of behavior may be found in the acid-catalyzed hydrolyses of an homologous series of acetate esters [143]. An all-sulfonate polystyrene colloid (diameter 162 nm, $\sigma_0 = 4.4\,\mu\text{C}\,\text{cm}^{-2}$) was used as the catalyst.

The effects may be expressed in terms of relative rates, i.e. the specific rate constant for the latex-catalyzed reaction (k) relative to that for simple acid catalysis, e.g. that for H_2SO_4 (k_0), as given in Table 6.3. It appears that the catalytic effect is optimal for a substrate of intermediate hydrophobicity (butyl) at this particular interface which had a surface area of $3.63\,\text{nm}^2$ per sulfonate group. Since a lone sulfonate group occupies no more than $0.50\,\text{nm}^2$, most of the surface – over $3.00\,\text{nm}^2$ per group – is hydrophobic polystyrene. Primary esters, both the *n*- and *i*-isomers, undergo hydrolysis through a slow step involving addition of a water molecule to the protonated intermediate. The tertiary esters, on the other hand, undergo a different mechanism in which very fast alkyl cleavage occurs. There may be a small

TABLE 6.3
Relative second-order rate constants for latex sulfonic acid-catalyzed hydrolysis of alkyl acetates

Acetate ester	k/k_0 at 50°C
Methyl	2.6
n-Butyl	7.7
i-Butyl	7.7
t-Butyl	0.91
n-Pentyl	3.7
i-Pentyl	5.5
t-Pentyl	0.57

disadvantage to surface catalysis in the latter cases because of a less favorable entropy of activation. Values of k/k_0 for the t-butyl and t-pentyl acetates are seen to be less than unity.

As an example of the effect of the nature of the catalyst surface, the values of k/k_0 were obtained for a single substrate, n-butyl acetate, on three different catalysts. These are shown in Table 6.4, and indicate that the catalyst with the highest surface acid concentration gives the greatest increase in rate.

The last column gives the relative second-order rate constant, k'', based upon the two-dimensional surface acid concentration. When looked at in this way, the differences in rate constants largely disappear, demonstrating that it is the differences in surface acid concentration and not the presence of –OH groups which are responsible for the observed effects. These results show that reaction rates almost eight times those in homogeneous conditions can be achieved with the simplest kind of latex catalyst. With additional catalytic groups and optimal spacing much higher rates may be possible.

6.3.2 Catalysis of other small molecule reactions

6.3.2.1 Decarboxylation

A more dramatic and sophisticated example of catalysis is the decarboxylation of 6-nitrobenzisoxazole-3-carboxylate (NBC) anion in the presence of a

TABLE 6.4
Relative rate constants for latex-catalyzed hydrolysis of n-butyl acetate

| Latex no. | Area/surface group (nm^2) | | [H$^+$] μmol $\times 10^5$ cm^{-2} | k/k_0 | k''/k_0 |
	RSO$_3^-$	ROH			
24	9.25	2.55	4.60	2.73	5.9
65	5.85		7.26	3.80	5.2
17	3.63		11.7	7.70	6.6

cationic polymer colloid. The reaction is:

$$\text{NBC} \longrightarrow + CO_2$$

(structure on left: benzene ring with COO⁻, C=N-O ring fused, and CH₂-NO₂ substituent, labeled NBC; structure on right: benzene ring with C≡N, O⁻, and CH₂-NO₂ substituent)

The catalysts were lightly crosslinked, cationic PS/co-chloromethyl styrene latexes, quaternized with different trialkyl amines. A cationic initiator (AIBA, Table 6.1) was used to ensure colloidal stability. It was found, by selective swelling experiments, that the core, not the shell, of the particles is the locus of catalysis, and that the most hydrophobic (tributyl ammonium) gave the greatest acceleration. Enhancements of the first-order rate constant of up to 21 000-fold at 25°C were obtained, due to:

- ion-exchange of the chloride counterions of the latex by the isoxazole carboxylate anions, which concentrates the substrate within the latex particles; and
- the lower dielectric constant within the tributyl ammonium particles as compared to that of water or in trimethyl-, triethyl- or tripropyl-ammonium particles. This leads to lower hydration and thus destabilization of the ground state of the reactant.

Added salt has two effects: deswelling of the particles, making them more hydrophobic (enhancing the rate), and screening of electrostatic interactions between substrate and catalytic sites, decreasing the degree of ion-exchange (decreasing the rate). Thus the apparent rate constant initially increases and then decreases as the ionic strength of the reaction mixture is increased [144].

6.3.2.2 Hydrolysis

In a different kind of reaction, a bimolecular process catalyzed by a small molecule, similar rate enhancements were observed in the presence of the same cationic polymer colloid. The reaction involved the hydrolysis of p-nitrophenyl diphenyl phosphate catalyzed by o-iodoso-benzoate (IBA). Again the catalysis took place within the interior of the particles and was greatest for the most hydrophobic polymer colloid (trimethyl < triethyl < tripropyl < tributyl). In this case the second-order rate constant was increased 10-fold, and the concentration of substrate within the particles, 630-fold for an overall rate enhancement of 6300 [145]. The mechanism involves concentration of the substrate in the particles by ion-exchange, as above, along with a small modification of the rate constant due to a change in the dielectric constant and solvency of the medium. Latex particles of $c.\ 0.2\,\mu m$ diameter have the advantage of their small size compared with that of conventional ion-exchange resin beads ($c.\ 20\,\mu m$ diameter) such that diffusion of substrate and (in this case, catalyst) and product in and out of the particle occurs rapidly. Both this reaction and the preceding one exhibit the highest rate increases found so far for any colloidal or polymeric catalytic system.

$$\text{pNO}_2\text{-C}_6\text{H}_4\text{-O-PO}(\text{O}\varnothing)_2 + \text{IBA-iodide} \longrightarrow (\varnothing\text{O})_2\text{PO}_2^- + \text{IBA}$$

$\varnothing$ represents phenyl

6.3.3 Practical applications

How would latex catalysts be used in practical situations? Obviously they must be stable to reaction conditions and then they must be removed easily at the end. If small amounts are involved and the cost is not great, the best procedure is to coagulate the latex by adding salt or (for weak electrolyte colloids) by changing the pH. Alternatively one can separate the latex catalyst from the reaction medium at the outset with a semi-permeable membrane through which reactants and products can pass. Such membranes exist in the form of hollow fiber bundles with very high specific surface areas, and are obtainable in industrial sizes and materials.

> Particles can become powerful reagents in chemical analysis, medical diagnostics, cell separation and drug delivery.

Most manufacturing engineers would be happy to achieve a 10% increase in throughput in a production process. The fact that 10^3- to 10^4-fold rate increases may be had by using polymer colloidal catalysts should provide an enormous inducement towards efforts in this direction. Even though the financial rewards are potentially great, essentially no industrial applications have as yet been explored. This is probably because of the need for a combination of colloid, polymer, and organic chemistry combined with creative engineering. An interdisciplinary team with these talents could possibly – with the right reaction – effect huge increases in process efficiency.

A review of 'Cationic Latexes as Catalytic Media' has been published by Ford and coworkers [146].

6.4 Biomedical applications

6.4.1 Binding biomolecules to particle surfaces

With appropriate surface functional groups, such as aldehyde (−CHO), amine (−NH$_2$) or thiol (−SH), it is possible to bind proteins, DNA fragments or antibody molecules to the surface of latex particles. Why do this? Because the particles can become powerful reagents in chemical analysis, medical diagnostics, cell separation and drug delivery. Even though many biological macromolecules such as proteins appear to adsorb onto polystyrene irreversibly, it is better to bind them chemically to the particles to ensure that no undesired desorption or adsorption (nonspecific binding of unwanted proteins) occurs, and to be able to further modify the surface when required. This should become clearer during the following discussions.

An easy way to introduce functional groups for biomolecule binding is to copolymerize glycidyl methacrylate, which is commercially available, and which provides epoxy groups along the polymer chains. The surface epoxy groups will readily react with thiol or amine groups in proteins. Alternatively, a comonomer with a short chain polyethylene glycol group which is −OH terminated can be used. This is converted to a sulfonate ester by reaction with an alkyl sulfonyl chloride. The latter, in turn, reacts with a thiol or amine group in a biomolecule to link it to the particle surface:

$$P(\diagup O \diagdown)_n OH + ClOSO_2R \longrightarrow P(\diagup O \diagdown)_n OSO_2R + HCl$$

$$P(\diagup O \diagdown)_n OSO_2R + HS-L \longrightarrow P(\diagup O \diagdown)_n S-L + HOSO_2R$$

$$P(\diagup O \diagdown)_n OSO_2R + H_2N-L \longrightarrow P(\diagup O \diagdown)_n NH-L + HOSO_2R$$

In this scheme, P represents the polymer, R is a suitable moiety such as tolyl (ClOSO$_2$R is then 'tosyl chloride'), and L is the ligand or biomolecule (such as an immunoglobulin like IgG or IgM). The sulfonate ester-containing latex particles are quite stable under room conditions, losing only about 10% of their reactive functionality in a year.

An alternative method employs a carbodiimide which 'activates' surface carboxyl groups. This is followed by a rearrangement reaction which results in a covalent bond between the particle and the antibody, as depicted in Fig. 110.

With an antibody attached ('conjugated') to its surface, and any bare areas treated so as to avoid nonspecific binding of biomolecules, the particle becomes a highly immunospecific reagent for the corresponding antigen which may reside on the surface of a cell or in solution. This is represented schematically in Fig. 111 in which a cancer cell, which contains tumor-specific antigens (TSA), is selected by the particle–antibody conjugate.

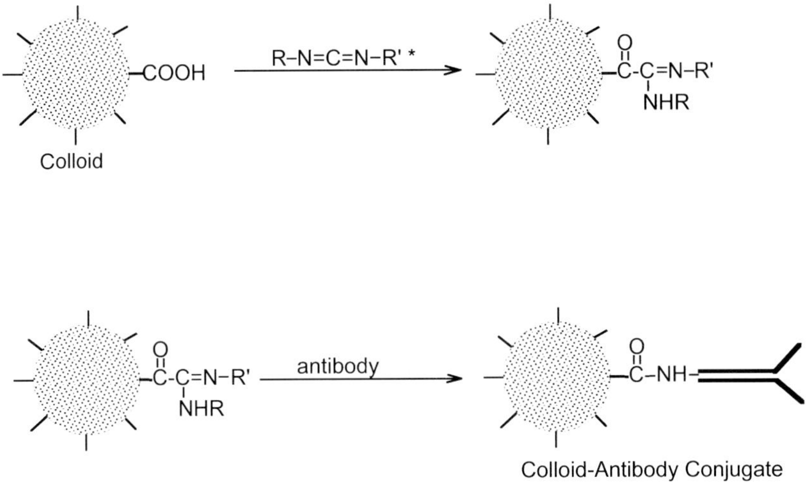

Fig. 110. Carbodiimide method for antibody conjugation. *R′ is 1-ethyl-2,3-dimethyl-aminopropyl carbodiimide.

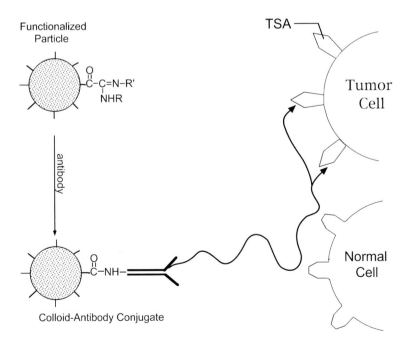

Fig. 111. Mechanism of immunoselectivity by particle–antibody conjugate [147].

6.4.2 Immunospecific cell separation

6.4.2.1 Immunomagnetic cell separation

When magnetic latex particles are covered with a shell of hydrophilic gel and carry surface-bound antibodies, they will attach themselves to the surfaces of cells only where the corresponding antigen resides and to no other. They then form a 'rosette' of particles surrounding the cell when there are a number of antigen sites on the cell surface. In peripheral blood or bone marrow, for example, red blood cells exist in suspension, and can be 'targeted' by this means. Experimentally a dispersion of the latex particles is simply mixed well with a slurry of the cells in an appropriate buffer solution. The rosetted target cells can then be removed readily by the application of a magnetic field. All other cells pass through without retention. This means that, for example, leucocytes – white blood cells – which all look approximately alike under the microscope, can be separated *preparatively* according their various functions. An illustration of how cells bind to particles is given in Fig. 112.

6.4.2.2 Immunofluorescent cell separation

Instead of using the magnetic properties of immunospecific particles, one can label them with fluorescent dye molecules. When these are attached to cells with the corresponding antigens, it is possible then to separate the cells in a flow cytometer using laser light scattering and/or fluorescence detection to distinguish cells carrying different surface antigens in the same mixture.

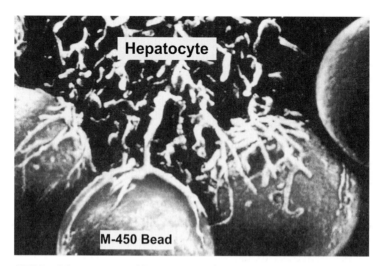

Fig. 112. SEM of immunospecific attachment of liver cell to 4.5 μm microspheres. From Danielsen, H. *et al.* (1986). *Scand. J. Immunol.* **24**, 179–187. Photo courtesy of the Radium Hospital, Oslo, Norway.

> It is possible to distinguish cells carrying different surface antigens in the same mixture.

A flow cytometer uses Rayleigh jet technology, the same as that in ink jet printers, to create microscopic droplets. If the liquid involved contains a dilute suspension of cells, one can regulate the conditions such that there is a high probability that any given droplet contains either one or no cells. The droplets are charged electrostatically by means of an externally applied field, and fall first through an exciting laser beam and fluorescence/scattering detectors and then between electrodes whose field can be turned rapidly on or off. It is important that the polymer colloidal particles are monodisperse so that their scattering signal is the same for all particles. Drops containing no cells or those containing unlabelled cells fall directly into a receiver for them. Those drops containing, say, green dye-labelled particles attached to 'T4' cells, when detected, cause a field to be applied to the electrodes with the result that these drops are deviated so as to fall into a different receptacle. On the other hand, red dye-labelled particle/cell rosettes may correspond to 'B' cells, which can be deviated with a different applied potential to fall into their own container. Counting rates as high as $10^5 \, min^{-1}$ are possible [148]. Although this may sound impressive, and is quite acceptable for analytical purposes, preparative quantities are small for ordinary collection times.

6.4.2.3 Cancer therapy

A dramatic example of the application of immunospecific cell separation technology is in the treatment of a cancer of the nerve cells, neuroblastoma. Many of the cancerous cells are found in the bone marrow. In clinical treatment, a large sample of bone marrow is taken from the patient and slurried with magnetic, monosized 4.5 μm particles which are immunospecific for the neuroblastoma cell surface. The slurry is passed through a series of strong permanent magnets which remove, in two passes, essentially all of the cancer cells. The patient is then treated with chemotherapy and radiation to kill all the remaining bone marrow cells in his body, both cancerous and normal, as well as neuroblastoma cells in other parts of the body.

> Some patients have survived without symptoms for as long as seven years.

Subsequently the 'cleaned' suspension of normal cells from the magnetic separation is reintroduced to the patient's bones. The cells proliferate rapidly, so that within about 30 days the patient is able to leave the hospital. Remissions are rare, with some patients having survived without symptoms for as long as seven years.

168   POLYMER COLLOIDS: A COMPREHENSIVE INTRODUCTION

In addition to neuroblastoma, clinical treatments of lymphoma and leukemia have been successful using the same immunomagnetic cell separation techniques. Similar treatments for other cancers such as myeloma, breast cancer and small cell lung cancer are currently under experimental investigation [149].

6.4.3 Clinical diagnostics

6.4.3.1 Detection of specific proteins

A large number of diseases are either caused by various proteins or are accompanied by the production of characteristic proteins. These may exist in the peripheral blood, various fluids within the body, or on cell surfaces. Antibodies specific to any particular protein can be produced in humans, mice, goats or other animals. Thus it is possible to 'manufacture' the desired antibodies and then bond them to polymer microspheres to create reagents which detect the desired proteins. The detection may be in the form of 'agglutination' – coagulation – which may be seen visually or by means of turbidimetry. Or it may be by fluorescence or methods derived from those used in the development of color films. Thus it is now possible to detect, for example, the sex of a fetus and have the result appear as words on a film saying in blue letters 'It's a boy!' or in pink, 'It's a girl!'

6.4.3.2 Detection of specific DNA fragments

Molecules other than proteins can also be detected, most importantly DNA fragments, which give information about inherited diseases, viruses and bacteria. The process is schematically represented in Fig. 113. The specific DNA sequence is cleaved out by restriction enzymes from a large DNA molecule which may exist within a complex mixture of various biomolecules. If the original DNA sample is very small it can be multiplied hugely by means of the polymerase chain reaction (PCR) technique. The desired piece of DNA is then combined with biotin 'primer' (which signals initiation to the polymerase enzyme), and is reproduced further by PCR about a million times to obtain enough material to be detected.

At this stage it is still mixed with many other undesired molecules. Binding to magnetic particles allows for isolation of the desired sequences from all others, greatly simplifying purification [149]. This is done by initially binding streptavidin ('Str' in Fig. 113) to the particles. The streptavidin, in turn, will combine specifically to the biotin end group on the desired DNA fragment. Magnetic separation then isolates the target DNAs from all the rest. Subsequently these fragments can be tagged with a color-producing end group called lac I-β galactosidase. Detection of the color intensity provides a measure of the amount of the particular DNA fragment in the original sample. The method also may provide enough pure material for sequencing of the DNA upon breaking the Str–biotin bond, releasing the pure DNA fragment.

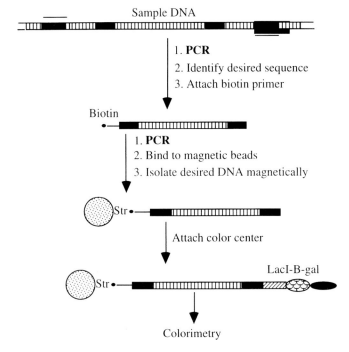

Fig. 113. Detection of immobilized amplified nucleic acid (DIANA).

> Bacteria can leave a DNA 'footprint,' say in cooked food, which epidemiologists can trace.

Bacteria can be detected by immunospecific DNA detection/separation in a similar manner since they will have DNA sequences characteristic of their species. Any small adaptive variant will be accompanied by changes in the genetic make-up, and therefore should be at least theoretically detectable by these methods. Interestingly, the bacteria do not have to be alive; they can leave a DNA 'footprint,' say in cooked food (i.e. the DNA is still present in the dead cells), which epidemiologists can trace in efforts to find out, for example, why some residents of a town died of food poisoning.

6.4.4 Other applications involving DNA

There are many other uses of this technology in molecular biology. For example, DNA-binding proteins can be isolated magnetically from a mixture of many different proteins by using magnetic particles with the appropriate complementary DNA already bound to the surface. Under the right conditions only proteins with the desired amino acid sequence ('fusion proteins') will associate with the surface DNA. After magnetic separation and washing, the protein is removed with a high salt buffer and it is ready for analysis, characterization and use. The particles are ready to be recycled.

> Specific messenger RNAs can be purified by binding to magnetic particles.

Specific messenger RNAs can be purified by modification of these methods, again involving binding to magnetic particles. Short sequences of synthetic polythymine are bound to the particle surface. The desired mRNA is labelled with a 'tail' sequence of polyadenine which binds with the oligo-T sequences. After magnetic separation the mRNA can be 'melted' off by elevating the temperature.

Other applications involve solid phase cloning and *in vitro* mutagenesis. A review of all of these applications has been given by Ugelstad *et al.* [149]. All of this is based upon the pioneering work of Rembaum who wrote a lucid review of the basic principles and applications involved [150].

6.4.5 Chemotherapy: targeted drug delivery

6.4.5.1 General considerations

Combination of immunospecificity and controlled release capabilities of functional polymer colloids leads to the possibility of sending particles to specific cell sites and of releasing drugs to them and to no others. The promise is manifold:

- highly toxic drugs may be employed if the entire amount can be directed to only the target site;
- very small amounts of drug may be sufficient since it is not diluted throughout the entire body; and
- release rates can be such that administration of the drug may involve a single dose which can last for long periods of time.

There are, however, problems associated with the use of polymer colloids within the body. The principal one is that particles in the normal colloidal size range can be rapidly cleared by the reticuloendothelial system (RES), and end up primarily in the liver by either phagocytosis or pinocytosis. Some success at avoiding this problem has been achieved by adsorption of polyethylene oxide (PEO)-containing surfactants onto the latex particles. Permanent bonding of PEO would presumably be a safer procedure. This material, however, may render the surface less able to recognize antigens because the particle surface would be covered with a gelatinous, relatively thick layer of PEO chains.

> Colloidal particles can be rapidly cleared by the reticuloendothelial system.

Furthermore the particles, once in the body, may remain permanently if they are not biodegradable. Thus efforts are being made to develop particles which will undergo quantitative breakdown and subsequent harmless elimination from the body after a reasonable time. Perhaps the most promising of these are certain poly(lactide-co-glycolide) latexes, which have been approved for human use by the US Food and Drug Administration. These have to be made, however, in nonaqueous media because of the polyesterification reactions involved (Section 4.2.5.2). The colloids must subsequently be transferred to an aqueous system.

6.4.5.2 Targeted drug delivery

The basic idea of targeted drug delivery is to immunospecifically attach a drug-containing latex particle to the target cell and to release the drug to that cell only in a manner similar to that depicted in Fig. 111. The drug initially may be absorbed by the particle, be chemically a part of the particle as a result of copolymerization of a drug-containing monomer, or it may be attached to the particle surface by chemical bonding post emulsion polymerization. The drug may be released by simple diffusion out of the particle or by, for example, surface hydrolysis (both with likelihood that some will go elsewhere in the body), or it may act by physical contact with the cell surface. In the last instance surface-attached drug may possibly remain on the latex particle without the need to be released at all – the perfect situation.

> A drug may remain on the particle surface without the need to be released – the perfect situation.

A specific example of targeted drug release, but not using polymer colloids, involves the drug arabino-furanosyl-adenine-monophosphate (ara-AMP). When this compound is conjugated with lactosaminated serum albumin (L-SA) it will specifically attach to hepatocytes (liver cells). The drug ara-AMP is active against *Ectromelia* virus hepatitis. The conjugated drug was found to act over four to five times as long in mice as the free drug. Its mechanism of action is apparently by inhibition of DNA synthesis in liver cells.

Thus there remains a huge opportunity for targeted drugs which act by way of conjugation to polymer colloids. Concomitantly there are also major challenges which must be met. These arise from the fact that colloidal particles are seen by the mammalian immune system as undesirable intruders and are quickly (sometimes in a matter of minutes) cleared from the blood stream. On the other hand, such drugs may be especially effective with individuals having weak immune systems.

A review of polymer-based drug delivery is given in the book by Tarcha [151]. Chapter 14 in that book, by P.J. Tarcha and R.S. Levinson, discusses targeted drug delivery.

References

126. Fitch, R.M. (1982). In *IUPAC Macromolecules* (H. Benoit and P. Rempp eds), p. 52, Pergamon Press, Oxford.
127. Goodwin, J.W., Hearn, J., Ho, C.C. and Ottewill, R.H. (1973). *Br. Polym. J.* **5**, 347.
128. Goodwin, J.W., Ottewill, R.H., Pelton, R., Vianello, G. and Yates, D.E. (1978). *Br. Polym. J.* **10**, 173.
129. Liu, L.J. and Krieger, I.M. (1978). In *Emulsions, Latices and Dispersions* (P. Becher and M.N. Yudenfreund eds), p. 41, Marcel Dekker, New York.
130. Traut, G.R. and Fitch, R.M. (1985). *J. Colloid Interface Sci.* **104**, 216.
131. Fitch, R.M. (1982). In *IUPAC Macromolecules* (H. Benoit and P. Rempp eds), p. 39, Pergamon Press, Oxford.
132. Tsaur, S.L. and Fitch, R.M. (1987). *J. Colloid Interface Sci.* **115**, 450.
133. Charreyre, M.T., Boulanger, P., Delair, Th., Mandrand, B. and Pichot, C. (1993). *Colloid Polym. Sci.* **271**, 668.
134. Fitch, R.M. and McCarvill, W.T. (1978). *J. Colloid Interface Sci.* **66**, 20.
135. Fifield, C.C. (1985). Ph.D. Thesis, University of Connecticut.
136. Vidal, F., Guillot, J. and Guyot, A. (1994). In *Proceedings of the 3rd International Symposium on Radical Copolymers in Dispersed Media*, p. 235, Réseau des Polymèristes Lyonnais et Groupe Français des Polymères, Lyon.
137. Guyot, A. and Tauer, K. (1994). *Adv. Polym. Sci.* **111**, 43.
138. Fitch, R.M., Gajria, C. and Tarcha, P.J. (1979). *J. Colloid Interface Sci.* **71**, 107.
139. Scholsky, K.M., Tarcha, P.J. and Hsu, A. (1988). In *Polymers for Controlled Drug Delivery* (P.J. Tarcha ed), p. 163, CRC Press, Boca Raton, FL.
140. Delair, T., Marguet, V., Pichot, C. and Mandrand, B. (1994). *Colloid Polym. Sci.* **272**, 962.
141. Ugelstad, J., Berge, A., Ellingsen, T., Schmid, R., Nilsen, T.-N., Mørk, P.C., Stenstad, P., Hornes, E. and Olsvik, O. (1992). *Prog. Polym. Sci.* **17**, 87.
142. Ugelstad, J., Berge, A., Ellingsen, T., Aune, O., Kilaas, L., Nilsen, T.-N., Schmid, R., Stenstad, P., Funderud, S., Kvalheim, G., Nustad, K., Lea, T., Vartdal, F. and Danielsen, H. (1988). *Makromol. Chem., Makromol. Symp.* **17**, 177.
143. Fitch, R.M., Mallya, P.K., McCarvill, W.T. and Miller, R.S. (1981). *Preprints*, Div. Colloid Surf. Chem., American Chemical Society 182nd Meeting, 23–28 August, Paper No. 64.
144. Lee, J.J. and Ford, W.T. (1993). *J. Organic Chem.* **58**, 4070.
145. Lee, J.J. and Ford, W.T. (1994). *J. Amer. Chem. Soc.* **116**, 3753.
146. Ford, W.T., Lee, J.J., Yu, H., Ackerson, B.J. and Davis, K.A. (1995). In *Macromolecular Symposia* (J. Guillot, A. Guyot and C. Pichot eds), p. 333, Hüthig and Wepf Verlag, Heidelberg.
147. Scholsky, K.M. (1982). Ph.D. Dissertation, University of Connecticut.
148. Steinkamp, J.A. (1984). *Rev. Sci. Instrum.* **55**, 1375.
149. Ugelstad, J., Berge, A., Ellingsen, T., Schmid, R., Nilsen, T.-N., Mørk, P.C., Stenstad, P., Hornes, E. and Olsvik, O. (1992). *Prog. Polym. Sci.* **17**, 87.
150. Rembaum, A., Yen, S.P.S. and Molday, R.S. (1979). *J. Macromol. Sci. – Chem.* **A13**, 603.
151. Tarcha, P.J. (ed) (1990). *Polymers for Controlled Drug Delivery*, CRC Press, Ann Arbor, MI.

Chapter 7

Latex Stability

7.1 Thermodynamic versus kinetic stability

When a piece of material in the bulk state, immersed in a fluid medium, is subdivided into many small colloidal particles, there is an accompanying change in the standard free energy of the system, ΔG_f°. This is represented schematically in Fig. 114 in which the $\Delta G_f^\circ = \gamma_{sl} \Delta A_{sl}$ can be either positive or negative. Here γ_{sl} is the solid/liquid interfacial surface tension or free energy, expressed in $J m^{-2}$, and ΔA_{sl} is the increase in interfacial area. If ΔG_f° is positive, the colloidal state is unstable relative to the bulk, and the colloid is referred to classically as lyophobic (where 'lyo-' refers to the surrounding medium). If, on the other hand, ΔG_f° is negative, the colloid is said to be lyophilic, and it is thermodynamically stable. Examples of both kinds, synthetic and natural, are given in Table 7.1.

Lyophobic colloids, even though they are thermodynamically unstable, can be made metastable for long periods of time if an energy barrier of sufficient height can be erected between the bulk and colloidal states. This is illustrated schematically in Fig. 115. The nature of this barrier, E_{act}, and the experimental factors which govern it in electrically charged and neutral colloids, have been the subject of a very large body of theory and experiment. When the barrier is absent or is very small relative to the thermal energy of the particles, kT – as in the lower curve in Fig. 115 – then the particles tend to revert to the bulk state by aggregating. This process is known as coagulation or flocculation, depending upon the detailed nature of the potential energy curves between particles, to be discussed below.

The tendency for colloidal particles to aggregate arises from the universal attractions among all uncharged molecules, usually known as van der Waals interactions. These are dipole–dipole interactions which act through space. Colloidal particles contain many molecules attracting each other, not through space, but through the fluid medium. In some cases the particle–medium attractions are stronger than those between particles, with the result that the colloidal state is preferred, i.e. the system is lyophilic. In this case the relative positions of the bulk and colloidal states in Fig. 115 would be reversed, and there would be no thermodynamic driving force for the particles to aggregate.

174 POLYMER COLLOIDS: A COMPREHENSIVE INTRODUCTION

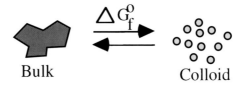

Fig. 114. Free energy change from bulk to colloidal states.

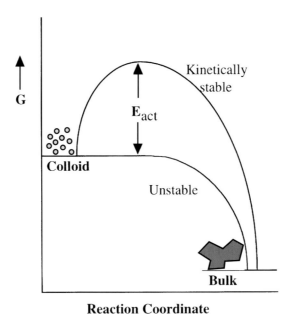

Fig. 115. Kinetic stability of a lyophobic colloid.

7.2 Kinetic stability of electrically charged hydrophobic colloids

7.2.1 The van der Waals attraction

Neutral molecules attract each other primarily because thermal oscillations of the nuclei relative to their electronic clouds result in transient dipoles. These, in turn, can induce dipoles in neighboring molecules. The latter always will be

TABLE 7.1
Examples of lyophobic and lyophilic colloids

Lyophobic	Lyophilic
Polystyrene in water	PVC (polyvinyl chloride) in plasticizer
AgI in water	Crosslinked polyacrylamide particles in water ('Microgel')
PMMA in hexane	Crosslinked PMMA particles in toluene ('Microgel')
Milk (butterfat in water)	Starch granules in water

oriented such that the two attract, as shown in Fig. 116. The frequencies of oscillation were originally considered to be in the ultraviolet, and from these considerations Hamaker calculated by summing all the microscopic interactions between two spheres that the attractive potential energy at close distances in a vacuum can be approximated by:

$$V_a = -\frac{Aa}{12H_o} \qquad (93)$$

where A, the 'Hamaker constant,' is characteristic of the material of the particles, a is their radius and H_o is the minimum distance of separation of the spheres. The corresponding relationship for two flat plates is

$$V_a = -\frac{A}{12\pi H_o^2} \qquad (93a)$$

When the particles are immersed in a dispersion medium their interaction potential is reduced by the self-attraction of the medium, A_{22}, and A is replaced by A_{121}:

$$A_{121} = (A_{11}^{1/2} - A_{22}^{1/2})^2 \qquad (94)$$

where A_{11} is the value of A in Eq. 93. For PS particles A_{11} is 9×10^{-20} J and A_{22} is 6×10^{-20} J, which gives by Eq. 94 $A_{121} = 3 \times 10^{-21}$ J, a 30-fold reduction in their attractive potential when immersed in water. These values are approximate at best, but one can say that generally for most aqueous polymer colloids A_{121} should lie in the range between 0.5×10^{-21} and 10×10^{-21} J.

A plot of the attractive potential as a function of the distance of separation is given in Fig. 117 for the two cases. The negative values for V_a indicate a favorable free energy change as the particles approach each other, leading to agglomeration.

7.2.1.1 Retardation

As the distance between interacting particles becomes greater, the inducing radiation becomes out of phase with the 'responding' radiation, thereby weakening the interaction. Thus the Hamaker constant should be viewed as a Hamaker *function*. This will change slightly the curvature of the V_a curve in water at higher values of H_0 in Fig. 117.

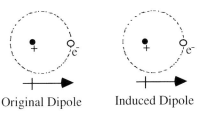

Fig. 116. Origin of the van der Waals attraction.

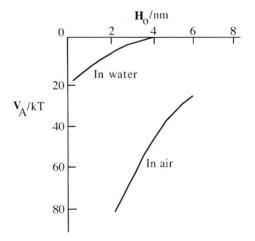

Fig. 117. Attractive potential between two PS 100 nm spheres in air and water.

7.2.2 The electrical double layer

7.2.2.1 For flat plates

To overcome the van der Waals attraction, the particles can be stabilized by giving them a surface charge. This may be in the form of surface-bound ionogenic groups, e.g. sulfonate chain ends, or of adsorbed ionic surfactant, either anionic or cationic. In the discussion below we shall arbitrarily choose negative ions to be fixed on the particle surface, but it should be understood that a surface of opposite charge would be treated identically, but with the signs reversed. The charge on two particles leads to a repulsive potential between them, but this is not a simple electrostatic repulsion because of the existence of ions in the surrounding medium.

Because of the requirement for electroneutrality, there must be an equal number of ions of opposite charge in the vicinity of the surface ions. These are usually relatively small, e.g. Na^+, K^+ or NH_4^+, and are thus able to move about rapidly, but always under the influence of the rather highly charged surface of opposite charge. The result is an *electrical double layer* composed of the more or less fixed surface ions and a diffuse layer of moving ions. The corresponding electrical potential, $\psi(x)$ (reduced by the surface potential, ψ_0) is plotted in Fig. 118 as a function of distance x from a charged, flat plate immersed in aqueous electrolyte. In this case the Debye–Hückel approximation (see below) for low surface potential, ψ_0, was used. The calculation of these and similar curves involves the simultaneous solution of two equations, one for the distribution of charged, Brownian particles in a potential field, and the other for the change in potential as a function of the charge density. The first is due to Boltzmann, and the second, to Poisson.

Boltzmann observed that the distribution of charges in a field is given by $n = n_0 \exp(ve\psi/kT)$, where n and n_0 are the concentrations of charges in the field and in zero potential, respectively. From this one may calculate that the

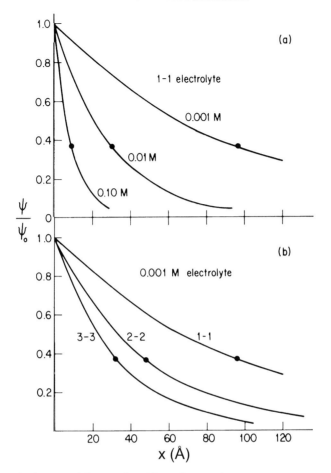

Fig. 118. Electrical potential vs x for (a) various salt concentrations and (b) various salt valences [152].

net density, ρ, of ions – here assumed to be acting as point charges – at any given potential, ψ, will be given by the hyperbolic sine function in Eq. 95:

$$\rho \equiv ve(n_+ - n_-) = -2nve \sinh\left(\frac{ve\psi}{kT}\right) \qquad (95)$$

where $\sinh y \equiv (e^y - e^{-y})/2$, and v is the valence of the ions in solution (assuming they are all the same), n is the number concentration of ions of the one valence in excess, and e is the negative of the charge of an electron. Since charge times voltage (potential) equals energy, the dimensionless term $(ve\psi/kT)$ is the electrical energy of an ion relative to its thermal energy.

Poisson showed that the charge density is a function of the curvature of the potential:

$$\nabla^2 \psi = \left(\frac{\partial^2}{\partial x^2} + \frac{\partial^2}{\partial y^2} + \frac{\partial^2}{\partial z^2}\right)\psi = -\frac{4\pi\rho}{\varepsilon} \qquad (96)$$

where ε is the dielectric constant of the medium. Since the field for a flat plate does not vary in the y and z directions, it is only the $(\partial^2/\partial x^2)$ term which is required in this case. Combining Eqs 95 and 96 gives the well-known Poisson–Boltzmann distribution of ions in an electrical field:

$$\nabla^2 \psi = \frac{8\pi n v e}{\varepsilon} \sinh\left(\frac{v e \psi}{kT}\right) \qquad (97)$$

For low potentials $\exp(ve\psi/kT) \approx 1 + (ve\psi/kT)$, to give the Debye–Hückel linear approximation for a flat plate:

$$\nabla^2 \psi = \frac{\partial^2 \psi}{\partial x^2} = \frac{8\pi n v^2 e^2}{\varepsilon kT}\psi = \kappa^2 \psi \qquad (98)$$

where
$$\kappa^2 \equiv \frac{8\pi n v^2 e^2}{\varepsilon kT}$$

The term κ has the dimensions of reciprocal distance, such that κ^{-1} is referred to as the Debye length, or the thickness of the diffuse part of the double layer. Kappa is also a measure of the ionic strength, μ, of the solution where

$$\mu \equiv \tfrac{1}{2} \sum_i n_i v_i^2$$

and where n is expressed in molality.

Integration of Eq. 98 gives the simple relationship for the potential curves shown in Fig. 118:

$$\psi = \psi_0 e^{-\kappa x} \qquad (99)$$

for small values of the surface potential, i.e. mathematically for $v\psi_0 \ll 25\,\mathrm{mV}$. It turns out that Eq. 99 holds quite well for surface potentials up to 50 mV, which covers the great majority of practical systems. For higher potentials the full integration of Eq. 97 due to Gouy and Chapman is used, the result of which can be expressed in terms of the dimensionless parameter γ:

$$\gamma = \gamma_0 e^{-\kappa x} \qquad (100)$$

where
$$\gamma \equiv \frac{\exp(ve\psi/2kT) - 1}{\exp(ve\psi/2kT) + 1}$$

When the potential $\psi(x)$ falls to $(\psi_0/e)^*$, then $x = \kappa^{-1}$, the Debye thickness. This is illustrated in Fig. 119, taken from Fig. 118a, in which the Debye thicknesses are 96, 30 and 9.6 Å for 0.001, 0.010 and 0.100 molar 1:1 electrolyte, respectively. Clearly the addition of salt has a profound effect on the double layer thickness!

---

* e here represents the basis of natural logs: $e = 2.7183$

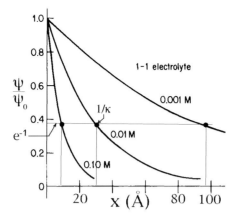

Fig. 119. Debye thickness at three salt concentrations.

7.2.2.2 For spheres

For spherical particles of radius a, the electrical potential is a function only of the radial distance, r, from the center of the particle, and Eq. 99 becomes

$$\psi = \psi_a \frac{a}{r} e^{\kappa(a-r)} \qquad (101)$$

for small values of ψ_a. However, in this case the potential falls off more rapidly than that for flat plates, so that the approximation for Eq. 101 is applicable for higher surface potentials than those to which Eq. 99 applies.

7.2.2.3 Counterion condensation and the Stern layer

The assumption that the counterions act as point charges appears to be valid in the outer parts of the diffuse layer, but calculations based on Eq. 97 of concentrations at the interface result in impossibly high values for relatively large values of ψ_a or ψ_0. For example, for $\psi_0 = 300$ mV (possible for some inorganic colloids such as AgI, but unlikely for polymer colloids), the Boltzmann equation, $n = n_0 \exp(ve\psi/kT)$, gives for 0.001 molar electrolyte $n = (0.001) e^{12} = 160$ molar! Clearly, counterion size will restrict the number of ions in the first layer.

Careful measurements of both colloids and polyelectrolytes by a variety of techniques have shown that some counterions are strongly associated with the surface even when the surface potential is only on the order of magnitude of kT. This so-called 'counterion condensation' creates a layer at the interface which neutralizes some of the surface charge. This layer is known as the Stern layer [153] after the man who first proposed its existence in 1924. Counterion condensation may therefore depend upon specific ion effects, such as their polarizability, hydrated ion size and charge, greatly complicating our ability to generalize. On the other hand, as shown schematically in Fig. 120, the Stern layer potential, ψ_δ, may be sufficiently lower than ψ_a or ψ_0 that the Debye–Hückel linear approximation holds over an even broader range, if we have a means for determining ψ_δ independently.

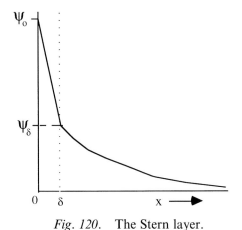

Fig. 120. The Stern layer.

7.2.3 Interactions between charged particles

7.2.3.1 DLVO theory

The theory which follows was formulated independently by two groups, Derjaguin and Landau in Russia [154], and Verwey and Overbeek in the Netherlands [155], and together is known as the 'DLVO theory.'

The repulsive free energy for flat plates. When two charged particles approach each other as a result of their thermal motion, the ion clouds in their diffuse electrical layers overlap, leading to an ionic concentration between the particles that is higher than elsewhere. As a consequence the free energy of the system, as measured by the electrochemical potential, is raised. This creates a driving force for separation of the particles, shown schematically in Fig. 121 for two flat plates. From another point of view, the higher concentration of ions between the particles leads to an osmotic pressure to bring solvent from outside, tending to force the particles apart. The magnitude of this free energy change was calculated by Verwey and Overbeek by

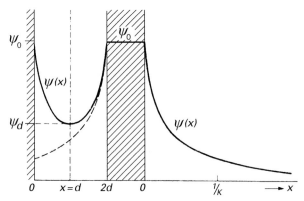

Fig. 121. Electrical potential between two plates compared with that for a single double layer [155].

determining the difference between $\psi(x)$ and ψ_δ at a point half way between the plates [155]. This was done by integration of Eq. 97 between the appropriate limits as defined in Fig. 121.

The fact that the halfway point gives the lowest potential (least negative; slope = 0) means that normally the linear approximation again may be used, except for very close approach. Under these conditions the repulsive free energy is:

$$V_R = \frac{32nkT\gamma_0^2}{\kappa}(1 - \tanh \kappa d) \qquad (102)$$

where
$$\gamma_0 \equiv \frac{\exp(ve\psi_0/2kT) - 1}{\exp(ve\psi_0/2kT) + 1}$$

The average concentration of salt, n, is in both the pre-exponential term and in the exponential function, $\tanh \kappa d$ – in κ – so that the repulsion is very sensitive to salt concentration. It is also a strong function of the surface potential, ψ_0, as illustrated in Fig. 122 in which $\log V_R$ has been calculated as a function of the dimensionless distance of separation κd for various values of ψ_0. It is interesting to note that (a) the curves are all mostly linear and parallel, differing only in magnitude, and (b) the approximate curves are just slightly higher than the rigorous ones, but are good down to a distance of separation of half the Debye thickness.

It may be helpful in understanding the origins of this repulsion at all distances of separation to perform a thermodynamic thought experiment: if one takes a charged plate in air with no counterions and immerses it into a salt solution, the double layer spontaneously forms. Equilibrium is reached and the system is at its lowest free energy condition. If this double layer is disturbed, e.g. by introduction of another, overlapping diffuse layer of ions, the free energy of the system must be increased, leading to a driving force for separation.

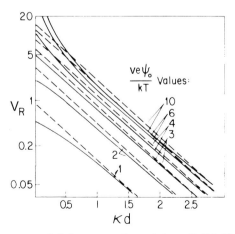

Fig. 122. Repulsive potential between flat plates. Solid lines, full theory; dashed lines, Eq. 102 [156].

The total free energy for flat plates. The combination of both the double layer repulsions and the van der Waals attractions, which act over about the same distances, leads to a rather complex set of curves. Even so, one additional energy should be included for completeness, i.e. the Born repulsion where the two plates or particles come into 'contact.' These are shown schematically in Fig. 123 in which the energy barrier, $V_{T_{max}}$, is the same as the 'E_{act}' in Fig. 115. There are two minima: (a) the primary, which results from the combination of the van der Waals attraction (very strong at short distances) and the Born repulsion and (b) a secondary minimum, which is usually quite shallow (often just a fraction of kT) if it exists at all. If we ignore the Born repulsion, the total interaction can be given, to a good approximation by the sum of Eqs 93a and 102:

$$V_T = \frac{32nkT\gamma_0^2}{\kappa}(1 - \tanh \kappa d) - \frac{A}{12\pi H^2} \qquad (103)$$

The primary minimum is generally so deep that the energy barrier for the reverse process of bringing two particles out of it is considered to be infinite. Thus the exact depth is unimportant, and the Born repulsion can be ignored. There are exceptions, however, in which coagulation can be reversed, but these are rare.

Figures 124 and 125 give curves calculated from Eq. 103. It should be noted that they are for values of the Hamaker constant which are higher than A_{121} given earlier for polystyrene in water, but the trends and shapes involved are still informative. Generally the Hamaker constant is not a variable, since it is governed by the materials comprising the particles and the medium in which they are immersed. Most commercial polymers fall within a narrow range of A-values, with the silicones and perfluoro-polymers having the lowest.

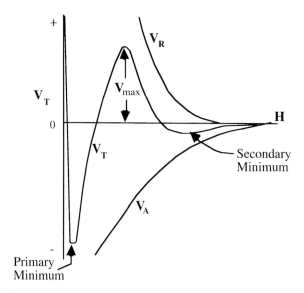

Fig. 123. Total interaction free energy between two particles (schematic).

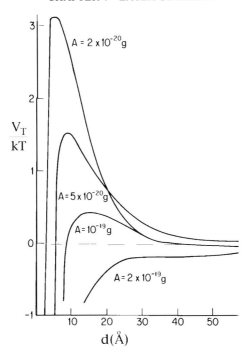

Fig. 124. Total interaction energy between flat plates; variable Hamaker constant, A_{121}. $\kappa = 10^7$ cm^{-1}; $\psi_0 = 100$ mV [157].

The most important experimental variable is ionic strength, and its effect on the interaction energy between two flat plates is shown Fig. 125. As salt is added to the colloid, the diffuse part of the electrical double layer contracts, as shown in Fig. 119, and this, when combined with the van der Waals attraction, leads to pronounced diminution of the barrier, allowing coagulation to occur. The barrier maximum always occurs at about the same distance of separation, c. 4–5 nm in these cases. An increase in the surface potential, e.g. by increasing the surface charge density, will increase the barrier height according to Eq. 103 (through the γ_0^2 term).

The total free energy for spheres. The DLVO theory for flat plates has been developed rather thoroughly above because it is used directly to obtain the interaction between spheres by means of the Derjaguin approximation [154]. Derjaguin took as the model for two interacting spheres two series of concentric, circular flat plates of infinitesimal width, dh, facing each other a distance H apart. The double layer repulsion of each pair of ring plates could then be calculated by means of integrating the derivative of Eq. 102 between appropriate limits as defined in Fig. 126.

From the geometry in Fig. 126, one may formulate the interaction as

$$V_R = 2\pi a \int_{H_0}^{\infty} (f_H - f_\infty) \, dH \qquad (104)$$

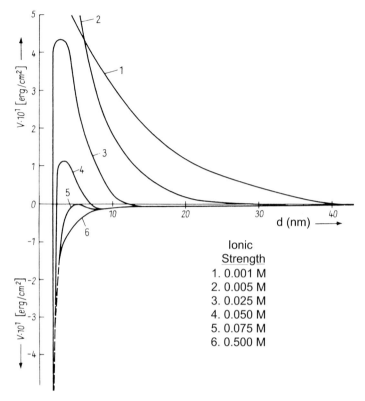

Fig. 125. Total interaction energy between flat plates; variable ionic strength. $A = 10^{-19}$ J; $\psi_0 = 25.6$ mV [158].

Verwey and Overbeek used Eq. 102 to obtain an expression for the term $(f_H - f_\infty)$ from which the approximate, low potential value for the repulsion is derived, according to Eq. 105.

$$V_R = \frac{\varepsilon a \psi_0^2}{2} \int_{H_0}^{\infty} \left(1 - \tanh \frac{\kappa H}{2}\right) d\left(\frac{\kappa H}{2}\right) = \frac{\varepsilon a \psi_0^2}{2} \ln(1 + e^{-\kappa H_0}) \qquad (105)$$

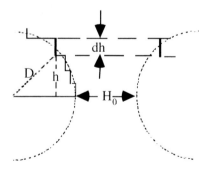

Fig. 126. Derjaguin model for calculating the interaction between two spheres.

The attraction between two spheres is obtained by integrating the dispersion forces (transient dipole-induced dipole) acting between two spheres, as given earlier in Eq. 93

$$V_A = -\frac{Aa}{12H_0} \tag{93}$$

This equation is an approximation which is good for relatively small distances of separation, i.e. on the order of $H_0 = 0.4a$. For 100 nm particles this corresponds to 40 nm.

Combination of Eqs. 105 and 93 gives the total free energy of interaction, V_T, in Eq. 106.

$$V_T = \frac{\varepsilon a \psi_0^2}{2} \ln(1 + e^{-\kappa H_0}) - \frac{Aa}{12H_0} \tag{106}$$

7.2.3.2 Lifshitz–Parsegian–Ninham (LPN) theory

Parsegian and Ninham have pointed out that the pairwise summation of individual molecular interactions to calculate A, which was derived for molecules in a vacuum, is not suitable for condensed media, and that the correction for the influence of the medium (Eq. 94) is inadequate [159]. They applied the Lifshitz theory, which employs the dielectric susceptibilities of the particles and the fluid medium to calculate what they call a 'Hamaker function,' $A(H_0, b)$, where two slabs of material of thickness b are separated by a distance H_0, such that the flat plate interaction energy would then be, in analogy to Eq. 93a:

$$V_A = \frac{A(H_0, b)}{12\pi H_0^2}$$

Lifshitz pointed out that, because the important radiation from oscillating dipoles lies in the microwave and infrared regions, the wavelengths involved are much longer than molecular dimensions, so that an ensemble of molecules in resonance can be treated as a continuum.

> Wavelengths are longer than molecular dimensions, so that an ensemble of molecules can be treated as a continuum.

The equations involved are somewhat complex, and are summarized below. The reader is encouraged to go to the original literature [160] or to the book of Russel, Saville and Schowalter [161] for detailed derivations. The total free energy of attraction for two plates of material 2 and thickness b, separated a distance H_0 by material 3, all immersed in material 1, is:

$$V_a(H_0, b) = \frac{kT}{2\pi c^2} \sum_{n=0}^{\infty}{}' \xi_n^3 \varepsilon_3 \int_1^{\infty} p\,dp$$
$$\times \{\ln[1 - \Delta_{31}^2(\text{eff})\,e^{-2w_3 H_0}] + \ln[1 - \bar\Delta_{31}^2(\text{eff})\,e^{-2w_3 H_0}]\} \quad (107)$$

where

$$\Delta = 1 - \left(\frac{1 + \Delta_{32}\Delta_{21}\,e^{-2w_2 b}}{\Delta_{32} + \Delta_{21}\,e^{-2w_2 b}}\right)^2 e^{-2w_3 H_0} \equiv 1 - \Delta_{31}^2(\text{eff})\,e^{2w_3 H_0}$$

and

$$\bar\Delta = 1 - \left(\frac{1 + \bar\Delta_{32}\bar\Delta_{21}\,e^{-2w_2 b}}{\bar\Delta_{32} + \bar\Delta_{21}\,e^{-2w_2 b}}\right)^2 e^{-2w_3 H_0} \equiv 1 - \bar\Delta_{31}^2(\text{eff})\,e^{2w_3 H_0}$$

where the prime on the summation symbol indicates that the $n = 0$ term is multiplied by 0.5, and where for $j, k = 1, 2, 3$:

$$w_j = \frac{\varepsilon_3^{1/2}\xi_n s_j}{c}; \quad s_j = \left(p^2 - 1 + \frac{\varepsilon_j}{\varepsilon_3}\right)^{1/2}; \quad \varepsilon_j = (i\xi_n);$$

$$\Delta_{jk} = \frac{s_k - s_j}{s_k + s_j}; \quad \bar\Delta_{jk} = \frac{s_k \varepsilon_j - s_j \varepsilon_k}{s_k \varepsilon_j + s_j \varepsilon_k}$$

in which p is a dummy integration variable. The $\varepsilon(i\xi)$ values are obtained from major peaks in the absorption spectra as a function of real frequencies by an appropriate algorithm [159].

Since it is the infrared and microwave frequencies which are more important than the ultraviolet, the result is that the attractions are much longer range, extending out to more than 1000 Å, as shown in Fig. 127, and that retardation is not very important. The relative contributions of the various frequencies involved are shown in Fig. 128 in which the 'zero frequency,' or microwave, component, i.e. the dielectric constant of the material, plays the major role in attraction. Because of this, there is a temperature-dependence (not shown) which was not taken into account in earlier theories.

Solutes can have significant influence on the dielectric properties of the medium, which in turn will also affect the interactions. An example is given in Fig. 129 where two flat plates of polystyrene of infinite thickness are separated, in one case, by pure water and, in the other, by 0.1 M NaCl solution [160]. The calculated Hamaker function, expressed in units of 10^{14} ergs (10^{21} J), is plotted against H_0 on a log scale. Since nearly all practical polymer colloids contain neutral electrolyte, this effect obviously cannot be ignored. It is entirely different from that in which salt decreases the thickness of the diffuse part of the double layer. These calculations indicate an attraction at

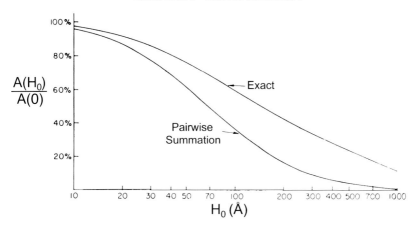

Fig. 127. Comparison of exact, Lifshitz–Hamaker function with pairwise addition values [159].

distances up to 10,000 Å in pure water, whereas in salt water it almost disappears at 1000 Å! The combination of both salt effects, on the dielectric susceptibility of the medium and on the collapse of the double layer, mean extreme sensitivity to 'inert' electrolyte.

Finally, in Fig. 130, we have the total LPN interaction energy between two polystyrene spheres where the salt concentration, in terms of $1/\kappa$, is varied, and where $a = 100$ nm and $\sigma_0 = 2.55\,\mu\text{C}\,\text{cm}^{-2}$ (5000 charges/sphere). It can be seen that when $1/\kappa = 10$ Å ($\kappa^2 \propto n$ – Eq. 98) the barrier height is greater than $25kT$, sufficient for long-term stability, although there is a small secondary minimum which may cause some weak flocs to form. Lowering the ionic strength four-fold ($1/\kappa = 20$ Å) moves the barrier out considerably, leaving no

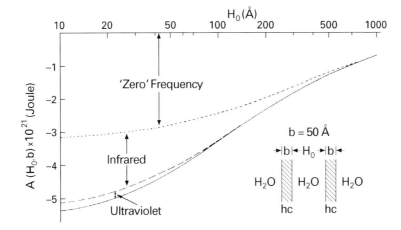

Fig. 128. Component contributions to the Hamaker function as a function of plate separation [159].

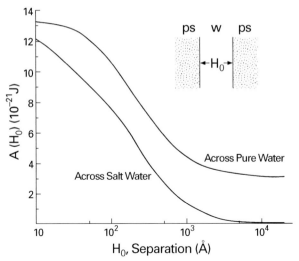

Fig. 129. The Hamaker function vs separation of two PS plates [160].

secondary minimum, whereas increasing it 25-fold ($1/\kappa = 2$ Å) eliminates the barrier, leading to fast coagulation (Section 7.5.1.1).

There are many general references on the DLVO theory; Hiemenz' book is quite thorough and lucid [162], although it does not cover the Lifshitz–Parsegian–Ninham (LPN) theory.

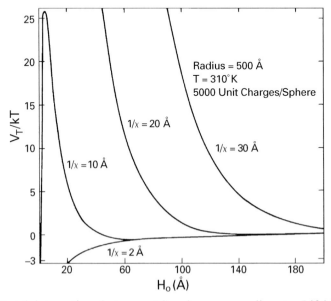

Fig. 130. Total interaction between PS spheres according to Lifshitz–Parsegian–Ninham theory. Variable ionic strength [160].

7.2.3.3 Langmuir–Ise–Sogami (LIS) theory

A look at Fig. 124 indicates that polystyrene particles in water, with a Hamaker constant of $A_{121} = 3 \times 10^{-21}$ J, would have an energy barrier, $V_{T_{max}}$, much higher than the highest shown, with the added realization that the existence of a secondary minimum, where the colloidal particles weakly attract each other over intermediate distances, would not be predicted. There is, however, a large body of experimental work which appears to show conclusively that long-range attractions exist and that they are on the order of one or more times kT in depth. For example, Figs 55 and 56 (Section 5.3) show attractions between negatively charged glass and negatively charged particles (Fig. 55) and among the particles themselves to form voids (Fig. 56).

In the former case measurements by laser confocal microscope of particle concentration as a function of distance from the glass surface and as a function of ionic strength are shown in Fig. 131. It is clear that in the absence of salt (rigorously 'deionized,' but still containing counterions necessary for electroneutrality) there is an attractive force operating at distances of up to 50 *micro*meters, i.e. 50 000 nanometers (Fig. 131A), far beyond the range of interactions discussed heretofore! When salt is added, the effect is diminished at $[\text{NaCl}] = 5 \times 10^{-5}$ M and disappears when that concentration is doubled to

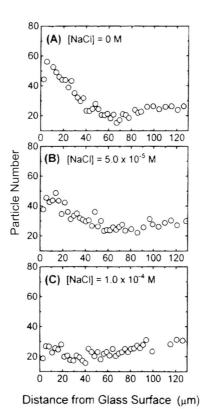

Fig. 131. Particle number concentration as a function of distance from cover glass [163].

0.10 mM. This is not at all what one would expect if this attraction were due to the dispersion (Hamaker) force.

It is, however, exactly what one expects if the effect were due to long-range *electrostatic* interactions as proposed by Irving Langmuir in 1938. Langmuir argued that if one looks at the colloid as a whole, the small counterions swimming among the particles attract the particles in a manner similar to that in which chloride ions act among the sodium ions in a crystal of salt [164]. He stated that the dispersion forces among colloidal particles in water were not strong enough to cause the attractions observed, e.g. phase separation into particle-rich and particle-poor domains, as in Fig. 56. The effect of the medium in reducing the van der Waals forces had not been taken into account by previous workers. Langmuir argued that the Debye–Hückel osmotic pressure, Π, of the system, which is composed of two parts:

$$\Pi = kT \sum n_i - \frac{e^2 \kappa}{6\varepsilon} \sum n_i v_i^2$$

will turn negative when the second term in this equation is sufficiently large. This occurs at relatively high surface charge, just what would not be predicted by DLVO. The first term in the equation is the combinatorial contribution of all particles, including ions, regardless of charge, while the second term reflects the clustering of counterions around the charged particles. Langmuir assumed that the approximation for low surface potentials could apply, an assumption that Verwey and Overbeek strongly criticized.

In the face of these arguments Ise and Sogami have shown that the electrostatic attraction between two spheres can be calculated if one takes into consideration the difference between the Gibbs free energy, G, and the Helmholtz free energy, F, i.e. the osmotic pressure-volume change. One then obtains

$$\Pi \Delta V = \Delta G - \Delta F = \frac{1}{2} \sum_{m \neq n} \frac{\partial U_{mn}^{E}}{\partial R_{mn}} \Delta R_{mn} \qquad (108)$$

where U_{mn}^{E} is the adiabatic pair potential between two charged particles resulting from their relative positions, and R is the center-to-center distance between them. It is important to distinguish between the volume of the colloidal dispersion and the volume V in Eq. 108 which is that occupied by the counterions. The volume V will change under the influence of the changes in electrical field as the two particles move towards or away from each other, a consideration which Ise and Sogami claim DLVO theory overlooked. These considerations lead ultimately to the Langmuir–Ise–Sogami (LIS) Gibbs potential, $U^{G,}$ which has two major differences from V_T of DLVO: (1) the pressure–volume effect is included and (2) the Hamaker attraction is ignored:

$$U^{G} - \frac{1}{\varepsilon} \left[\frac{Ze \sinh \kappa a}{\kappa a} \right]^2 \left[\frac{1 + \kappa a \coth \kappa a}{R} - \frac{\kappa}{2} \right] e^{-\kappa R} \qquad (109)$$

where Z is the *number of surface charges* on the particles [165]. This contrasts with the DLVO energy (Eq. 106) which depends upon the surface *potential*, ψ_0. Ise and Sogami are aware that the net surface charge will vary with the nature and concentration of the small ions because of counterion condensation. Plots of Eq. 109 for various values of ionic strength as given by the dimensionless variable κa are shown in Fig. 132. In this example $a = 60$ nm, $Z = 1400$ and $T = 23°C$.

It is clear from these plots that the secondary minimum is on the order of 5 to 10 times kT, where $kT/e = 25.6$ mV, much deeper than the depths of the DLVO minima in Fig. 130. But all depends upon the degree to which the Stern layer screens the surface charge. Recall that the value of Z is that at the surface of the Stern layer and can easily be less than one tenth that at the surface of the particle (see Section 5.7.4).

One of the strongest arguments in favor of the LIS theory comes from studies of the thermal motions of polymer particles in systems in which there is equilibrium between ordered and disordered domains. X-ray diffraction studies of dilute polystyrene sulfonates in solution have shown that the average spacing between particles in crystalline domains is often on the order of one half the average in the system as a whole. Thus there must be an equilibrium between highly ordered (crystalline) and disordered (gas-like) domains, in which the long-range LIS attractions act to stabilize the ordered regions. Ito and coworkers photographed the motions of particles in latexes and plotted their trajectories over a period of 11–15 s to produce Fig. 133.

The interparticle spacings, the distinct boundary at the 'surface' of the crystalline domain, and the constrained motions within the secondary minimum within the ordered region as contrasted to those in the disordered region, all support the concept of a long-range electrostatic attraction outside of the tenets of the DLVO theory. A further compelling experimental

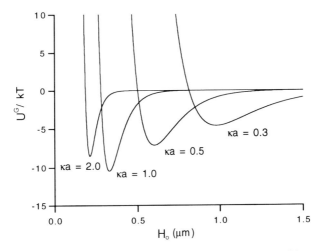

Fig. 132. Sogami potential between two spheres as a function of interparticle separation. Variable κa [165].

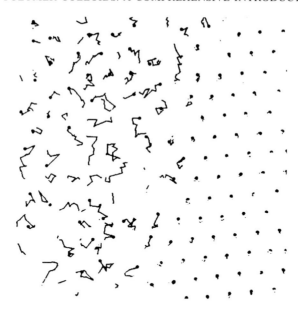

Fig. 133. Trajectories of PS latex particles in 2% colloid. Order–disorder equilibrium [166].

observation is that the addition of inert electrolyte to such a system causes an increase in the interparticle spacings in the ordered regions in accordance with LIS, whereas the DLVO theory predicts a decrease.

What follows in the next section is a discussion of the interactions among particles carrying no electrical charge, so-called sterically stabilized colloids.

7.3 Kinetic stability of electrically neutral hydrophobic colloids

Historically many colloids were stabilized by nonionic materials which were low molecular weight, hydrophilic polymers or nonionic surface active agents, usually referred to as 'protective colloids' and nonionic surfactants, respectively. Typically pigments were ground in the presence of aqueous solutions of egg white, milk casein or gum arabic or in certain vegetable oils to give stable dispersions either in aqueous or organic media. In this section we shall focus on nonionic, lyophilic polymers adsorbed at the particle/medium interface which impart what is known as steric stabilization.

7.3.1 Polymeric stabilizers

Almost all polymers will adsorb at an interface because the chain molecule will have certain regions which are lyophilic and others which are lyophobic (Fig. 134). Deliberate synthesis of amphiphilic polymers can greatly enhance

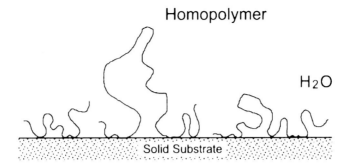

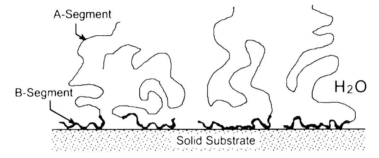

Fig. 134. Polymer conformations at a solid/liquid interface.

the strength of adsorption and allow control of polymer dimensions in the two phases. A simple example would be an 'AB' block copolymer in which the A-block is hydrophilic and the B-block is hydrophobic. This molecule will sit at a particle/water interface with the B-block serving as an 'anchor' and the A-block snaking out into the aqueous medium, as shown in Fig. 134. Other interfacial polymer conformations can be obtained by making, for example, ABA triblock copolymers or graft (commonly known as 'comb' or 'brush') copolymers. Thus not only the thickness of the adsorbed layer, but also the nature and density of the lyophilic portions are determined. Some examples are shown schematically in Fig. 135. Both figures involve the solid/aqueous interface, but they could apply to a solid/organic medium interface as well, in which case the A-segments would be soluble in the organic phase, while the B-segments would be insoluble.

The mechanism of stabilization involves the interaction of two surfaces which are covered in this manner. Clearly, there can be no interaction until the outermost segments of the two layers start to overlap. As the two surfaces are brought closer together the concentration of polymer units increases in the overlap region with a resulting increase in the osmotic pressure. This tends to bring in solvent from the surrounding medium, with a consequent force to separate the particles in direct analogy to the overlap of two diffuse parts of

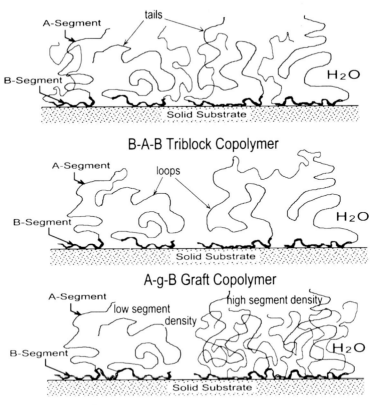

Fig. 135. Interfacial chain conformations and dimensions depend upon block- or graft-copolymer structures.

electrical double layers (Section 7.2.3.1, *The repulsive free energy for flat plates*).

This mechanism of stabilization clearly depends upon the free energy of mixing, ΔG_m, between the A-segments of the steric stabilizer and the solvent medium. If ΔG_m is negative, then there is favorable mixing and the osmotic pressure mechanism will operate to separate the particles. On the other hand if ΔG_m is positive, mixing is unfavorable, and there will be a tendency for solvent to be expelled from a region of high polymer concentration, and the particles will tend to coagulate under the influence of the van der Waals attractions. The Gibbs–Helmholtz equation tells us that there are two components to the Gibbs free energy change upon mixing:

$$\Delta G_m = \Delta H_m - T\Delta S_m \tag{110}$$

For some polymers the enthalpy of mixing is negligible (so-called athermal solutions), so that the free energy of mixing is entirely entropy-driven, and thus becomes more favorable as T is increased. For others, the enthalpy term

TABLE 7.2
Factors affecting the Gibbs free energy of mixing

ΔS_m	ΔH_m	ΔG_m	Effect on colloid stability
+	+	Depends on T	Unstable at low T; stable at high T
+	−	−	Stable at all T values
−	−	Depends on T	Stable at low T; unstable at high T
−	+	+	Unstable at all T values

dominates, either negatively or positively. These effects are summarized in Table 7.2.

The Flory–Huggins theory of polymer–solvent interactions employs these considerations as a basis, using lattice theory to calculate the entropy of mixing. Flory introduced the concept of the temperature, θ, to specify the conditions under which the free energy of mixing just goes to zero, and beyond which phase separation occurs.

When the polymers are attached to colloidal particles, one might expect that as the θ-condition is reached the driving force for stabilization disappears and the system will coagulate. This assumes that the Flory–Huggins lattice theory applies to polymers which are attached to a surface. In fact, experiments show a remarkable coincidence of θ and the critical coagulation temperature, CCT, as shown in Table 7.3.

The θ-condition is defined not just by the temperature, but by the nature of the solvent as well, as Eq. 110 suggests. Thus addition of a nonsolvent for the stabilizer polymer can lead to coagulation just as readily. Unlike electrostatic stabilization, however, coagulation can usually be reversed by adding good solvent or by changing the temperature. The surface polymer, upon re-solvation, will expand and tend to separate the particles.

TABLE 7.3
Comparison of critical coagulation temperature (CCT) and Flory θ for some stabilizing polymers

Stabilizer polymer	Molecular weight (Da)	Medium	CCT (K)	θ (K)
Polyethylene oxide	10 000	Aq. 0.39 M MgSO$_4$	318	315
Polyethylene oxide	96 000	Aq. 0.39 M MgSO$_4$	316	315
Polyethylene oxide	1 000 000	Aq. 0.39 M MgSO$_4$	317	315
Polyacrylic acid	9 800	Aq. 0.20 M HCl	287	287
Polyacrylic acid	51 900	Aq. 0.20 M HCl	283	287
Polyacrylic acid	89 700	Aq. 0.20 M HCl	281	287
Polyvinyl alcohol	26 000	Aq. 2.0 M NaCl	302	300
Polyvinyl alcohol	57 000	Aq. 2.0 M NaCl	301	300
Polyvinyl alcohol	270 000	Aq. 2.0 M NaCl	312	300
Poly-iso-butene	23 000	2-Methyl butane	325	325
Poly-iso-butene	150 000	2-Methyl butane	325	325

7.3.2 Interaction between two particles

7.3.2.1 Smitham–Evans–Napper (SEN) theory

When two surfaces, which are covered with a well solvated polymer layer of thickness L, approach each other at a distance of separation d, they will form an overlap region of thickness δ, and *interpenetration* volume, dV, as shown in Fig. 136. It is conceivable that the plates could come so close together that $d < L$, a situation known as *compression*. Experiments and calculations both show that the repulsion between the two plates due to interpenetration alone is so great that compression is not observed, assuming 'normal' coverage of the surface by the stabilizing polymer and normal thermal energies of the particles.

Calculation of the magnitude of the repulsion comes from the work of Smitham, Evans and Napper [167]. The generally close correlation between the θ-temperature for polymers in dilute solution and the CCT shown in Table 7.3 indicates that solution theory is applicable and that upon interpenetration, the thermodynamic considerations dominate, and any reconformations of the chains must be small enough to neglect. The starting point is the Flory–Krigbaum free energy for the interpenetration of two polymer chains in solution [168]:

$$\Delta G = 2kT \left(\frac{V_s^2}{V_1} \right) (\tfrac{1}{2} - \chi_1) \int \rho_d \rho_d' \, dV \qquad (111)$$

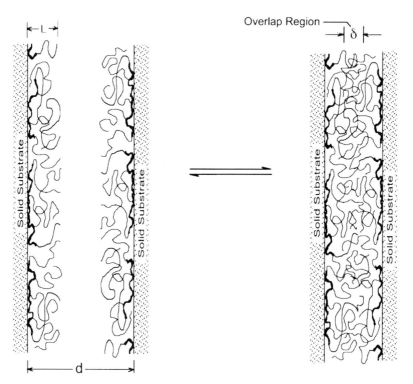

Fig. 136. Schematic representation of two plates with steric stabilizer on approach.

in which V_s is the volume of a chain segment, V_1 is that of a solvent molecule, χ_1 is the Flory–Huggins interaction parameter, and ρ_d and ρ'_d are the segment density distribution functions of the two chains, respectively. For flat plates the volume dV can be given by Adx, where x is the distance normal to the plate surface; and for unit area $dV = dx$. Taking the normalized distribution functions, $\hat{\rho}_d$ and $\hat{\rho}'_d$, these substitutions into Eq. 111 give the Gibbs free energy per unit area for interpenetration between two flat plates:

$$\Delta G^{FP} = 2kT \left(\frac{V_s^2}{V_1} \right) (\tfrac{1}{2} - \chi_1) \nu^2 i^2 \int_{d-L}^{L} \hat{\rho}_d \hat{\rho}'_d \, dx \qquad (112)$$

where ν is the number of chains containing i segments each within dx.

Generally the concentration of polymer chain segments in the interpenetration domain is not great, so that the distribution functions may be approximated by that of the plates at infinite separation, $\hat{\rho}_d \approx \hat{\rho}_\infty$.

For spheres, the Derjaguin approximation which was employed for ionic systems may again be used. The geometrical considerations are the same as those depicted in Fig. 126. For equal size spheres the free energy of interpenetration is obtained by integrating ΔG^{FP} between the limits of H_0 and infinite separation:

$$\Delta G^S = \pi a \int_0^\infty \Delta G^{FP} \, dH_0$$

which gives upon substitution of Eq. 112:

$$\Delta G_i^S = 2\pi a kT \left(\frac{V_s^2}{V_1} \right) (\tfrac{1}{2} - \chi_1) \nu^2 i^2 \int_0^\infty \int_{d-L}^{L} \hat{\rho}_\infty \hat{\rho}'_\infty \, dx \, dH_0 \qquad (113)$$

The pre-integral parameters in Eq. 113 determine the polymer/solvent interactions, and these are independent of the shape of the distribution functions. Experimentally it turns out that many steric stabilizers are so strongly adsorbed at the particle surface that their lyophilic moieties experience sufficient lateral crowding that the chains stand practically rigid, i.e. their shapes can be described by a uniform segment density distribution. For example, in PMMA organosols stabilized with polyhydroxy stearic acid (PHA) chains (A-segments in Fig. 135) anchored by PMMA moieties (B-segments), the surface area occupied by a single PHA chain was found to be 80–100 Å^2, not much more room than that occupied by an extended chain standing on end [169]. This makes solution of Eq. 113 straightforward, assuming the segment density, χ and molar volumes are known. Other distribution functions, ρ_d, e.g. Gaussian and exponential, have also been calculated.

Many steric stabilizers are so strongly adsorbed that the chains stand practically rigid.

7.3.2.2 Experimental results

Two distribution functions are shown against the experimental results of Dorowsowski and Lambourne in Fig. 137. The curve for the uniform segment density distribution is not shown, but essentially lies on the Gaussian curve in the range $1 < H_0/L < 2$. When H_0, the distance between particle surfaces, is less than L, the stabilizer layer thickness, the stabilizer chains come in contact with the opposing particle surface, a region known as the 'compression' region where theoretical treatment is less reliable. All theories predict an interaction similar to that of a 'brick wall' when the particles come into the compression domain.

Dorowsowski and Lambourne obtained their data on polyacrylonitrile particles stabilized with polystyrene chains suspended in toluene. They managed to float the particles on the surface of a Langmuir trough and then measured the force required to compress the monolayer, an elegant experiment. This two-dimensional experiment, however, does not mimic very well what happens in three dimensions in ordinary colloids. Under such conditions, it is highly unlikely that any given pair of particles would possess much more than $5kT$ of energy, a region on the curves in Fig. 137 where $H_0/L > 1$, where there is interpenetration only.

Another series of ingenious experiments by Ottewill and coworkers [171] allows us not only to measure the repulsive forces in a three-dimensional

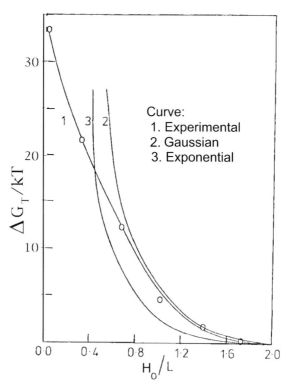

Fig. 137. Total repulsive potential energy between spheres for steric stabilizers. Theoretical and experimental curves [170].

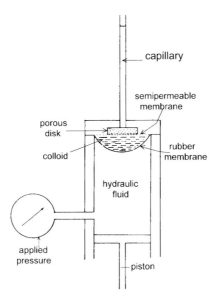

Fig. 138. Compression cell [171].

system, but to compare the results of steric and ionic systems. The apparatus is composed of a compression cell (Fig. 138) in which the latex sits upon a piston and below a semi-permeable membrane, which in turn is connected to a capillary tube partly filled with the fluid medium. As the piston is raised, the latex is compressed against the membrane, and fluid medium flows very slowly into the capillary, thereby concentrating the colloid. The change in height of the fluid column directly measures the volume change in the latex. A sensitive gauge measures the change in pressure. Some experimental results are shown in Fig. 139. The nonaqueous system was a PMMA colloid whose particles were covered with polyhydroxy stearic acid (PHA) chains (average DP = 5) in a medium of dodecane. The number-average particle size was 155 nm (dry particles including stabilizer, by TEM). The aqueous system was a polystyrene colloid in 10^{-5} M NaCl with a particle size of 182 nm.

The onset of repulsion in the sterically stabilized colloid was first measured at a volume fraction, ϕ, of 0.555 (Fig. 139 inset). Extrapolation to zero pressure suggests that 'first contact' should be made at about $0.53 < \phi < 0.48$, from which the chain dimensions could be estimated to be 7.3 nm, not quite the extended chain length of 9 nm for PHA of DP = 5. The 'brick wall' is very clear here, and occurs at $\phi = 0.566$, closely following the curves in Fig. 137 although with different coordinates.

In the aqueous system the behavior is entirely different, with a repulsive interaction observed already at $\phi = 0.2$, and a very gradual increase in pressure until the volume fraction reaches about 0.6 when it rises sharply. This follows the curves in Fig. 130 for larger values of $1/\kappa$, again with different coordinates. There is no secondary minimum attraction in either case; in the former, because the medium and the particles are not very different in composition, and in the latter, evidently because the surface charge was large.

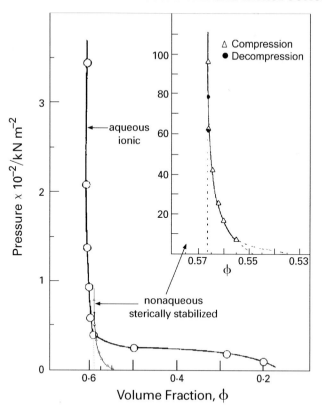

Fig. 139. Pressure vs volume fraction for nonqueous PMMA (inset) and aqueous PS colloids [171].

7.3.2.3 Practical considerations

The bottom line of all this is that steric repulsions are generally so strong that stabilization is essentially permanent, given that the stabilizer molecules are well designed and that the particles are in an appropriate solvent for the A-segments. Changes in temperature or solvency can move the system to θ-conditions for the stabilizer moieties, whereupon coagulation occurs. This is often desirable when the polymer is wanted in bulk form, and this mechanism can be environmentally friendly compared with the coagulation of ionically stabilized colloids where acid or salts must be added to obtain the coagulum and the supernatant, often also containing surfactants, initiator residues, etc., is ordinarily then dumped.

> The bottom line: steric repulsions are generally so strong that stabilization is essentially permanent.

In nonaqueous systems there can be intermediate regions of stability where the 'brick wall' effect is not so pronounced, which may obtain when:

- the A-segments are short, so that close approach may occur to such an extent that the van der Waals attractions are important, leading to secondary minimum flocculation (as opposed to coagulation); or
- the solvency of the medium is close to the θ-condition, in which case the chain conformations become more like an exponential segment density distribution with greater crowding near the particle surface, allowing for closer approach.

> Steric stabilizer leaves an incompatible polymeric layer among the particles in a film, potentially providing superior dynamic-mechanical properties.

Another practical consideration is that because of the polymeric nature of steric stabilizers, the amount of stabilizer required is usually considerably greater than that for ionic stabilizers – up to ten or more times as much. Because the anchoring is generally strong, there is no migration of stabilizer, e.g. when a film is formed by evaporation of the medium. This leaves an incompatible polymeric layer among the particles, such that the particles remain uncoalesced, which may or may not be desirable for the end-use intended. For example, in Section 2.4 phase-separated systems were seen to have superior dynamic-mechanical properties, so that appropriate engineering of the molecular architecture of the system could lead to improved performance of films and bulk materials derived from these polymer colloids.

An additional concern relates to the potential lack of mobility of the lyophobic segments of the stabilizers in solution prior to adsorption onto the particle surface. The stabilizers form micelles in aqueous as well as organic media, and because of their polymeric nature, the lyophobes can form highly viscous or glassy domains which constitute the core of the micelles. These are commonly in equilibrium with extremely low concentrations of free stabilizer molecules in solution. To form an efficient stabilizer layer requires initial adsorption of a micelle and subsequent rearrangement to form the monolayer. The latter process is often *slow*, and may take minutes, hours or even days [172, 173]. To enhance the mobility, one can add a good solvent for the lyophobe, raise the temperature or choose a stabilizer molecule with very short lyophobic chains.

Finally it can be said that steric stabilizers provide excellent colloidal stability, they are relatively insensitive to ionic strength, and they are necessary if freeze-thaw stability is required, apparently preventing ice crystals from forming in a way that would otherwise force particles together as water expands upon freezing.

A good, non-mathematical review of steric stabilization has been given by Napper in the book *Colloidal Dispersions* edited by Goodwin [174].

7.4 Electrosteric stabilization

There is also the whole domain of electrosteric stabilization [175] in which the stabilizing moieties are polyelectrolytes – polymer chains containing either

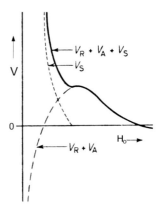

Fig. 140. Electrosteric interaction between two particles (schematic).

weak or strong electrolyte ionic groups. These can combine the best properties of both ionic and steric stabilization, i.e. long-range repulsion combined with a short-range 'brick wall' effect! This is shown schematically in Fig. 140 in which V_R represents the purely electrostatic repulsion.

Many electrosteric stabilizers are available commercially either as graft or block copolymers. The simplest of these are random copolymers containing ionizable groups, such as $-COOH$. They are used during emulsion polymerization to control particle size as well as to stabilize the latex. The hydrophobic groups, randomly distributed along the polymer chains, will adsorb onto the particle surfaces, and often will become grafted onto the particles as a result of chain transfer reactions. Any of the structures discussed above, e.g. those represented in Figs 134 and 135, could contain ionizable groups in the hydrophilic domains and would thus act as electrosteric stabilizers.

7.5 Kinetics of coagulation

The effect of coagulation on particle formation in emulsion polymerization was discussed briefly in Section 2.1.1.1 and Fig. 6. We now look at the subject in greater depth and from the point of view of colloid stability more generally.

7.5.1 A second-order rate process

Coagulation is considered to occur stepwise between pairs of particles, so that it is the collision frequency and the pair interaction potential which govern the overall rate. The analogy to chemical kinetics is close, so that the concept of an energy of activation (Fig. 115) can be invoked (an oversimplification – see next section), and the form of the expression will be:

$$\frac{dN}{dt} = -kN^2 \qquad (114)$$

where

$$k = A' e^{E_{act}/kT}$$

N is the particle number-concentration and A' is the pre-exponential factor, which for spheres will be just the collision frequency. As the barrier height is reduced to zero, the coagulation rate increases to a maximum which depends upon the rate of diffusion of particles towards each other.

The application of diffusion theory to this problem was first successfully made by von Smoluchowski [176] in 1917. He applied Fick's first law of diffusion to the case in which there is a concentration gradient, dN/dR, around a central particle due to the disappearance of other particles as they diffuse to, and stick to, this central particle. R is the radial distance from the center of the particle. In this model there is no barrier, and there is a deep energy well of attraction upon contact, so that every collision is irreversible. The corresponding geometry is shown in Fig. 141.

Fick's first law states that the flux of particles (particles s^{-1} m^2) towards the central one is directly proportional to the concentration gradient of particles. Then the total collision rate under steady state conditions is the flux, J, multiplied by the area, A, of the sphere:

$$AJ = -4\pi R^2 D \frac{dN}{dR} \quad (115)$$

where D is the Fickian diffusion coefficient for a particle. Since the central particle also diffuses, the *relative* diffusion coefficient is twice that for a single particle, assuming both are the same size, i.e. $D = 2D_1$ (otherwise $D = D_a + D_b$). Rearranging and integrating Eq. 115, noting the boundary conditions that $N = N_0$ ('bulk' value) when $R = \infty$, and $N = 0$ at $R = 2a$ gives:

$$\frac{AJ}{8\pi D_1} \int_{2a}^{\infty} \frac{dR}{R^2} = -\int_0^N dN \quad (116)$$

which shows that the collision frequency *on a single particle* is

$$AJ = -16\pi a D_1 N \quad (117)$$

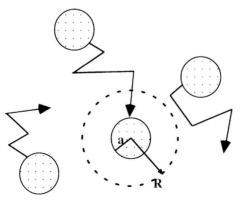

Fig. 141. Coagulation by diffusion to a central particle of radius a.

7.5.1.1 Fast coagulation

The collision rate for all particles is simply $A'N$ which, when compared with Eq. 114, indicates that the rate constant, k_{fast}, for coagulation in the absence of an energy barrier ($E_{\text{act}} = 0$, achieved by adding sufficient salt or heating to the θ-temperature or beyond), for electrostatically or sterically stabilized colloids, respectively, is

$$k_{\text{fast}} = A' = -16\pi a D_1 \quad \text{(no potential barrier)} \tag{118}$$

The Stokes–Einstein equation for D assumes that we have hard spheres (no solvent is carried with the particle as it diffuses), that the particles are large compared with the molecules of the surrounding medium, and that there are no wall effects:

$$D = \frac{k_B T}{6\pi \eta a} \tag{119}$$

where η is the viscosity of the medium. Combination of Eqs 118 and 119 gives

$$k_{\text{fast}} = \frac{8k_B T}{3\eta} \tag{120}$$

where the subscript B has been used to distinguish Boltzmann's constant, and the subscript fast indicates that this is the diffusion-controlled rate of coagulation and thus the fastest possible without agitation. Thus, for equal sized particles, the remarkable result is that the rate of coagulation is independent of particle size! The value of k_{fast} at 20°C is $1.08 \times 10^{-14}\,\text{dm}^3\,\text{s}^{-1}$ in water ($\eta = 0.010$ poise; in SI units this is $1.0\,\text{mPa s}$).

> The remarkable result is that the rate of fast coagulation is independent of particle size.

If we take an ordinary latex with a particle concentration of, say, $10^{16}\,\text{dm}^{-3}$, and dilute it one hundred times ($N = 10^{14}\,\text{dm}^{-3}$) and calculate the half-time for fast coagulation, i.e. the time it takes for the number of particles to decrease to one half of its original value, we proceed as follows:

Integration of Eq. 114 gives the standard second-order equation:

$$\frac{1}{N} - \frac{1}{N_0} = k_{\text{fast}} t \tag{121}$$

At the half-life time, $t_{1/2}$, $N = N_0/2$, which upon appropriate substitution into Eq. 121 and rearrangement gives

$$t_{1/2} = 1/k_{\text{fast}} N_0 \tag{122}$$

From this the half-life for the example above is calculated to be about 1 s; for the undiluted latex it would be a mere 10 ms – fast coagulation indeed! Equation 122 can only be used as a rough approximation, however, because we have so far ignored the differences between second-order molecular reactions and coagulation. In the latter, there will surely be some distribution of particle sizes since a doublet particle could coagulate with another single particle to form a triplet, etc. while some singles remain uncoagulated by the time $t_{1/2}$ is reached. But the arguments from which the equations above are derived, stemming from the model in Fig. 128, require only singlet–singlet collisions. For one thing, the diffusion constants for multiplet particles will be smaller. A treatment of the full distribution is quite possible and can be found in many references; see, for example, the book by Russell, Saville and Schowalter [161]. Nevertheless, the *rate* equations, Eqs 114–120, are perfectly applicable, as long as they are seen as *initial* rates, i.e. for the formation of doublets only.

> Flocculation is reversible; coagulation is not.

Incidentally there is some confusion in the literature concerning the terms 'flocculation' and 'coagulation.' Overbeek has clarified the issue by stating that the former term refers to reversible processes, and the latter to irreversible aggregation. Thus flocculation will deal with particles 'falling into' the secondary minimum, from which the energy barrier for disaggregation is not very great, whereas coagulation involves primary minimum aggregation with a large energy barrier for reversal. All of these involve non-zero energy barriers and thus fall into the overall class of 'slow coagulation.'

7.5.1.2 Slow coagulation

The 'energy of activation' in the coagulation process arises, as we have seen, from the pair interaction potentials from overlap of the diffuse parts of electrical double layers or steric, polymeric layers. Under these circumstances, not only the concentration gradient dN/dR in Eq. 115, but also the potential gradient serves as a driving force, either to accelerate or (usually) to retard coagulation. Thus to Eq. 115 must be added the potential gradient dV_T/dR as follows:

$$AJ = 8\pi R^2 D_1 \left[\frac{dN}{dR} + \frac{N}{kT} \frac{dV_T}{dR} \right] \qquad (123)$$

In the second term N appears because the flux of particles, AJ, will be proportional to it, and kT is in the denominator so that the energy term (dV_T/kT) remains dimensionless. This is the equation of Fuchs [177]. It can be integrated, with the boundary conditions as follows:

$$N = 0 \quad \text{when} \quad R = 2a, \quad \text{and} \quad N = N_0 \quad \text{when} \quad R = \infty$$

The result is

$$AJ = \frac{8\pi D_1 N_0}{\int_{2a}^{\infty} \exp\left(\frac{V_T}{kT}\right) \frac{dR}{R^2}} \tag{124}$$

From Eqs 117, 118 and 124 it is clear that the rate constant is:

$$k_{\text{slow}} = \frac{8\pi D_1}{\int_{2a}^{\infty} \exp\left(\frac{V_T}{kT}\right) \frac{dR}{R^2}} = \frac{k_{\text{fast}}}{2a \int_{2a}^{\infty} \exp\left(\frac{V_T}{kT}\right) \frac{dR}{R^2}} \tag{125}$$

$$k_{\text{slow}} = \frac{k_{\text{fast}}}{W}$$

in which W is called the Fuchs stability ratio, i.e.

$$W \equiv \frac{k_{\text{fast}}}{k_{\text{slow}}} = 2a \int_{2a}^{\infty} \exp\left(\frac{V_T}{kT}\right) \frac{dR}{R^2} \tag{126}$$

This neglects a hydrodynamic interaction when particles collide, discussed in Section 7.5.1.3 below.

As an arbitrary criterion of latex stability, let us say – following Verwey and Overbeek – that $W > 10^5$ for dilute, and 10^9 for very concentrated systems. The half life when $W = 10^5$ would be, according to Eq. 122 – but now using k_{slow} – on the order of 10^6 s or approximately 12 days, not very stable, but on the borderline. Because V_T enters W as an exponential in Eq. 125, it is $V_{T_{\text{max}}}$ which has a major influence on stability. For example, a small, 15% change in $V_{T_{\text{max}}}$ causes a ten-fold variation in W (at $V_{T_{\text{max}}} =$ approx. $15kT$), which, in turn, may result from a miniscule 7.5% change in ψ_a!

This suggests a useful approximation for the Fuchs stability factor, based upon expansion of the exponential in V_T in Eq. 126:

$$W \approx \frac{1}{2\kappa a} \exp(V_{T_{\text{max}}}/kT) \tag{127}$$

recognizing that $V_{T_{\text{max}}}$ is a function of κ, and that κ, in turn, is a measure of the inert electrolyte concentration, n (Eq. 98). A plot of Eq. 127 in the form of $\log W$ as a function of $\log n$ (Fig. 142) shows a very simple relationship in which there are two linear domains, one with a negative slope where $W > 1$, and one with a zero slope where $W = 1$ (but see next section). Clearly, where $W = 1$ the coagulation proceeds at the diffusion-controlled limit, so that higher concentrations of salt do not effect faster coagulation. In the other domain, 'slow' coagulation occurs. The intersection between the two is where $V_{T_{\text{max}}} \rightarrow 0$, and is called the critical coagulation concentration or ccc.

The logarithmic relationship of Eq. 127 suggests that a plot of $\log W$ against $V_{T_{\text{max}}}$ should give a straight line when κ is controlled. Prieve and Ruckenstein

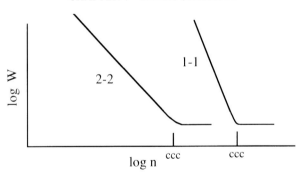

Fig. 142. Fuchs stability ratio vs. Electrolyte concentration. Variable valences (schematic).

[178] showed that this is indeed the case over a limited range of interaction potentials, i.e. when $12kT > V_{T_{max}} > 2kT$, as shown in Fig. 143. All of the points in this figure are calculated from their model involving weak acid surface groups. Under the conditions they chose, they found that the following approximative equation holds:

$$W \approx W_{\text{fast}} + 0.25\left[\exp\left(\frac{V_{\text{max}}}{kT}\right) - 1\right] \quad (128)$$

7.5.1.3 Hydrodynamic interactions

It can be seen from Fig. 143 that W_{fast} is greater than unity in the limit where $V_{T_{max}} \to 0$. There is a sizable body of experimental evidence to support

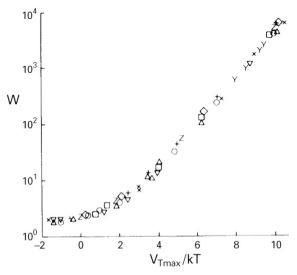

Fig. 143. Correlations of Eq. 127 with experimental data [178].

TABLE 7.4
Experimental values for the Fuchs stability ratio under fast coagulation conditions of polystyrene colloids

Particle radius (nm)	W_{fast}	Researchers
104	1.8	Lips and Willis [179]
179	1.9	Lips and Willis [179]
250	1.8	Lips and Willis [179]
46	1.7	Lichtenbelt et al. [180]
55	1.8	Lichtenbelt et al. [180]
88	1.6	Lichtenbelt et al. [180]
117	2.0	Lichtenbelt et al. [180]
338	1.9	Feke and Schowalter [181]

Note: It is not known how thoroughly these PS colloids were cleaned of polyelectrolyte.

this result, some of which is given in Table 7.4. The reason appears to come from the hydrodynamic interaction between particles upon close approach: molecules of the medium, with a finite viscosity, must flow out from between the particles before contact is made. The average experimental value obtained for W_{fast} is slightly under the theoretical value of 2.0. The latter was calculated by Spielman by including into Eq. 126 the change in diffusion coefficient of the particles as they approach each other, and as draining of the fluid medium becomes important [182]. This is given in Eq. 129:

$$W = 2a \int_{2a}^{\infty} \frac{D(\infty)}{D(R)} \exp\left(\frac{V_T}{kT}\right) \frac{dR}{R^2} \qquad (129)$$

where $D(\infty)/D(R)$ is the ratio of the relative diffusion coefficient at infinite separation and at the close distance R [182].

7.5.1.4 Specific ion effects

The classical theory enunciated above treats all ions as point charges with no specific interactions (with the exception of Prieve and Ruckenstein) with the lyophobic surface or any functional groups thereon. Yet the colloid literature has many examples of the so-called Hofmeister, or lyotropic, series in which different ions of the same valence exhibit different ccc values. Thus, for certain colloids monovalent cations exhibit relative ccc values in the order: $Li^+ > Na^+ > K^+ > H^+$; and divalent cations: $Mg^{2+} > Ca^{2+} > Sr^{2+}$. It appears that hydrated counterion size has a direct influence on the hydrodynamic drag of the particles upon close approach [183]. Furthermore, the dielectric data cited earlier (Section 5.7.5) also suggest that there must be specific ion interactions, such that different ions of the same valence would produce different Stern potentials. A few examples are given in Table 7.5. These effects are not trivial, but so far there seems to be no simple way to deal

TABLE 7.5
Critical coagulation concentrations for some polystyrene colloids

Colloid	Counterion	ccc (mol dm^{-3})	Researchers
PS-COOH	H$^+$	1.3	Ottewill et al. [185]
	Na$^+$	160	Storer [186]
PS/B	Na$^+$	200	Neimann and Lyashenko [187]
	K$^+$	320	Neimann and Lyashenko [187]
PS=NH$_2^+$	Cl$^-$	150	Pelton [188]
	Br$^-$	90	Pelton [188]
	I$^-$	43	Pelton [188]

with them. As Israelachvili said in 1981, 'What is clear so far, is that all non-specific theories, such as van der Waals' and double-layer force theories, cannot be extended down to contact separations' [184].

Additionally, some ions react with water to produce complex ions whose structures usually depend strongly upon pH. Aluminum ion, for example, forms not only aquo-ligands, but also polymeric species with valences of +4 and even higher. Phosphate is known to polymerize to form high-molecular weight polymers with valences equal to the degree of polymerization. Such ionic polymers can reduce the surface charge of colloidal particles to zero, thereby destabilizing them, and even reverse the charge, thereby re-stabilizing them.

7.5.2 Experimental methods

7.5.2.1 Light scattering: integrated method

The change in light scattering intensity with time can be used to follow the progress of a coagulating system. The advantage of using a polymer colloid is that the particles can all be spherical and of the same size, greatly simplifying interpretation of the experimental results. From the discussion in Section 5.1 one can derive the following equation which relates the scattering intensity, $I_k(\theta)$, of a k-fold aggregate at an angle θ to the degree of aggregation [179]:

$$I_k(\theta) = kI_1(\theta)\left[1 + \left(\frac{2}{k}\right)A_k(\theta)\right] \tag{130}$$

Equation 130 applies to uniform, monodisperse spheres. The geometrical factor $A_k(\theta)$ is given by the Rayleigh–Gans–Debye theory to be:

$$A_k(\theta) = \sum_{i=1}^{k-1}\sum_{j=i+1}^{k}\frac{\sin G_{ij}(\theta)}{G_{ij}(\theta)} \tag{131}$$

where

$$G_{ij}(\theta) \equiv \frac{4\pi n_2 h_{ij}}{\lambda} \sin\frac{\theta}{2}$$

and in which h_{ij} is the center-to-center distance between the ith and jth particles in an aggregate, λ is the wavelength of the incident light *in vacuo*, and n_2 is the refractive index of the dispersion medium. In the example described below, θ was chosen at 30°, and the colloid was a monodisperse PS latex with surface sulfonate groups only. Equation 130 indicates that as doublets are formed (k increases from 1 to 2) during coagulation, the scattering intensity will increase, even though the total number of particles is decreasing (recall that Rayleigh scattering intensity is proportional to the particle size raised to the sixth power, but to the number concentration of particles to the first power (Eq. 61)). It is best to look at initial rates so that only doublets have been formed; larger aggregates, which would complicate the angular dependence-of-intensity pattern, can be ignored. Typical light scattering intensity curves are shown in which the highly diluted (0.00100% solids) latex is rapidly mixed with $MgSO_4$ solution to the millimolarities shown in Fig. 144. All solutions were scrupulously filtered through a Millipore[TM] membrane to remove dust particles.

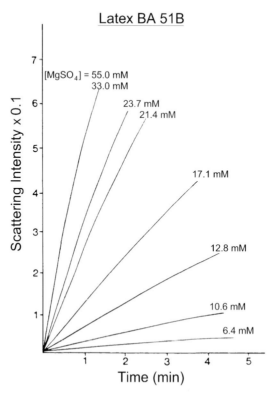

Fig. 144. Initial coagulation rates of a PS colloid. R_{30} vs time; variable 2:2 salt concentration [189].

It can be seen from Fig. 144 that above 33 mM electrolyte there is no further increase in the rate of coagulation. The initial slopes, extrapolated to $t \to 0$, can be shown to be linear [179] by calculation of the Rayleigh ratio at an angle θ from Eqs. 121 (for fast and slow coagulation), 129 and 130, as given in Eq. 132. The slopes are then converted simply by taking the ratios of

$$\text{slope}_{\text{fast}}/\text{slope}_{\text{slow}} = R_{\text{fast}}/R_{\text{slow}} = k_{\text{fast}}/k_{\text{slow}} = W$$

so that one can produce a so-called Reerink–Overbeek plot of log W versus log (concentration of coagulating electrolyte), as shown in Fig. 145.

$$R_\theta(t) \equiv \frac{I_\theta(t)}{I_\theta(0)} = 1 + 2 \sum_{k=2}^{\infty} \frac{N_k(t)}{N_0} A(k) \qquad (132)$$

Taking ratios of rates in this manner forces W_{fast} to be equal to unity, so that only the values of the slope and the intercept (ccc) are valid. But, given the approximations for low surface potential of the Reerink–Overbeek theory, it is possible to calculate Stern potentials from the slopes (Eq. 133) and Hamaker constants from a combination of the slopes and the ccc intercepts (Eq. 134):

$$-\frac{d \log W}{d \log n} = C \frac{a \gamma_\delta^2}{v^2} \qquad (133)$$

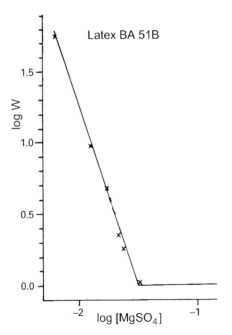

Fig. 145. Plot of Fuchs stability ratio vs salt concentration from data in Fig. 144 [189].

where C is a constant, γ_δ is defined as in Eq. 100, but using the Stern potential, ψ_δ, rather than the surface potential, ψ_0 and v is the valence of the electrolyte.

Understanding that the Hamaker constant is a fiction, nevertheless, we get a measure of the magnitude of the attractive potential by this means:

$$A = C' \frac{\text{slope}}{av} \text{ccc}^{-1/2} \tag{134}$$

Some results for the model colloids depicted in Fig. 145 and others are given in Table 7.6. There are seen to be strong deviations from theoretical values of surface potentials except at low surface charge densities, i.e. the Stern potentials, ψ_δ, fall in the range of 7–13 mV whereas the *calculated* surface potentials, ψ_0, are in the range of 9–62 mV. This may be explained by invoking 'counter-ion condensation,' the same phenomenon as that observed in the dielectric measurements of Sections 5.7.4 and 5.7.5, although to some extent it may be due to the approximations used for low surface potentials by Reerink and Overbeek. But when the Stern potential is calculated from the 'effective degree of dissociation,' f, obtained from dielectric measurements using the Gouy–Chapman equation (Eq. 135), one obtains good agreement with the values in Table 7.6 [190].

$$\sigma_0 = \varepsilon k T \frac{\sinh(ve\psi_0/2kT)}{2\pi ve} \tag{135}$$

It is strong testimony to the validity of these theories that measurements taken by such different methods as dielectric spectroscopy and coagulation kinetics are in close agreement. The slopes in Table 7.6 for 1:1 electrolyte are somewhat steeper than those for 2:2, as predicted by the DLVO theory and shown in Fig. 142. The Hamaker function, calculated here as a constant, is seen to be greater for higher surface charge densities, as predicted by LIS theory.

TABLE 7.6
Theoretical and experimental values of surface potentials and Hamaker constants for a series of polystyrene colloids [189] ($\bar{D}_n = 196$ nm)

Latex	σ_0 (μC cm^{-2})	$-d \log W / d \log n$	ccc (mol dm^{-3})	A (10^{-21} J)	ψ_δ (mV)	ψ_0 (mV)
1:1 Electrolyte						
51-I	10.2	3.63	1350	1.4	14	38
51-D	9.1	2.83	1220	1.3	12	34
51-C	5.0	2.06	800	1.1	10	26
51-H	1.0	1.31	325	1.0	8.1	8.6
2:2 Electrolyte						
51-I	10.2	3.18	29.5	4.1	13	62
51-D	9.1	2.34	34.8	2.8	11	55
51-C	5.0	1.80	61.0	1.5	9.8	36
51-H	1.0	1.07	60.0	1.0	7.5	9.8

The work cited above dealt with *relative* rates of coagulation. A method for obtaining *absolute* rates involves stopped flow turbidimetry in which the mixing of latex and coagulating electrolyte take only a few milliseconds, as carried out by Lichtenbelt, Pathmamanoharan and Wiersema (LPW) [180]. The turbidity, τ, of the system was obtained as a function of time, t, over periods of 40 ms to 3 s (depending on $t_{1/2}$) and then extrapolated to zero time, $(d\tau/dt)_0$, to ensure that relatively few doublets, and no higher multiplets, had formed. The light scattering behavior of the multiplets greatly complicates interpretation of the results. All experiments were carried out under conditions of fast coagulation by using concentrations of salt well above the ccc. The experimental rate constant for the formation of doublets, k_{11}, was calculated from Eq. 136:

$$k_{11} = \frac{1}{\tau_0} \left(\frac{d\tau}{dt}\right)_0 \bigg/ \left(\frac{C_2}{2C_1} - 1\right) N_1 \qquad (136)$$

where C_1 and C_2 are the extinction cross-sections of singlets and doublets, respectively. These were calculated from the Rayleigh–Gans–Debye theory, based upon Eq. 130: the turbidity is the amount of light transmitted, and therefore not scattered. The total amount scattered is found by integrating Eq. 130 over all angles, θ. The Smoluchowski fast coagulation rate constant is given by Eq. 120, and is independent of particle size. But LPW found that their rate constants were on the average 0.56 of the Smoluchowski value. The inverse of this, $1/0.56 = 1.8$, is the average of their 'W_{fast}' values given in Table 7.4 above. These results are considered to be in strong support of the existence of the hydrodynamic interaction, expressed in Eq. 129.

7.5.2.2 Other methods

Single particle optical sizing (SPOS). The more information one obtains about a system, the greater the precision and confidence level can be. Thus, being able to count individually the number and size of particles during the course of coagulation should represent the ultimate in this direction. This can be done by making a stream of colloidal particles pass through a laser beam. If both the beam and stream are of sufficiently small dimensions, and if the colloid is sufficiently dilute, then one will see flashes as single particles pass through the beam. The scattering pattern, I_θ, will provide information about particle size. This can be done by means of hydrodynamic focusing of a stream of latex particles in an apparatus like that shown in Fig. 146 [191]. A 10 μL, motor-driven syringe (a) creates a stream of the latex which enters a surrounding stream, the inner water flow, which in turn enters the outer water flow stream, all of which are in the laminar flow regime (see Eq. 79 and Fig. 62). The inner water flow rate (5.3×10^{-3} mL s^{-1}) is much faster than that of the latex stream (3.1×10^{-6} mL s^{-1}) so that the resulting extensional and shear forces lead to hydrodynamic focusing of the dispersion stream to a radius of 2.7 μm. With a laser beam thickness of 30 μm, this gives a sampling volume of 7×10^{-10} mL.

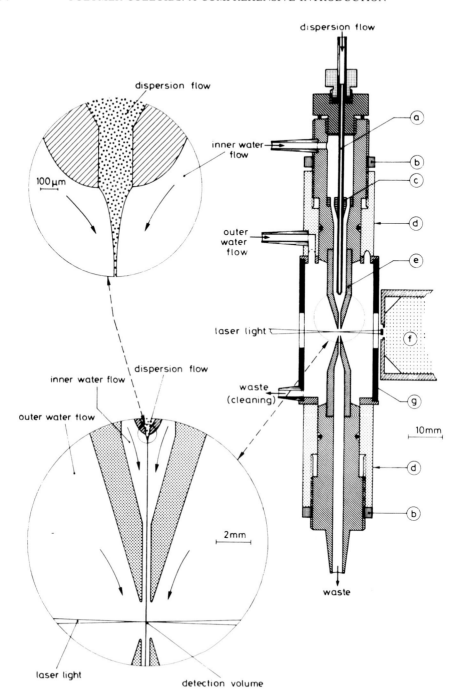

Fig. 146. Single particle optical sizing apparatus [191].

Calculation of the theoretical maximum particle concentration using coincidence statistics indicates that no more than 1.5×10^7 particles mL^{-1} should be present in order to ensure that two particles will almost never appear in the sampling volume at the same time. Up to 5000 particles per minute can be counted and sized by this means.

The water used in the flow streams must be degassed to remove any possible bubble formation, and filtered to remove dust particles. The coagulation is conducted outside of the apparatus by rapid mixing of latex and salt, waiting for a given time and then rapidly diluting so as to 'freeze' the degree of aggregation. Dilution reduces the salt concentration to well below the ccc and separates particles as well. The diluted colloid is then introduced into the flow apparatus, and the particle size distribution measured. Typical results are shown in Fig. 147 for a polystyrene latex with diameter of 696 nm and surface charge density of 4.7 μC cm^{-2}. From a series of such particle size distributions (PSDs) at different times, one can follow in great detail the coagulation process, as shown in Fig. 148. Aggregates larger than triplets were seen, but not counted in these experiments. The size distributions follow closely those calculated from DLVO theory.

Fraunhofer scattering. It is possible, with appropriate optics, to obtain light scattering patterns of polymer colloids in both the forward direction and at wider angles from which one can interpret the PSD over a broad size range. A schematic diagram of a commercial apparatus which employs this technique is shown in Plate 3. The combination of two frequencies of incident light allows one to obtain unequivocally the size distribution, since with a single frequency multiple solutions of the scattering equations for a given pattern may result. By reference to Figs 44 and 47, it can be seen that large particles

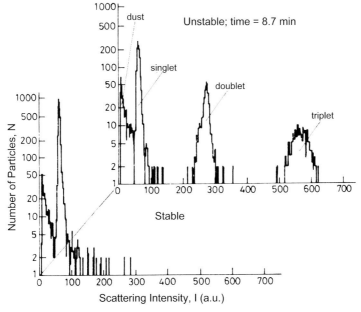

Fig. 147. Latex size distribution initially and after 8.7 min in 0.5 M KNO$_3$ [191].

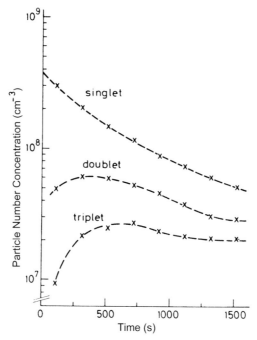

Fig. 148. PS latex size distributions vs time after mixing [180].

scatter light in the forward direction, whereas smaller ones scatter at wider angles. Furthermore, Eq. 62 tells us that the intensity of scattering is inversely proportional to the fourth power of the wavelength and to the number of particles times the square of their volume, Nv^2. Thus small particles scatter with little intensity, but this can be greatly offset by using incident radiation of short wavelengths.

The apparatus depicted in Plate 3 therefore employs two wavelengths, a blue one from an incandescent tungsten source and a red one from an He–Ne laser, and several angles of observation, θ, so as to obtain the broadest possible range of particle sizes. With this strategy the system can be fully automated to obtain particle size distributions between 0.02 and 1000 μm as differential and integral curves and in digital, spreadsheet form in a matter of 20 s! The progress of slowly coagulating systems may by this means be followed in great detail.

7.6 Practical applications

7.6.1 Latex paints

Latex paints probably consume the largest amounts of polymer colloids produced commercially. They are made by simply blending a pigment dispersion with a synthetic latex or, in rare cases, by grinding the pigment in the presence of a latex. There are two major sources of instability in latex paints:

(1) Dissolution of salts from the pigment into the medium, For example, calcium carbonate is used as a 'filler' pigment to lower the cost of the paint. It has little hiding power compared with, say, titanium dioxide. Even though calcium carbonate is considered insoluble in water, enough Ca^{2+} ions dissolve to destabilize a polymer colloid, especially since the ion is divalent.
(2) If the pigment is insufficiently covered with surfactant (in an effort to keep costs down, for instance), it will 'steal' surfactants from the polymer colloid, destabilizing it. This will be especially true if the dry pigment is ground in the presence of the latex.

To obviate these difficulties, as much as possible of the stabilizer should be chemically bound to the surface of the colloidal particles. Ionic stabilizing groups have the advantage of being required in relatively small amounts, especially when bound to the surface, but they have the disadvantage of sensitivity to salts. Nonionic, steric stabilizers impart much greater stability to the addition of pigments, but have the disadvantage of being required in relatively large amounts, and thereby imparting water-sensitivity to the final paint film. Electrosteric stabilizers are likely to provide the best compromise.

Another concern for paints arises in the process of film formation. During application a liquid film is spread onto the surface to be painted, water evaporates, and any inert salt in the system becomes more concentrated, threatening the stability of the paint. The way latex and pigment particles pack as the film forms is crucial to its ultimate integrity and therefore its ability to protect the substrate. If flocculation or coagulation occur prior to coalecence of the particles, an open structure can result, which would be especially important for corrosion barriers over steel, for example. This will be discussed in greater detail in Chapter 9 below.

Finally, adsorbed surfactants can become desorbed during film formation as the latex particles fuse together, reducing their total surface area. The surfactants will tend to migrate to new interfaces, in particular the film/air and film substrate interfaces. The latter, in particular, may lead to poor wet adhesion of the paint.

7.6.2 Molding resins

The second largest industrial use of polymer colloids may be as molding resins where the dry polymer is used to form molded objects. Here emulsion polymerization is employed because of its manufacturing advantages. In these cases the colloids are deliberately destabilized, coagulated, washed and dried. The supernatant liquid is discarded along with whatever coagulating reagents, initiator fragments, surfactants, buffers, etc. may be present. These all can constitute environmental hazards, thus creating costly clean-up problems.

There are two ways to overcome these problems:

(1) Again, by using surface-bound ionic groups, a minimum of stabilizing moieties is required, and none will be released into the waste water. As a rule of thumb, approximately one tenth as many surface-bound ionic

groups are required as adsorbed surface-active ones to get the same level of kinetic stability.

(2) Steric stabilizers can be made *in situ* during emulsion polymerization so that they are chemically bound to the particle surface. By changing the solvency of the medium or the temperature, the θ-condition can be passed to coagulate the latex. The resulting water-sensitivity of the molding resin may or may not be a concern.

7.6.3 Fibers

Some polymer colloids are coagulated – as in the case of molding resins – but are then either melted or dissolved in a solvent and spun into fibers. The environmental considerations for molding resins and ways to address them apply equally here. In these cases surface-bound ionic groups can be used to enhance dye-substantivity.

7.6.4 'Rubberized' concrete and stucco

If a synthetic latex of a rubbery polymer replaces all or most of the water used in making cement, concrete or stucco, a material is formed upon drying which has greatly improved flexibility. The latex particles apparently bond to the calcium phosphate crystals in such a way as to impart elasticity without greatly harming the strength of the structures formed. A major drawback may be the added cost of the latex.

Rubberized stucco has a reduced tendency to crack, so that it needs less maintenance than the regular material. Reduced maintenance costs are said to outweigh the initially higher cost of materials. Rubberized concrete is used on airport runways, for example, where the increased cost can be justified, thereby eliminating the need for asphalt strips otherwise required to allow for thermal expansion.

7.6.5 Controlled heterocoagulation for novel particle morphologies

Two dissimilar polymer colloids may be mixed under appropriate conditions of destabilization so that the particles of one heterocoagulate onto the surface of the other, as shown in Fig. 149.

The conditions of heterocoagulation will depend very specifically on the surface chemistries of the colloids involved. For example the anionic latex may contain RCOOH surface groups and the nonionic one, polyethylene oxide (PEO). It is well known that strong hydrogen bonds are formed between the two if the pH is low enough so that most of the carboxyls are in the acid form. Under such conditions, Thyebault and Riess have shown that slow, controlled agglomeration, such as that shown in Fig. 149, can occur to form core/shell morphologies [192]. Multiple layers in the shell are also claimed in the patent.

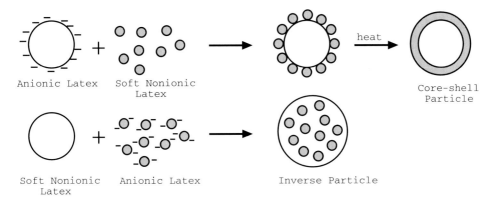

Fig. 149. Heterocoagulation of two dissimilar polymer colloids to form different particle morphologies (schematic).

A somewhat different, and perhaps more sophisticated, approach was used by Okubo, He and Ichikawa [193], in which a carboxylated latex with a diameter of 640 nm was mixed at pH 3 (again to suppress ionization) with a smaller sized, 140 nm, cationic latex, the latter formed by copolymerization of styrene and META (Table 6.2). Both colloids are sterically stabilized by a nonionic surfactant, PEO-sorbitan mono-oleate. There is essentially no coagulation upon gentle agitation at pH 3, unlike the previous example, probably because the PEO chains are not chemically bound to the small particles and thus bind to the surface RCOOH groups as small molecules. When the acid groups are changed by raising the pH to 9, heterocoagulation proceeds in a controlled manner to form framboidal core/shell particles like those shown in Fig. 1c. The average diameter of the resultant particles is 910 nm, within experimental error equal to the theoretical value of 920 nm for monolayer coverage of small ones on the big particles (640 + 2 × 140 nm). At this stage the heterocoagulation is reversible (see below) by reducing the pH back to 3. However, upon heating to 70°C for 4 h, two things occur: (1) the cloud point of the surfactant is neared (approach to the θ-temperature of PEO, combined with hydrophobic interactions), destabilizing the steric layer between particles in the agglomerate, and (2) the glass transition temperature, T_g, of the larger particles is exceeded, allowing for coalescence of the core and shell particles to occur. Subsequent lowering of the pH does not change the particle size. Because the small particles have a $T_g > 100°C$, they will not fuse to a smooth layer like those shown in Fig. 149.

Heterogeneous morphologies of these kinds can be extremely valuable in determining the dynamic mechanical properties of the derived films, fibers or molded objects. This will be dealt with in greater detail in Chapter 9 below.

7.6.6 Reversible coagulation

This term appears to be an internal contradiction, since coagulation is considered to be irreversible. Under a given set of conditions this definition still holds. But some colloids, especially sterically stabilized colloids that have been coagulated by crossing the θ-condition can, under favorable circumstances, be redispersed by returning to good solvation conditions for the stabilizing chains on the particle surface. This is a largely unexplored domain, but practical applications can be envisioned: because polymer colloids have high specific surface areas, they may be useful as heterogeneous catalysts, scavengers for heavy metal ions, etc.

After a catalytic reaction, it is often desirable to remove the catalyst. A change in temperature which caused coagulation, and thus precipitation, of the catalyst is the simplest way for its removal. By redispersion, the colloid is rendered ready and re-usable for the next batch.

Adding a polymer colloid containing appropriate chelating surface groups to waste water, for example, can be an efficient way to scavenge heavy metal ions. Subsequent coagulation removes the metal ions and polymer colloid together. Treatment of the concentrated coagulum to remove the metal ions, and redispersion of the colloids, allows for re-use of the relatively costly latex.

7.7 Lyophilic polymer colloids[*]

How is it possible to have a polymer which is lyophilic, i.e. soluble in the surrounding medium, and yet in the form of discrete colloidal particles? This is accomplished by forming crosslinked networks within the particles such that, even though the polymer undergoes thermodynamically favorable mixing with the medium, the polymer molecules are not free to escape the confines of the particle. Flory's theory for the swelling of gels applies here [194]. The swelling of the crosslinked polymer is favored by thermal interactions combined with an increase in entropy due to the larger number of possible configurations the molecule can occupy upon swelling. The swelling is limited by the tendency of the elastic network to contract. Swelling equilibrium is reached when the corresponding free energies are equal:

$$q_m^{5/3} \cong \bar{\nu} M_c \left(1 - \frac{2M_c}{M}\right) \left(\frac{\frac{1}{2} - \chi_1}{V_1}\right) \quad (137)$$

where q_m is the swelling ratio, i.e. the volume of the swollen gel divided by the volume of the unswollen polymer; $\bar{\nu}$ is the specific volume of the polymer; M_c is the molecular weight between crosslinks; M is the primary molecular weight (ordinarily very large in networks, so that the second term in Eq. 137 can be approximated by unity); V_1 is the molar volume of the solvent; and χ_1 is the Flory–Huggins interaction parameter (cf. Eqs 34 and 111). The first term, $\bar{\nu}M_c$, which represents the purely configurational entropy of mixing, becomes smaller as the density of crosslinks increases to the point where swelling is negligible. This can occur even though $\chi_1 < \frac{1}{2}$. Thus, to form a microgel

---

[*] The classical literature defines lyophilic colloids as solutions or colloidal dispersions of polymers in thermodynamically good solvents. We take the narrower definition involving the latter only. The former are now referred to as polymer solutions.

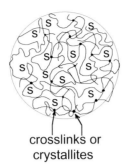

Fig. 150. Schematic of microgel particle. S represents the solvent.

particle, M_c should be on the order of 10 000 or greater, i.e. the amount of crosslinking comonomer should ordinarily be less than 10% of the total monomer charge.

> To form a microgel particle, M_c should be on the order of 10 000 or greater.

There are two different fundamental ways to achieve the 'crosslinking.' One is by interchain chemical bonds, usually introduced by copolymerizing multifunctional monomers with functionality greater than 2, e.g. divinyl benzene or ethylene dimethylacrylate. The other method is to have the polymer undergo partial crystallization. The crystalline domains must have sufficient lattice energy to be insoluble in the medium. The crosslinking must not be so great that the polymer as a whole becomes insoluble. The result can be considered a 'microgel' particle, as schematically represented in Fig. 150 in which 'S' represents solvent. We shall define the two types as *crosslinked microgels* and *microcrystalline microgels*. Some examples of lyophilic polymer colloids were given in Table 7.1.

7.7.1 Synthesis of microgels

Microgel particles can be synthesized in a variety of ways. Crosslinked microgels will form spontaneously in emulsion polymerization when a monofunctional monomer is copolymerized with a difunctional one or a difunctional one is polymerized alone.[*] For example, Obrecht, Seitz and Funke have compared the emulsion polymerization of pure 1,4-divinyl benzene (DVB) with that of styrene (S) [195]. They found that in recipes in which only the monomers were different:

- particle sizes of the DVB colloids were roughly half those of the S colloids;

---

[*] Note that formally speaking a monomer with a single double bond has a functionality, f, of 2, so that upon polymerization it will form linear chains. A monomer with two double bonds – 'difunctional', as used above – has $f = 4$ and is capable of forming a three-dimensional network.

- the number of particles formed by DVB was therefore about 6–9 times that formed by the styrene;
- the rate of emulsion polymerization for DVB was very fast initially and then slowed to a conversion of double bonds of about 50% of theoretical, whereas S gave conventional conversion-time curves and went to high conversions;
- the surface concentration of unreacted double bonds in the DVB colloids was about 1.6 vinyl groups per $1.0\,\text{nm}^2$, equivalent to $c.\,0.63\,\text{nm}^2$ per functional group;
- the amount of coagulum in the DVB colloids was large, from 22 to 37% of the total polymer, whereas the S colloids were essentially coagulum-free.

These results are quite typical, if somewhat extreme, given that pure difunctional monomer was used. Small particle sizes with considerable coagulum are likely when crosslinking is introduced into polymer colloids. The reasons for this have not been researched fully, but limited particle growth in highly crosslinked systems is likely to occur simply because of crowding of functional groups, even though they can be detected at the particle surface. The same phenomenon occurs in the synthesis of dendrimers in which growth stops after a few generations, even though functional groups available for further polymerization cover the surface [196]. An additional factor is that crosslinking may reduce the value of the critical degree of polymerization for nucleation of particles, j_{cr}. This is known to increase particle number and therefore decrease particle size (Fig. 10, Section 2.1.1.3).

> Small particle sizes with considerable coagulum are likely when crosslinking is introduced.

The large amount of coagulum produced when crosslinking is involved is not well understood, and offers an opportunity for practical research investigation. It may be associated with the interaction between surfactant and the particle surface, in which the hydrophobic tails of the surfactant, buried to some extent in the particles, are relatively immobile in a crosslinked matrix and cannot move freely when two particles coagulate in the early stages of emulsion polymerization. Thus limited coagulation, which almost *always* occurs as depicted in Fig. 6 and which requires an increasing concentration of surface charge by lateral movement of surface groups, cannot happen. And thus coagulation in crosslinked systems continues to very large sizes, i.e. coagulum.

Water-borne lyophilic colloids have been formed by polymerization directly in an aqueous environment, in some cases with a water-miscible organic solvent. For example, Rembaum and coworkers carried out the emulsion copolymerization of water-soluble 4-vinyl pyridine (4-VP) and the crosslinking monomer methylene bisacrylamide (BAM) under a variety of conditions [197]. Initiating free radicals were formed *in situ* by means of ^{60}Co gamma radiation.

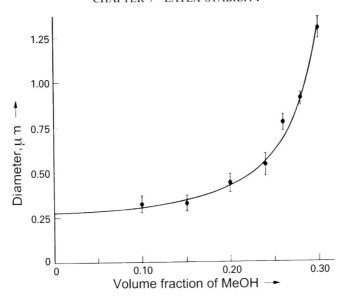

Fig. 151. Emulsion copolymerization of 4-VP/BAM (90/10); 4% polymer, 0.4% PEG. Effect of methanol concentration on particle size [197].

A small amount of polyethylene glycol (PEG) was added to control particle size. Apparently the PEG was grafted onto the main polymer in the presence of ionizing radiation. Quite narrow particle size distributions were obtained by this means. The dramatic effect on particle size of organic solvent in the form of methanol (MeOH) is shown in Fig. 151, presumably because it causes an increase in the value of j_{cr} during particle formation.

That this is directly related to solubility of the oligomers is borne out by the dependence of particle size on the solubility parameter of the medium. In Fig. 152 it is shown that with two different solvents in water, particle sizes over a four-fold range in monomer concentration were approximately the same when the solubility parameter of the two media were matched. The dependence of particle size on monomer concentration, i.e. the shapes of the curves, is a function of the solvency of the medium as well (not shown here, but see ref 197). These results were taken from a study of biocompatible polymer colloids which were to be used for protein binding and subsequent immunospecific binding to cells for clinical diagnostics, targeted drug delivery and chemotherapy.

7.7.2 Transfer to good solvents

Styrene/divinyl benzene copolymer colloids formed in water are obviously hydrophobic and thus not swollen by this medium. To form the desired lyophilic colloids it is relatively easy to transfer the particles to a good solvent for the polymer, such as toluene, whereupon they become swollen with solvent

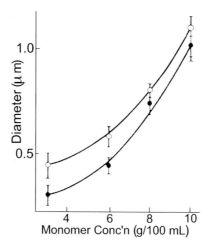

Fig. 152. Effect of aqueous solvents of the same solubility parameter ($\delta = 21.6$) on particle size. –o–, 13% acetone; –•–, 20% methanol [197].

to an extent that depends upon the degree of crosslinking. Transfer can be accomplished in several ways.

7.7.2.1 Destabilization and mixing with solvent

In this method the aqueous latex is destabilized by adding salt or acid (the latter if there are weak acid groups stabilizing the latex), not necessarily to the point of coagulation. The hydrophobic PS/DVB particles may be regarded as macro-amphiphiles, in that they possess a hydrophilic surface and a hydrophobic core. They thus have a tendency, especially under mildly destabilizing conditions, to accumulate at the water interface. If the organic solvent is on the other side of the interface, then the latex particles can migrate across the boundary into a region in which their free energy is considerably lower. An extreme example of this method is coagulation and drying of the polymer particles, after which they are stirred into good solvent. This works only for rather highly crosslinked particles which have no tendency to fuse together in the dry state.

7.7.2.2 Gradual solvent transfer through semi-permeable membrane

This second method involves the gradual change in the nature of the medium by diffusion of organic solvents through a semi-permeable membrane on one side of which is the latex, and on the other, the solvent. The organic solvent must at all times be soluble in the medium of the latex. So, for example, one could start with ethanol, which is miscible with water, to convert the medium to an ethanol/water mixture. The gradual exchange of water and ethanol through the membrane under constant stirring avoids the high local concentrations of ethanol that would arise from direct addition of this solvent. This tends to avoid coagulation. Subsequently a more hydrophobic solvent is introduced, perhaps toluene or through an intermediate stage involving butanol [198]. By this means it is possible to form lyophilic colloids in a great variety of solvents.

7.7.3 Applications of lyophilic polymer colloids

The application of biocompatible hydrophilic colloids to clinical and medical work has already been mentioned above. Crosslinked polyacrylic acid (PAA) particles have found large markets as 'superabsorbents' primarily in babies' diapers. Starch grains have many applications in the bakery and kitchen. They are hydrophilic particles, swellable by water, but are really larger than colloidal in size, being many micrometers in diameter.

> A mixture of microgel particles with soluble polymer and pigment has exhibited superior application and performance properties.

Organophilic, crosslinked polymethyl methacrylate (PMMA) colloids have been found to be an important component in automotive lacquers. A mixture of such particles with soluble polymer and pigment has exhibited application and performance properties superior to lacquers made from soluble polymer alone. Perhaps the largest application of organophilic polymer colloids is polyvinyl chloride (PVC), in which the particles are 'crosslinked' through crystallites. When these are swollen in a good solvent, they are often called 'organosols'; when they are swollen with a plasticizer for the PVC, they are called 'plastisols.' Organosols, which have a fairly low viscosity, are used as protective and decorative coatings, whereas the plastisols, which are high-viscosity pastes, are used to form molded objects. In the latter case, for example, the plastisol is poured into a mold; the mold is then heated to above the melting temperature of the PVC and then cooled. The PVC recrystallizes throughout the molded object in a rigid or rubbery form, depending upon the plasticizer level. Wire insulation, tool handles, dustpans, wall covering and rubber dolls, to mention only a few examples, are all made in this manner.

References

152. Hiemenz, P.C. (1986). In *Principles of Colloid and Surface Chemistry*, 2nd edn, p. 690, Marcel Dekker, New York.
153. Stern, O. (1924). *Z. Elektrochem.* **30**, 508.
154. Derjaguin, B. (1940). *Trans. Faraday Soc.* **36**, 203.
155. Verwey, E.J.W. and Overbeek, J.Th.G. (1948). *Theory of the Stability of Lyophobic Colloids*, Elsevier, Amsterdam.

156. Hiemenz, P.C. (1977). In *Principles of Colloid and Surface Chemistry*, p. 386, Marcel Dekker, New York.
157. Hiemenz, P.C. (1986). In *Principles of Colloid and Surface Chemistry*, 2nd edn, p. 712, Marcel Dekker, New York.
158. Sonntag, H. and Strenge, K. (1969). *Coagulation and Stability of Disperse Systems*, p. 42, Halsted Press, John Wiley, New York.
159. Parsegian, V.A. and Ninham, B.W. (1971). *J. Colloid Interface Sci.* **37**, 332.
160. Parsegian, V.A. (1975). In *Physical Chemistry: Enriching Topics from Colloid and Surface Science* (H. van Olphen and K.J. Mysels eds), p. 27, Theorex, La Jolla, CA.
161. Russell, W.B., Saville, D.A. and Schowalter, W.R. (1989). *Colloidal Dispersions*, p. 279, Cambridge University Press, Cambridge.
162. Hiemenz, P.C. (1986). *Principles of Colloid and Surface Chemistry*, 2nd edn, Marcel Dekker, New York.
163. Ito, K., Muramoto, T. and Kitano, H. (1995). *J. Amer. Chem. Soc.* **117**, 5005.
164. Langmuir, I. (1938). *J. Chem. Phys.* **6**, 873.
165. Sogami, I. and Ise, N. (1984). *J. Chem. Phys.* **81**, 6320.
166. Ito, K., Nakamura, H., Yoshida, H. and Ise, N. (1988). *J. Amer. Chem. Soc.* **110**, 6955.
167. Smitham, J.B, Evans, R. and Napper, D.H. (1975). *J. Chem. Soc., Faraday Trans. I*, **71**, 285.
168. Flory, P.J. and Krigbaum, W.R. (1950). *J. Chem. Phys.* **18**, 1086.
169. Fitch, R.M. and Kamath, Y.K. (1972). *J. Indian Chem. Soc.* **49**, 1209.
170. Smitham, J.B. (1976). Ph.D. Thesis, p. 82, The University of Sydney.
171. Ottewill, R.H. (1977). In *Colloid and Interface Science* (M. Kerker, R.L. Rowell and A.C. Zettlemoyer eds), Vol. I, p. 379, Academic Press, New York.
172. Liang, S.J. and Fitch, R.M. (1982). *J. Colloid Interface Sci.* **90**, 51.
173. d'Oliveira, J.M.R., Xu, R., Jensma, T., Winnik, M.A., Hruska, Z., Hurtrez, G., Riess, G., Martinho J.M.G. and Croucher, M.D. (1993). *Langmuir*, **9**, 1092.
174. Napper, D.H. (1982). In *Colloidal Dispersions* (J.W. Goodwin ed), The Royal Society of Chemistry, London.
175. Bassett. D.R. (1981). In *Emulsion Polymers and Emulsion Polymerization* (D.R. Bassett and A.E. Hamielec eds), ACS Symposium Series 165, p. 276, American Chemical Society, Washington, DC.
176. von Smoluchowski, M. (1917). *Z. Physik Chem.* **92**, 129.
177. Fuchs, N. (1934). *Z. Physik* **89**, 736.
178. Prieve, D.C. and Ruckenstein, E. (1980). *J. Colloid Interface Sci.* **73**, 539.
179. Lips, A. and Willis, E. (1971). *Trans. Faraday Soc.* **67**, 2979.
180. Lichtenbelt, J.W.Th. Pathmamanoharan, C. and Wiersema, P.H. (1974). *J. Colloid Interface Sci.* **49**, 281.
181. Feke, D.L. and Schowalter, W.R. (1983). *J. Fluid Mech.* **133**, 17.
182. Spielman, L.A. (1970). *J. Colloid Interface Sci.* **33**, 562.
183. van de Ven, T.G.M. (1988). *J. Colloid Interface Sci.* **124**, 138.
184. Israelachvili, J.N. (1981). *Phil. Mag. A* **43**, 753.
185. Ottewill, R.H. and Walker, T. (1968). *Kolloid Z. Z. Polym.* **227**, 108.
186. Storer, C.S. (1968). Ph.D. Thesis, University of Bristol, UK.
187. Nieman, R.E. and Lyashenko, O.A. (1962). *Colloid J. USSR* (English trans.) **24**, 433.
188. Pelton, R. (1976). Ph.D. Thesis, University of Bristol, UK.
189. Tsaur, S.L. and Fitch, R.M. (1987). *J. Colloid Interface Sci.* **115**, 463.
190. Fitch, R.M., Su, L.S. and Tsaur, S.L. (1990). In *Scientific Methods for the Study of Polymer Colloids and their Applications* (F. Candau and R.H. Ottewill eds), p. 388, Kluwer, Dordrecht.
191. Pelssers, E. (1988). *Single Particle Optical Sizing; Aggregation of Polystyrene Latices by Salt and Polymer*, Doctoral Dissertation Landbouwuniversiteit te Wageningen.
192. Thyebault, H. and Riess, G. (1990). European Patent 0249554/B1.
193. Okubo, M., He, Y. and Ichikawa, K. (1991). *Colloid Polym. Sci.* **269**, 125.
194. Flory, P.J. (1953). *Principles of Polymer Chemistry*, pp. 576 ff., Cornell University Press, Ithaca.

195. Obrecht, W., Seitz, U. and Funke, W. (1974). In. *Emulsion Polymerization* (I. Piirma and J.L. Gardon eds), ACS Symposium Series 24, p. 92, American Chemical Society, Washington, DC.
196. de Brabander-van den Berg, E.M.M. and Meijer, E.W. (1993). *Angew. Chem. Int. Ed. Engl.* **32**, 1308.
197. Rembaum, A., Yen, S.P.S. and Molday, R.S. (1979). *J. Macromol. Sci. – Chem.* **A13**, 603.
198. Krieger, I.M. (1972). *Adv. Colloid Interface Sci.* **3**, 111.

Chapter 8

Electrokinetics

8.1 Flow past the electrical double layer

The flat electrical double layer is described in Section 7.2.2.1 in terms of a negatively charged surface with small, positive counterions (or the reverse), either bound to the surface or in a Poisson–Boltzmann distribution in the diffuse part. A schematic representation of this is given in Fig. 153. An externally imposed tangential flow of the medium over the surface leads to a distortion of the diffuse ions to create a 'streaming potential.' Because the process is reversible, flow can be created by application of an external electrical field. This phenomenon is known as 'electro-osmotic flow.' The fluid velocity gradient depends upon the choice of coordinate axes. In Fig. 153 the origin of the Cartesian coordinates is at the surface, and the surface (such as a glass plate) does not move relative to the observer. If the origin remains at the surface, but the surface moves (such as that of a colloidal particle), then the maximum fluid velocity would be at the boundary of the diffuse layer.

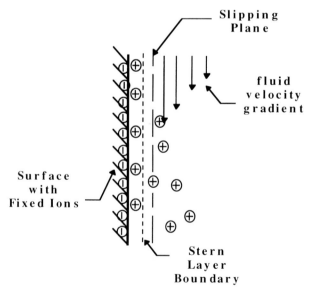

Fig. 153. Flow of medium over electrical double layer.

As we have seen earlier, some of the counterions may be strongly adsorbed to the surface, leading to an effective decrease in the surface electrical potential. In the case of the above electrokinetic effects, other counterions may be sufficiently strongly attracted to the surface that they will not move under the influence of the shear gradient, whereas those further out in the diffuse layer will. Thus we must define the various regions and their corresponding electrical potentials as follows: the surface and the surface potential, ψ_0, the Stern boundary and the Stern potential, ψ_δ, and the 'slipping plane' and the corresponding zeta potential, δ. These are represented schematically in Fig. 154, in analogy to the curves in Fig. 120, Section 7.2.3.1. The position of the slipping plane may or may not correspond to that of the Stern layer boundary. This will be a function of the surface potential, the magnitude of the applied field, E, and the magnitude of the shear gradient. When ψ_0 is small, say less than $c.$ 50 mV, then the Stern layer may not exist for simple counterions such as Na^+ (complex or organic ions may still adsorb because of their polarizability, etc.). Under these conditions there is evidence that all three potentials are identical [199].

8.2 Electrophoresis

8.2.1 Basic concepts

When a colloidal particle with its surrounding double layer is subjected to an external electrical field, E, it will tend to move with a velocity, v, towards the electrode with charge opposite to that on the particle surface. If one considers the double layer to be a condenser with a dielectric constant $\varepsilon\varepsilon_0$ and a thickness κ^{-1}, then the charge per unit area at the slipping plane, σ_ζ, will be given by Eq. 138

$$\sigma_\zeta = \varepsilon\varepsilon_0 \zeta \kappa \tag{138}$$

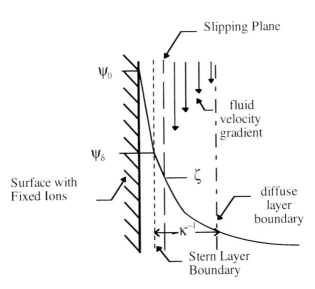

Fig. 154. Various electrokinetic potentials.

The electrical force acting on unit surface area of a particle is, at steady velocity, just counterbalanced by the viscous resistance:

$$\sigma_\zeta E = \eta v \kappa$$

which, from Eq. 138 and rearrangement, gives for the particle velocity, v:

$$v = \frac{\varepsilon \varepsilon_0 \zeta E}{\eta}$$

The electrophoretic mobility, u, is defined as the particle velocity per unit applied electrical field, with units usually given as $\mu m\, s^{-1}/V\, cm^{-1}$:

$$u \equiv \frac{v}{E} = \frac{\varepsilon \varepsilon_0 \zeta}{\eta} \tag{139}$$

Equation 139, due to Smoluchowski, involves a rather naïve approach in that the action of the applied field upon the *counterions* and the solvent that moves with them, known as retardation, has not been considered. Furthermore, the distortion of the diffuse layer in the applied field leads to a polarization such that the particle and counterions tend to be drawn back together, known as relaxation. And yet another problem concerns the distortion of the local electrical field by the particle. It turns out that when the particle is small relative to the double layer thickness ($\kappa a \ll 1$), this effect is negligible (Hückel's theory), but when $\kappa a \gg 1$, it needs to be taken into account (Smoluchowski's theory), as shown in Fig. 155. Hückel had calculated that the mobility should be 2/3 that of Smoluchowski.

8.2.2 Quantitative theory

The Debye thickness of the diffuse part of the double layer, κ^{-1}, was given by Eq. 98. It is actually the distance from the surface at which the potential falls to ψ_δ/e, where e is the base of the natural logarithms. Table 8.1 gives dimensions of the double layer at different ionic strengths.

It is important to note that within the double layer the ions rearrange themselves rapidly as a result of any perturbations due to an externally applied

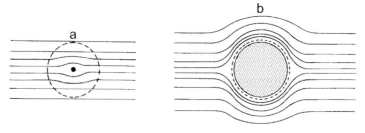

Fig. 155. Distortion of applied field by particle and double layer. (a) $\kappa a \ll 1$; (b) $\kappa a \gg 1$. Dashed line represents the boundary of diffuse part of the double layer.

TABLE 8.1
Debye length as a function of ionic strength

Molarity (1:1 Electrolyte)	κ (cm^{-1})	κ^{-1} (nm)
0.1	10^7	1.0
0.001	10^6	10
10^{-5}	10^5	100

field. Einstein's Law of Brownian Motion states that the time required for an ion to travel the distance κ^{-1} in 1 mM salt solution, i.e. 10 nm, will be on average

$$t = \frac{\kappa^{-1}}{2D} \tag{140}$$

which, for $D = 10^{-9}$ m^2 s^{-1}, will be about 0.05 ms, quite a bit faster than the electrophoretic processes. This means that the particle moves because of small rearrangements in ion concentrations *outside of* the double layer, with correspondingly 'instantaneous' reorganization *within* the double layer.

The steady state flow of ions may be expressed in terms of the forces acting upon them:

$$\text{div}[\pm(n_\pm v_\pm e)\,\text{grad}\,\psi - kT\,\text{grad}\,n_\pm + n_\pm F_{\text{res}_\pm}\mathbf{v}] = 0 \tag{141}$$

where n is the concentration of ions, ψ is the electrical potential at any point, F_{res} is the friction coefficient of the ions and $\mathbf{v}$ is the velocity of the liquid. Equation 141 is known as the Poisson–Boltzmann (cf. Eq. 97)–Nernst–Planck equation, the first two terms of which give the electrochemical potentials of the ions. The friction coefficient is related to the diffusion coefficient, D, for the ions through Einstein's Law of Diffusion:

$$F_{\text{res}} = \frac{kT}{D} \tag{142}$$

where a is the radius. The third term in Eq. 141 gives the force of convective transport, and is described by the Navier–Stokes equation for fluid motion:

$$\nabla^2 \mathbf{v} = \frac{\nabla p + \rho \nabla \psi}{\eta} \tag{143}$$

where ∇ is the differential operator, p is pressure, and ρ is volume charge density. The first term on the right in Eq. 143 gives the force due to a pressure gradient and the second, that from the electrical potential acting on the ions within the liquid. These equations are quite general and apply to particles of any shape.

8.2.2.1 Theory of O'Brien and White

A widely used numerical solution to these equations was devised by O'Brien and White [200] who proposed that the problem could be subdivided into two forces which are counterbalanced under steady flow conditions: αv, the force to move the particle at velocity v, and βE, that to hold the particle stationary in the applied electrical field, E. Then

$$\alpha v = \beta E \tag{144}$$

and the electrophoretic mobility, $u = v/E$, is simply

$$u = \alpha/\beta \tag{145}$$

The dimensionless mobility, $\mathscr{E}$, can be plotted against the dimensionless zeta potential, $\tilde{\zeta}$, as a function of the dimensionless double layer thickness, κa, where

$$\mathscr{E} \equiv \frac{3\eta e}{2\varepsilon\varepsilon_0 kT} u \quad \text{and} \quad \tilde{\zeta} \equiv \frac{e\zeta}{kT} \tag{146}$$

Plots are presented for two ranges of κa in Figs 156 and 157. In the limit of $\kappa a \ll 1$, the Hückel equation is achieved, i.e.

$$\mathscr{E} = \tilde{\zeta} \tag{147}$$

as shown in Fig. 156. At the other limit in which $\kappa a \gg 1$, the Smoluchowski equation is reached, i.e.

$$\mathscr{E} = \tfrac{3}{2}\tilde{\zeta} \tag{148}$$

which is represented by the limiting slope in Fig. 157. At low zeta potentials, when $\tilde{\zeta} \leq 2$ ($\zeta \leq 50\,\text{mV}$) all of the curves converge to the 'ideal' Hückel limit, especially when the diffuse layer is greatly expanded (Fig. 155a). This condition is reached experimentally by deionization of the polymer colloid, such as by using mixed bed ion-exchange resins. At higher salt concentrations, where $\kappa a > c.3$, more than one value of the mobility is obtained for certain values of $\tilde{\zeta}$. This is only a problem when $\tilde{\zeta} \geq c.4$ or when the value of ζ is greater than about 100 mV, a situation which is rarely achieved in 'hard sphere' polymer colloids. It apparently is common in inorganic colloids, e.g. metal oxides and clays.

8.2.2.2 The meaning of the zeta potential

Before going further, it is important to understand exactly what is being measured. To gain some perspective, we look at the potential gradient in the

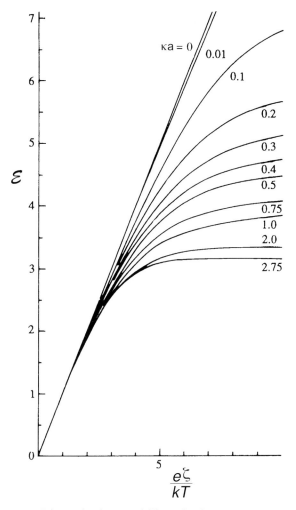

Fig. 156. Dimensionless mobility of spheres vs zeta potential [200].

region of interest, and take for example a particle with a surface potential of 50 mV and a slipping plane at a distance of 1 nm from the surface. The local electrical field within this region is then roughly

$$\frac{d\psi}{dx} \approx \frac{50 \text{ mV}}{1 \text{ nm}} \approx 5 \times 10^5 \text{ V cm}^{-1}$$

such that the slightest variation in the position of the slipping plane leads to an enormous variation in ζ! Under ideal Hückel conditions it turns out, as we shall see below, that the slipping plane is at the surface, and this consideration is moot. But as soon as the value of ζ becomes greater than about 50 mV, and/or κa becomes greater than about 3, the slipping plane apparently moves

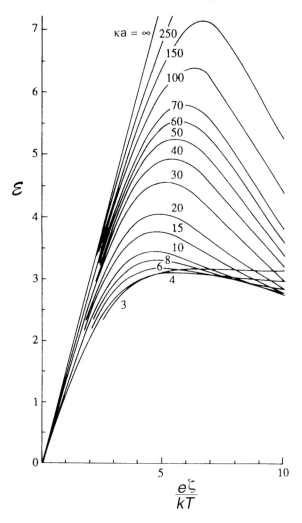

Fig. 157. Dimensionless mobility of spheres vs zeta potential [200].

off the surface and its position may then vary with the experimental conditions, in particular, the applied field. *Because there is no direct measure of the distance of the slipping plane from the surface, it is impossible to relate ζ directly to ψ_0.* Lyophilic polymer adsorbed at the surface can move the slipping plane out to any distance from the surface. When the thickness of the polymer layer exceeds κ^{-1}, the mobility should fall. It may not fall to zero because of ion mobility within this surface layer (see Section 8.3.2 below).

Thus the zeta potential and the electrophoretic mobility from which it is derived, although valuable in providing information about the magnitude of the surface potential and its sign, must be viewed within its limitations. In some instances, it is these electrokinetic parameters which are the most important, especially where diffusion of the particles plays a determining role such as in coagulation kinetics.

8.2.3 Experimental results

8.2.3.1 Electrophoretic mobilities of low charge polymer colloids

Bagchi and coworkers synthesized a series of polyvinyl toluene (PVT) latexes whose diameters (monodisperse) ranged from 82 to 150 nm and whose surface charge densities varied from 0.50 to 0.63 $\mu C\,cm^{-2}$ – small particles with low charge densities [199]. They used two methods for measuring electrophoretic mobilities – microelectrophoresis and the moving boundary method. The latter was found to be inaccurate and was abandoned. The former technique involves measuring the velocities of individual particles in a capillary tube with blackened platinum electrodes at either end. The applied electrical field along the axis of the capillary induces electro-osmotic flow of the solvent near the inner wall of the capillary and, because this is a closed system, there is flow in the opposite direction in the middle of the tube. Between these two flow regions there exits a stationary layer in which the measurements are made.

> Between the two flow regions there exits a stationary layer in which the measurements are made.

Intense illumination and a microscope are required to see, not the resolved particle, but the spot of light it reflects. Even so, Bagchi's particles were so small that they could not be observed. So they coated AgCl monodisperse crystals with their latex particles by means of heterocoagulation, and then measured the velocities of the complex particles which would have the same zeta *potential* (not surface charge) as the individual latex particles alone. Their mobility results, taken as a function of pH, are shown in Fig. 158. The fact that the mobilities are independent of pH indicates that the surface ionic groups are strong electrolytes, in this case $-SO_4^-$. This corroborated conductometric and potentiometric titration results which showed essentially only strong acid groups.

The mobility (u) data from Fig. 158 were combined with ionic strength, particle size and surface charge calculations to produce the results in Table 8.2. In these results it is clear that the zeta potentials and the surface potentials calculated from the surface charge densities are almost equal, and that therefore the slipping plane is at or very near to the surface of the particles.

The microelectrophoretic method has serious limitations:

- the individual motions of a statistically significant number of particles must be followed;
- the stationary layer must be found precisely, or the radial velocity profile of the medium must be known precisely and mobilities taken at various radial distances;
- the capillary walls must be scrupulously clean; no contamination by adsorption of previous colloids can be tolerated;

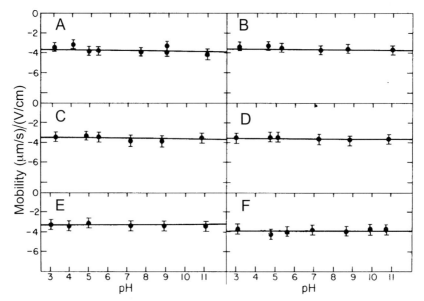

Fig. 158. Microelectrophoresis of PVT latexes [199].

- eye strain and tedious labor are a problem; and
- Joule heating as a result of the electrical resistance of the colloid can lead to density gradients, which in turn disturb the velocity profiles.

Because they are so easily made with narrow particle size distributions and with controlled surface chemistry, polymer colloids have been viewed as the ideal model colloids for studies such as these. However, polymer colloids with higher surface potentials – and which are therefore more stable – generally do not produce results which agree with the theories outlined above, with a few notable exceptions.

8.2.3.2 Electrophoretic mobilities of highly charged polymer colloids

To get some idea of the complexities involved, the work of Paulke and coworkers [201] is instructive. They synthesized five latexes of different surface

TABLE 8.2
Electrophoretic mobility results for selected polyvinyl toluene latexes

Latex	σ_0 ($\mu C\,cm^{-2}$)	ψ_0 (calculated)[a] (mV)	u ($\mu m\,s^{-1}/V\,cm^{-1}$)	ζ (mV)
A	0.50	43	3.7	50
B	0.63	55	3.4	47
C	0.51	45	3.5	48
D	0.51	45	3.4	47

[a] Calculated using Gouy–Chapman equations.

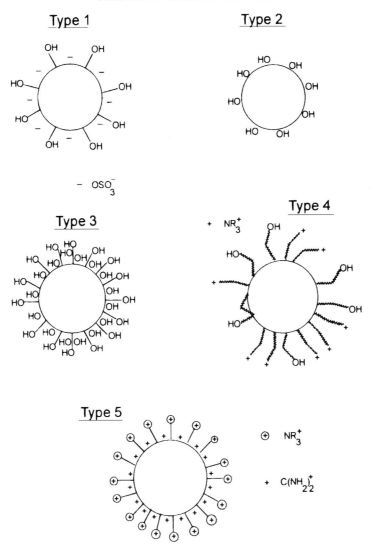

Fig. 159. PS colloids of different surface chemistries (schematic) [201].

configurations represented schematically in Fig. 159, using the synthetic strategies given in Chapter 6. The colloids were purified and then subjected to electrophoresis at various pHs and ionic strengths, represented by pλ, the positive, decadic logarithm of the conductivity in units of $S\,m^{-1}$.

The mobility data are represented by three-dimensional plots, as shown in Fig. 160. The complexity of the graphs, even in any of the two-dimensional 'slices,' indicates that there is little correspondence to theory. Some of the 'bumps' on the graphs may be artifacts of the SURFER spline approximation algorithm employed to interpolate among experimental points, and some may be due to inadequate purification or even contamination of the colloids during sample preparation.

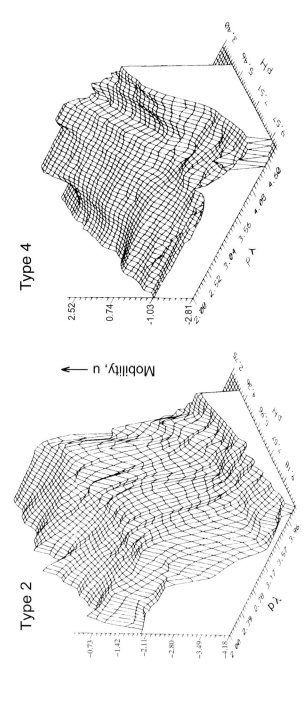

Fig. 160. Typical electrophoretic mobility profiles of PS colloids of various surface chemistries shown in Fig. 159 [201].

Only two of the graphs are presented here, those for the Type 2 and Type 4 latexes, which represent the two broad categories found by Paulke, i.e. one in which the dependence of mobility on conductance is strong and that on pH, is weak (Types 2, 3 and 5), and the other in which pH-dependence is strong, and that on pλ is weak (Types 1 and 4). The former class is comprised of colloids with strong electrolyte groups, and the latter, those containing polyethylene glycol-like moieties which are very weak acids capable of protonation at low pH and perhaps anion adsorption at high pH. Perhaps of equal utility is that these mobility profiles can serve as 'fingerprints' of colloids which are highly characteristic of their surface chemistries. Rowell and coworkers, who coined the term, have written extensively on the subject including theoretical interpretations of such three-dimensional plots based on the classical electrochemical theory given above, and the problems associated with the variability of the position of the slipping plane [202, 203].

> In certain rare instances polystyrene colloids behave ideally.

In certain rare instances polystyrene colloids have been found to behave ideally. One example is that of a commercially available latex (Interfacial Dynamics Corporation) described by Gittings and Saville [204]. The particles had a diameter of 156 nm and a surface charge density $\sigma_0 = -1.08\,\mu\text{C}\,\text{cm}^{-2}$ due to surface sulfate groups. The colloid was purified by 12 successive centrifugations/washings. Gloved hands were always used to handle samples because the authors found that negatively charged oligomers exist in the oil on human skin, and that they can lead to measurable contamination. Microelectrophoresis gave a mobility at $T = 298\,\text{K}$ and ionic strength at 1 mM NaCl ($\kappa a = 8.1$) of $u = -4.24\,\mu\text{m}\,\text{s}^{-1}/\text{V}\,\text{cm}^{-1}$ ($\tilde{\mathscr{E}} = 3.25$) corresponding to *two* possible zeta potentials, according to the calculations of O'Brien and White [200] as represented in Fig. 157, of $\zeta = -100\,\text{mV}$ and $-140\,\text{mV}$ ($\tilde{\zeta} = 3.9$ and 5.5).

8.3 Dielectric spectroscopy

8.3.1 An ideal polymer colloid?

To choose between the two possible zeta potentials Gittings and Saville resorted to dielectric spectroscopy, noting that the frequency-dependent dielectric response is a function of the volume fraction of the colloid. Furthermore, by taking the dielectric *increment*, the permittivity compared with that at a reference frequency at zero volume fraction, they could eliminate most instrumental errors. Gittings and Saville assumed that the dielectric increment could be expanded in a series as given in Eq. 149:

$$\frac{\varepsilon(\omega,\phi) - \varepsilon(\omega_r,\phi)}{\phi} = [\Delta\varepsilon(\omega) - \Delta\varepsilon(\omega_r)] + \tfrac{1}{2}[\Delta^2\varepsilon(\omega) - \Delta^2\varepsilon(\omega_r)]\phi + \cdots \quad (149)$$

where $\Delta\varepsilon$ and $\Delta^2\varepsilon$ represent the first and second derivatives and ω is the angular frequency (subscript r denotes the reference frequency). The dielectric increments at 50 kHz relative to the response at 100 kHz (the upper limit of the response

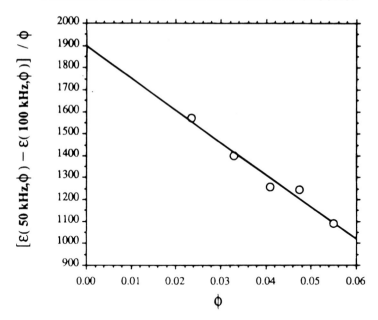

Fig. 161. Dielectric increment as a function of volume fraction for a PS colloid [204].

range in 1 mM salt solution) are plotted versus volume fraction in Fig. 161. The linearity of the plot justifies the dropping of cubic and higher terms in Eq. 149.

From a series of such graphs at various relative frequencies, the authors could construct a plot of the zero-concentration dielectric increment as a function of the applied frequency, the results of which are shown in Fig. 162. The experimental points are compared to theoretical curves for $\zeta = 100\,\text{mV}$ and $\zeta = 140\,\text{mV}$. The fit clearly favors the higher potential, although there is some doubt at the higher frequencies.

The extraordinary success of these results is attributed to several key factors:

- Close agreement between particle sizes obtained by electron microscopy and by photon correlation spectroscopy (PCS) (Section 5.1.6) indicate the absence of a 'hairy layer' on the particle surface;
- Extreme care in purification and avoidance of contamination.

Since the theory deals with particles in isolation, extrapolation to zero volume fraction is important.

8.3.2 Hairy or porous particle surfaces?

Surface 'hairiness' has been a nagging concern for years, and may distinguish *polymer* colloids from others such as silica, iron oxide or gold sols. Presumably the surface hairs are chain ends made hydrophilic by polar or ionic end groups and which extend into the aqueous medium. Experimental evidence for this was first given by van den Hul and Vanderhoff [205] who heated polystyrene latexes under pressure to 120°C, considerably above the T_g of the

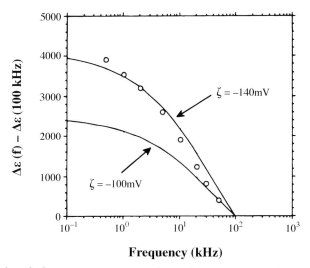

Fig. 162. Dielectric increment at zero volume fraction versus frequency. –o–, experimental points. Curves calculated from theory [204].

polymer. They measured the surface areas before and after the heat treatment by electron microscopy and by nitrogen gas adsorption (BET method) and found in nearly every case, a pronounced decrease in surface area, as shown in Table 8.3. The authors speak more in terms of surface porosity rather than hairiness, which may be the better way to look at such results.

Coagulation during particle formation in emulsion polymerization (Section 2.1.1.1) may lead to latex particles which have internal structure arising from the imperfect fusion of the material which was on the surface of the primary particles. Independent evidence for this kind of structure has been given by Winnik and coworkers using fluorescence methods to measure diffusion of solvent molecules into the particles [206]. Heating above T_g could lead to a more complete fusion of such 'interstitial' material with the polystyrene matrix within the particles with a consequent decrease in porosity, as shown schematically in Fig. 163.

Rosen and Saville looked at the effect of heat treatment on the dielectric response of two latexes, one anionic and one amphoteric, which they prepared [207]. Their methodology and theoretical treatment was much the same as

TABLE 8.3
Surface areas of PS colloids before and after heat treatment [205]

Latex	Surface area before heating (m² g⁻¹)		Surface area after heating (m² g⁻¹)	
	S_{TEM}	S_{BET}	S_{TEM}	S_{BET}
A-2	65	78	42	55
B-1	36	51	26	29
A-6	5.5	10	5.6	6.9
C-1	22	37	26	32

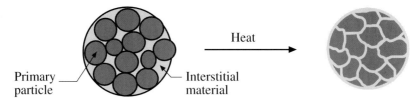

Fig. 163. Effect of heating above T_g: reduction in particle porosity (schematic).

that of Gittings and Saville described above. They found that the dielectric increment for both latexes was profoundly affected by the heat treatment even though the microelectrophoretic mobilities were only slightly changed. In the case of the amphoteric colloid, heat treatment resulted in the dielectric increment as a function of the applied frequency coming into agreement with the calculations of DeLacey and White [208] based upon the same classical theory as that of O'Brien and White [200]. This is shown in Fig. 164 in which it can be seen that the high-frequency increment is enormously reduced by the heating, and agreement with the theoretical curve (for $\zeta = +81\,\text{mV}$ in 1 mM HCl) is achieved after some 4 h of heating. The reference frequency was 0.5 kHz. Although similar results were obtained with the anionic latex, its dielectric increment never reached the theoretical curve.

> Heat treatment resulted in the dielectric increment coming into agreement with the calculations.

It is possible that these changes are due at least in part to hydrolysis of the surface functional groups under the harsh conditions involved during the heat treatment. Counterion condensation could also be affected by changes in

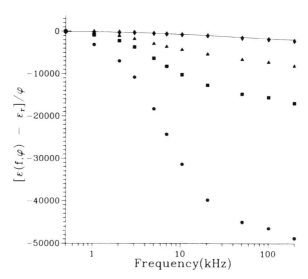

Fig. 164. Effect of heating PS amphoteric colloid at 120°C. –●– untreated, –■– 1 h, –▲– 2 h, –◆– 4 h, — theory [207].

TABLE 8.4
Experimental zeta potentials and surface charge densities [209]

Latex	u (μm s^{-1}/V cm^{-1})	ζ^a (mV)	ζ^b (mV)	σ_0 (μC cm^{-2})
Bare	−4.5	−80	−127	−0.61
Bare/heated	−4.1	−69	−71	−0.48
Hairy	−3.0	−46	−169	−0.28
Hairy/heated	−3.9	−64	−87	−0.44

[a] Calculated from measured mobilities and [b] from experimental dielectric difference spectra. [KCl] = 0.0005 M.

surface structure (see Section 5.7.4). A further concern is that the heating was conducted in glass bottles. It is known that polysilicates can be leached from glass, and become adsorbed onto the latex particles (Section 5.7.2). No checks were made to determine by titration how much the surface charge density had changed. In such experiments surface groups should be chosen which do not hydrolyze readily (see Section 6.1). In later work Rosen and Saville did titrate before and after heating a PMMA latex and found a 68% increase in surface acid groups, apparently due to weak acid, suggesting that surface methacrylate groups were converted to RCOOH [209].

In that same study Rosen and Saville took the anionic PMMA colloid which had surface-copolymerized aldehyde (RCHO) groups derived from acrolein and had polyacrylamide (PAm) grafted to it to produce a synthetically hairy latex which was otherwise the same. Electrophoretic mobilities and dielectric difference spectra of the bare and hairy latexes before and after heating tended to confirm their earlier hypotheses (Table 8.4).

For example, mobilities and measured zeta potentials decreased as a result of surface polymer, decreased upon heating bare particles but increased on heating hairy particles. The bare latex was brought to near-ideal behavior by heating, just like the amphoteric PS colloid described above. Many details, however, could not be resolved because of the large number of uncontrolled experimental variables including possible hydrolysis of the PMMA and PAm, hydrolysis of the chemical bond for grafting of the PAm to the particle surface, and oxidation of aldehyde to carboxylic acid.

8.4 Electrokinetic instrumentation

8.4.1 Electrophoretic light scattering (ELS)

Dynamic light scattering, which measures the rate of diffusion of particles though the medium (Section 5.1.6), is conducted in the presence of an externally applied electric field to obtain the electrophoretic mobility. Because the particles move in the field in addition to their Brownian motion, it has been found that by use of a reference light beam in heterodyne mode, one can measure the Doppler shifts due to electrophoresis. This in turn gives a *mobility*

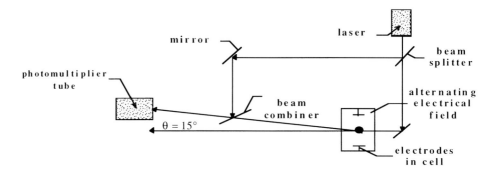

Fig. 165. Schematic of electrophoretic light scattering apparatus.

spectrum, from which the zeta potential is calculated. The schematic diagram of commercially available apparatus, the Brookhaven* Zeta Plus or the Photal 800 from Otsuka Electronics, is shown in Fig. 165. The low scattering angle of $\theta = 15°$ reduces stray scattering from the cell walls and minimizes the effect of Brownian motion on the mobility distribution.

The same apparatus, with the electrical field turned off, can also be used to obtain particle sizes using an autocorrelation analysis like that represented by Eq. 68. The open cell design with closely spaced electrodes placed far from the cell walls essentially eliminates the problem of electro-osmotic flow, so important in microelectrophoresis. The cells are made of acrylic plastic and are disposable. It takes about 1 min to make a measurement.

The apparatus is based upon a design described by Novotny and Hair whose theory is outlined below [210]. Because there are both diffusional and translational motions of the scatterers, the autocorrelation function Eq. 68 is modified as follows:

$$g(\tau) = \langle I_S \rangle \exp[-i\omega_S \tau - DQ^2 \tau] \tag{150}$$

where

$$Q \equiv \frac{4\pi n_0}{\lambda_0} \sin \frac{\theta}{2}$$

and in which $\langle I_S \rangle$ is the average intensity of scattered light and ω_S is its frequency. The scattering wave vector, Q, is the same as that in Eq. 68. In order to separate the Doppler-shifted translational scattering from the rest, the scattered radiation is mixed with a beam-split part of the incident wave. The heterodyne autocorrelation function is then

$$g_{Ht}(\tau) = \langle I \rangle^2 + \langle I_S \rangle^2 \exp(-2DQ^2 \tau)$$
$$+ 2\langle I_R \rangle \langle I_S \rangle \cos[(\omega_S - \omega_R)\tau] \exp(-DQ^2 \tau) \tag{151}$$

---

* Brookhaven Instruments Worldwide, Holtsville, New York, USA; Otsuka Electronics USA, Fort Collins, Colorado, USA.

where $\langle I_S \rangle$ and $\langle I_R \rangle$ are average total and reference intensities and ω_R is the frequency of the reference beam. The second term in Eq. 151 describes the homodyne interaction, and is negligible when $\langle I_S \rangle \ll \langle I_R \rangle$. The electrophoretic velocity, v, is then related to the Doppler shift, $\omega_S - \omega_R$, in the third term in Eq. 151 by:

$$v = (\omega_S - \omega_R)/Q \qquad (152)$$

from which the electrophoretic mobility is obtained through Eq. 139. Novotny and Hair provide solutions to these equations under conditions where the (square wave) frequency of the applied field is small such that v is constant, as well as where the applied frequency is high enough so that particle accelerations are observed [210].

8.4.2 Dielectric spectroscopy

8.4.2.1 The Solartron frequency response analyzer (FRA)

This dielectric spectrometer consists of three parts, the Solartron (Solartron Instruments, Houston, USA) frequency response analyzer (FRA), an electrochemical interface (ECI) and a computer system. A schematic layout of the instrumentation is given in Fig. 166. The FRA has three main functions: a programmable generator to provide the input signal; a correlator to filter out harmonic oscillations and random noise; and a data-storage and display device to present data points in Cartesian coordinates.

The ECI, in conjunction with the FRA, controls the potential difference between the electrodes and carries out AC-characteristic measurements. The whole experiment is operated from the computer keyboard. Frequency response analysis has the advantage over other methods of signal processing

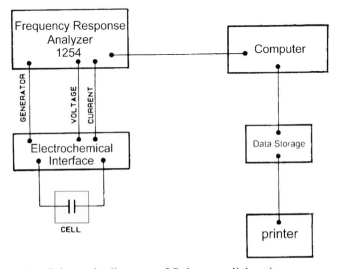

Fig. 166. Schematic diagram of Solartron dielectric spectrometer.

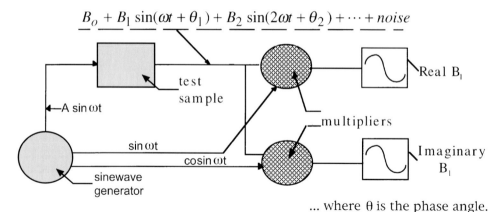

Fig. 167. Schematic diagram for frequency response analysis.

in that random noise and unrelated harmonics are readily eliminated. A schematic representation of the process is shown in Fig. 167. A sinewave generator provides the oscillating electrical field to the sample. The output from the sample is comprised of an in-phase component (due to pure resistances) and an out-of-phase component (due to capacitances) (see Section 5.7.2), corresponding to the real and imaginary parts of the output signal. The output is divided into two parts, one of which is multiplied by the sine function from the sinewave generator, and the other, by the cosine function. Subsequent integration averages out noise and eliminates harmonics to give the real and imaginary parts.

The instrumentation is capable of operating at frequencies from 10^{-3} Hz to several MHz. At the low-frequency end of the spectrum it is important to avoid electrode polarization. This is accomplished through a control mechanism in the electrochemical interface. The theory of the instrumentation and data analysis, along with some experimental results, have been given in, Section 5.7 [211].

8.4.2.2 The four-electrode dielectric spectrometer

The dielectric difference spectra discussed in an earlier section were obtained on a four-electrode instrument which was designed to minimize effects of electrode polarization and other instrument-derived errors [212]. The unique cell design is shown schematically in Fig. 168. The advantage of this design is that the sensing electrodes are separate from source and sink electrodes which supply the external field. Current flow and thus electrode polarization are minimized by a high impedance circuit combined with the buffer amplifiers. The measured (complex) impedance then will be simply the voltage drop between the sensing electrodes and the current: $Z = \Delta V/I$.

Both the driving and reference frequencies are fed to the source electrode from separate signal generators through a summing amplifier as shown in Fig. 169. By converting the input potential to a current through the V/I converter,

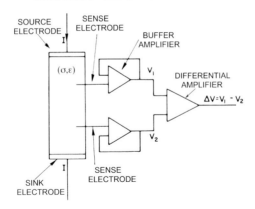

Fig. 168. Four-electrode dielectric spectrometer cell design [212].

changes in the source electrode impedance due to polarization will not affect the measurement, since the current is not affected by polarization:

$$I = I(\omega) + I(\omega_R) \qquad (153)$$

In the detector part of the system, a feedback voltage is generated which is proportional to the current within the cell to further enhance signal-to-noise ratio. This is done by means of an I/V converter linked with a feedback network comprised of a capacitance, C_F and a variable resistance, R_F, the

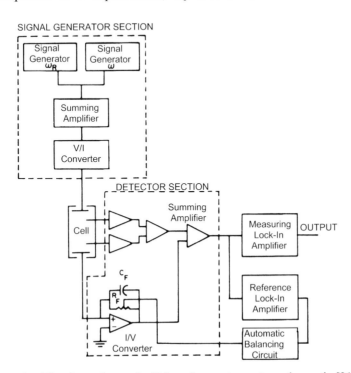

Fig. 169. The four-electrode dielectric spectrometer schematic [212].

output of which is

$$V_F = -IZ_F \tag{154}$$

The combined cell and feedback voltages are fed through a summing amplifier to give the balance voltage:

$$V_0 = -\Delta V - V_F \tag{155}$$

Given the cell impedance, Z_C, i.e. that between the sensing electrodes, the voltage drop is:

$$\Delta V = IZ_C \tag{156}$$

Combination of Eqs 154, 155 and 156 gives the final output from the detector, namely a voltage which is a function of the difference in impedances between the sample and reference:

$$V_0 = -I(Z_C - Z_F) \tag{157}$$

The lock-in amplifier measures the in-phase and out-of-phase components of this signal.

Typical results have been shown above in Figs 161–164 and in Table 8.4.

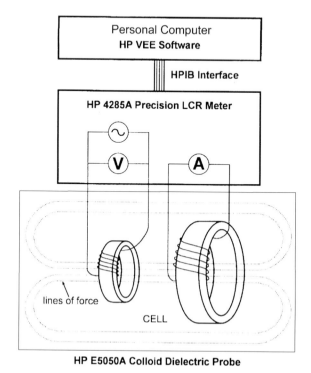

Fig. 170. Hewlett-Packard electrodeless dielectric probe [214].

8.4.2.3 Electrodeless dielectric spectrometer

Perhaps the best way to get rid of electrode polarization is to get rid of the electrodes. This has been accomplished in the following highly creative manner, best described by reference to the schematic diagram in Fig. 170. The probe is comprised of two ferrite rings placed close together, each of which is wound with a coil of wire. The oscillating input voltage is put through one of the coils, which then produces a corresponding, oscillating magnetic field. This in turn induces an electric field within the sample, the magnitude of which depends upon the dielectric properties of the system. The other solenoid senses this oscillating field, thereby generating its own electrical response, which is fed to an inductance-capacitance-resistance (LCR) meter. The entire apparatus thus acts to sense the complex impedance of the polymer colloid. The real part of the impedance depends on the conductivity of the sample, while the imaginary part depends upon its permittivity. This instrument was developed by the Kobe Instruments Division of the Hewlett-Packard Company [213, 214].

For more detailed discussions of electrokinetic phenomena the reader is referred to the textbook by Russel, Saville and Schowalter [215]. For a classical introduction to dielectric spectroscopy of colloids, the book by Dukhin and Shilov may be useful [216].

References

199. Bagchi, P., Gray, B.V. and Birnbaum, S.M. (1980). In *Polymer Colloids II* (R.M. Fitch ed), p. 225, Plenum Press, New York.
200. O'Brien, R.W. and White, L.R. (1978). *J. Chem. Soc., Faraday Trans. II*, **74**, 1607,
201. Paulke, B.-R. Möglich, P.-M. Knipper, E. Budde, A. Nitzsche, R. and Müller, R.H. (1995). *Langmuir*, **11**, 70.
202. Prescott, J.H., Shiau, S.J. and Rowell, R.L. (1993). *Langmuir* **9**, 2071.
203. Marlow, B.J. and Rowell, R.L. (1991). *Langmuir*, **7**, 2970.
204. Gittings, M.R. and Saville, D.A. (1995). *Langmuir* **11**, 798.
205. van den Hul, H.J. and Vanderhoff, J.W. (1971). In *Polymer Colloids* (R.M. Fitch ed), pp. 14–16, Plenum Press, New York.
206. Winnik, M.A. and Pekcan, Ö. (1990). In *Scientific Methods for the Study of Polymer Colloids and Their Applications* (F. Candau and R.H. Ottewill eds), NATO ASI Series C **303**, p. 236, Kluwer, Dordrecht.
207. Rosen, L.A. and Saville, D.A. (1990). *J. Colloid Interface Sci.* **140**, 82.
208. DeLacey, E.H.B. and White, L.R. (1981). *J. Chem. Soc., Faraday Trans. II* **77**, 2007.
209. Rosen, L.A. and Saville, D.A. (1992). *J. Colloid Interface Sci.* **149**, 542.
210. Novotny, V. and Hair, M.L. (1980). Iin *Polymer Colloids II* (R.M. Fitch ed), p. 37, Plenum Press, New York.
211. Fitch, R.M., Su, L.S. and Tsaur, S.L. (1990). In *Scientific Methods for the Study of Polymer Colloids and Their Applications* (F. Candau and R.H. Ottewill eds), NATO ASI Series C **303**, p. 373, Kluwer, Dordrecht.
212. Myers, D.F. and Saville, D.A. (1989). *J. Colloid Interface Sci.* **131**, 448.
213. Siano, S.A. (1996). *R&D Magazine* **Feb.**, 45.
214. HP E5050A Colloid Dielectric Probe, Hewlett-Packard Co., Santa Clara, CA.
215. Russel, W.B., Saville, D.A. and Schowalter, W.R. (1989). *Colloidal Dispersions*, p. 211, Cambridge University Press, Cambridge.
216. Dukhin, S.S. and Shilov, V.N. (1974). *Dielectric Phenomena and the Double Layer in Disperse Systems and Polyelectrolytes*, Halsted/John Wiley, New York.

Chapter 9

Order–Disorder Phenomena

9.1 Origins of ordering in polymer colloids

9.1.1 Bragg diffraction

Plate 2 in this book displays the beautiful colors which arise from ordering of the particles in a colloid which is nevertheless completely fluid. The colloid in the beaker is being stirred with a magnet bar, and yet the colors are due to Bragg diffraction because the particles are aligned in regular arrays, at least within a short range. Krieger and Hiltner investigated such dispersions by constructing a diffraction cell [217] which could hold a fluid, shown in Fig. 171. They found, as others had before them, that only under certain conditions of colloid concentration and ionic strength would *monodisperse* colloids exhibit this phenomenon.

The Bragg diffraction law gives the angle at which constructive interference of light beams reflected from regularly spaced rows of particles reaches a maximum, the so-called Bragg angle. For colloids, with a refractive index n_c behind a glass plate of refractive index n_g, the Bragg law is modified by the application of Snell's law to the form given in Eq. 158 in which m is the diffraction order, λ_0 is the wavelength in air, and θ (Fig. 171) and θ_0 are the Bragg angle and the goniometer angle, respectively.

$$m^2 \lambda_0^2 = \tfrac{8}{3} D^2 (n_c^2 - n_g^2 \cos^2 \theta_0) \tag{158}$$

Typical results are shown in Fig. 172 for a single latex with a particle diameter $D_0 = 171$ nm, in which the Bragg angle is seen to be a function of the wavelength of the incident light, and quantitatively in accordance with Eq. 158. Thus as one observes the colloid at different angles one sees different colors, as so beautifully illustrated in Plate 2.

The explanation for this behavior is based upon the observation that the ionic strength has a large influence. Deionization of monodisperse polymer colloids as polymerized brings about the appearance of colors. Addition of relatively small amounts of salt, which collapses the electrical double layer, completely destroys the opalescence. So it appears that if all the particles are the same size and they all carry the same charge, and therefore the same electrical potential, they all must repel each other to the same extent. They therefore will find equilibrium positions in which they 'stand' apart to the same distance in an array which is equivalent to hexagonal close packing. If this is the case, then the interparticle spacing, D, must obey Eq. 159 because

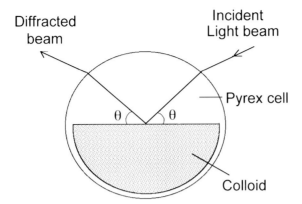

Fig. 171. Apparatus for light diffraction from a colloid.

the volume fraction, ϕ, which in these circumstances includes the surrounding double layer of each particle, must be 0.74.

$$\phi\left(\frac{D}{D_0}\right)^3 = 0.74 \qquad (159)$$

A test of this equation was conducted with a great variety of colloids, both in aqueous and in organic media (charged particles will still be ionized to some extent in media of low dielectric constant). By obtaining Bragg reflections at various dilutions, such as those shown in Fig. 173, and applying Eqs 158 and 159, Krieger and Hiltner showed that the interparticle spacing was indeed hexagonally 'close packed' at all volume fractions down to 1%!

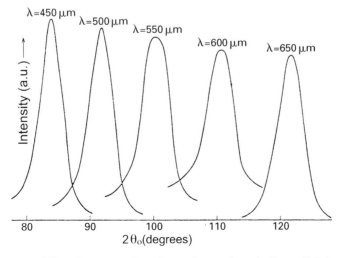

Fig. 172. Bragg diffraction as a function of wavelength for a PS latex. $\phi = 0.092$ [217].

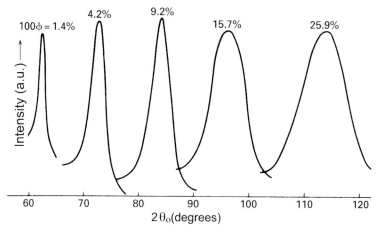

Fig. 173. Bragg diffraction at various volume fractions for a PS colloid. $D_0 = 171$ nm, $\lambda = 450$ nm [217].

The nonaqueous colloids were formed from slightly crosslinked particles which were transferred into dimethyl formamide, benzonitrile, cyclohexanone, acetophenone, dioxane, methyl benzoate, ethyl acetate, benzyl alcohol or m-cresol. All of these systems, in a plot of $(D_0/D)^3$ versus ϕ, fell on the same line with a slope of 0.74. So it would appear that the simple theory of mutual repulsions of double layers is sufficient to explain all of these results.

9.1.2 Ionic strength effects

But then Krieger and Hiltner made the 'mistake' of doing one more set of experiments in which they looked very carefully at the influence of ionic strength on interparticle spacings. These results are given in Fig. 174 in which the dimensionless inverse interparticle spacing cubed is plotted as a function of volume fraction, according to Eq. 158, at various concentrations of NaCl and $CaCl_2$. Both 1:1 and 2:1 electrolyte gave the same results at equal ionic strengths. At a given ionic strength, as the colloid is diluted the spacing, D, increased until a critical volume fraction ϕ^* known as the Krieger point was reached, after which the value of D remained unchanged (horizontal lines). At somewhat greater dilution, ϕ', opalescence disappeared and no further Bragg diffraction could be observed, indicating the loss of order. These changes are indicated in Fig. 174 for the NaCl concentration of 6.1×10^{-3}M, but they were observed for all salt concentrations except for the completely deionized system.

9.1.3 Equilibrium three-phase systems

How is it that the spacing, upon dilution beyond ϕ^*, remains fixed? In other words, why do the particles continue to retain a closest packed array with the same dimensions even though there is now room to spread out further? The

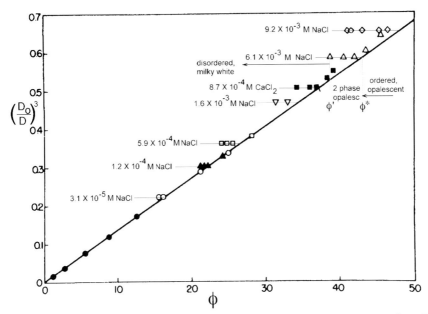

Fig. 174. Interparticle spacing vs volume fraction at various salt concentrations [217].

conclusion appears to be that the system, in the interval between ϕ^*, and ϕ', must exist as two phases, one ordered and close packed, and the other disordered. This interpretation has been supported by many subsequent experiments by other workers which clearly have shown the coexistence of ordered and disordered regions in the same colloid. An example is shown in Plate 4 in which a polymethyl acrylate (PMA) colloid of uniform particle size has been allowed to reach phase equilibrium after standing for several months. There is a sharp boundary between the crystalline and milky regions, and a diffuse boundary between the latter and a zone where there are almost no latex particles. (N.B. The upper region is colored yellow by a free radical inhibitor used in certain experiments.) The regions seem to correspond on a macroscopic scale with the triple point of a substance in which solid, liquid and vapor phases are in equilibrium. Even though there is room for the particles to expand into the vapor phase, it appears that there must be an attractive force, such as the Sogami potential (Section 7.2.3.3, especially Figs 132 and 133) holding the particles together in the ordered domain. This would be analogous to the van der Waals attraction among molecules in a molecular crystal.

9.1.4 Hard sphere model

There are alternative explanations for what is happening here, however. For example, Hachisu and Takano argue that all can be explained in terms of the interactions of hard spheres, i.e. where the electrical potentials of the particles play an insignificant role [218]. They argue that the *particles with their expanded double layers* occupy somewhere between 0.50 and 0.55 volume fraction, a condition under which the Alder–Wainwright transition occurs to form an equilibrium mixture of ordered phase with $\phi = 0.55$, and a disordered one with $\phi = 0.50$. This occurs for purely entropic reasons: by separating into two phases the system gains entropy by creating more space within the disordered domain. Under the conditions shown in Plate 3 sedimentation pressure is just compensated by the osmotic pressure of the colloid.

This is schematically represented in Fig. 175 in which a sketch of the system depicted in Plate 3 is drawn alongside of a schematic, inverted volume-vs-pressure isotherm plot of a hard sphere system undergoing a first-order phase transition. Here 'M' designates the melting point and 'F' designates the freezing point. Alder and Wainright calculated by computer modelling that the pressure at this transition, P_t, was 11.6 times that for an ideal gas, P_i, at the same volume [219]. To obtain values of the pressures, latexes were allowed to reach equilibrium over several months at constant temperature, and then Hachisu and Takano siphoned off samples with extreme care above the phase boundary. At equilibrium

$$P_t = P_{osm} = P_{sed} = \frac{\rho_p - \rho_m}{\rho_p} \frac{W}{A} \qquad (160)$$

where ρ_p and ρ_m are the densities of the polymer and medium, W is the dry weight of polymer above the phase boundary, and A is the cross-sectional area of the container. To get the corresponding ideal gas pressure, $P_i = nkT$, where n is the particle number density, they took a small sample of latex from immediately above the phase boundary. From its dry weight, W', its volume, V, and the known particle diameter, D, they could calculate n:

$$n = \left(\frac{W'}{\rho_p V}\right) \Big/ \left(\frac{\pi D^3}{6}\right) = \frac{\phi \pi D^3}{6} \qquad (161)$$

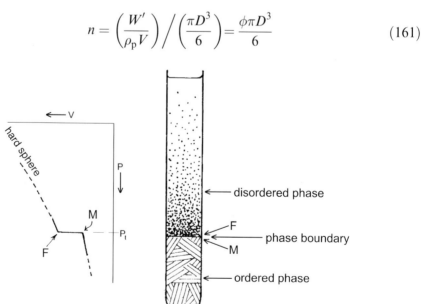

Fig. 175. Balance between sedimentation and osmotic pressures in a two-phase system [218].

CHAPTER 9 ORDER–DISORDER PHENOMENA

TABLE 9.1
Sedimentation pressures of colloids showing three-phase equilibrium behavior [218]

	Salt concentration (M)				
	0.01	0.001	0.01	0.005	0.008
D (nm)	563	563	500	500	260
P_t (Pa)	0.42	1.2	0.39	0.41	2.7
P_i (Pa)	0.028	0.018	0.031	0.03	0.192
P_t/P_i	15	68	12.6	14	14
T (K)	307	307	293	293	293

Their results for three different PS colloids at three ionic strengths and two temperatures are given in Table 9.1.

In most cases the value of P_t/P_i was close to that of the Alder hard sphere model, i.e. 11.6, even though these 'spheres' were more like compressed dodecahedra of soft, diffuse ionic layers! Deviations occurred at lower ionic strengths, e.g. 0.001 M above, where the softness of the ionic layers would be much greater.

9.1.5 The Yoshino quantized oscillator model

Yoshino has observed order–disorder behavior between domains in the *horizontal plane* like those found by Ito and coworkers, as shown in Fig. 133. A computer-enhanced photomicrograph taken in a horizontal plane of a 400 nm diameter PS latex with surface sulfate groups is displayed in Fig. 176 [220]. The particles in the disordered domain cannot be seen because of their Brownian motion. The interparticle spacing was 2.4 μm. The colloid was very dilute, yet the phase boundary was stable for over an hour.

The interparticle distances are too great and the attractive forces are too strong to be explained by a secondary minimum in the DLVO theory. Yoshino proposes a theory quite different from that of Ise and Sogami and analogous to the theory of dispersion forces (see, for example, the

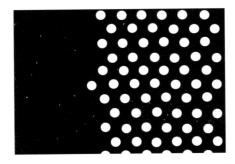

Fig. 176. Interphase between ordered domain and the medium in a PS latex [220].

Lifshitz–Parsegian–Ninham theory in Section 7.2.3.2). During Brownian motion a latex particle will move with respect to the center of its spherically symmetrical, diffuse electrical double layer, in a manner similar to that during electrophoresis and to that of an atomic nucleus within its electronic cloud. This produces fluctuating polar moments which can induce corresponding polar moments in neighboring particles, a result which always leads to attraction (cf. Fig. 116). Yoshino models these motions as electrical multipoles with energies in quantized states, as schematically shown in Fig. 177.

Let R_0 be the time-averaged radius of the Debye sphere ($R_0 = 1/\kappa$). Then its deformation in spherical coordinates is given by

$$R(\theta, \phi, \tau, t) = R_0 \left[1 + \sum_l \sum_m (\alpha_{lm(\tau,t)} + \beta_{lm} E) Y^*_{lm(\theta,\phi)}\right] \qquad (162)$$

where $Y^*_{lm(\theta,\phi)}$ are spherical harmonics and $\alpha_{lm(\tau,t)}$ are the deformation parameters which determine the shape of the double layer, β_{lm} is the resonance factor and E is the electrical field imposed by a neighboring particle.

The eigenvalues of m can be from $-l$ to $+l$ so that there can be a total of $2l+1$ values. The first mode, with $l=1$, is then a dipole; the second mode has the value of $l=2$ which gives a quadrupole deformation. With $m=0$, the relaxation time for a deformation, τ, is on the order of 1–10 s. This is just what was found by Fitch, Su and Tsaur [221] using dielectric spectroscopy for several polystyrene colloids, i.e. characteristic dielectric loss relaxation times on the order of seconds (Fig. 178), which they attributed to the ensemble motions of the diffuse layer. Yoshino has not calculated pair potentials based upon this model, although he asserts that his van der Waals-like forces should act over several micrometers distance.

> DLVO theory is most successful with closed systems in which repulsions among particles force the colloid against the walls.

To date there is no resolution of these and other theories. It is probable that in many cases the DLVO picture is satisfactory. It is sometimes referred to in

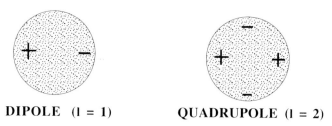

DIPOLE (l = 1)   QUADRUPOLE (l = 2)

Fig. 177. Multipole configurations of the diffuse part of the electrical double layer surrounding a colloidal particle [220].

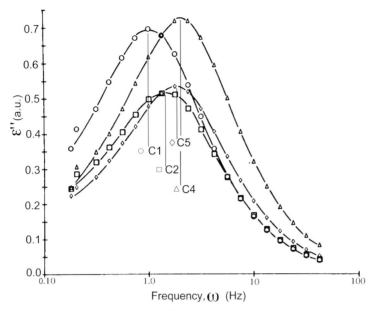

Fig. 178. Dielectric loss, ε'' spectra for four PS colloids [221].

the literature as the 'repulsion only' theory, referring to the double layer interaction only, which is an incorrect characterization since it does comprehend a secondary minimum arising from dispersion forces; there are also many weakly flocculated systems which are well described by DLVO. It is certainly most successful in dealing with closed systems in which repulsions among particles force the colloid against the walls of the container such that the interparticle spacing is governed strictly by the volume fraction occupied by the particles. This is described by Eq. 159 for hexagonal close packing. Incidentally, there are other close packing geometries which have been observed, most notably face-centered cubic (fcc).

But when there is the possibility for expansion into a more dilute, liquid-like region, and still the close packed structures remain stable as in Figs 133 and 176, then clearly there is no expansion pressure against the walls. Thus interparticle *attractions* must be operational.

9.2 Melting–freezing behavior of colloidal crystals

9.2.1 The role of Brownian motion

9.2.1.1 Direct microscopic observation

It is clear from the discussion above that polymer colloids can serve in many ways as models for molecular systems including phase transition behavior. Under conditions where repulsions dominate and Eq. 159 holds true, i.e. above the Krieger point ($\phi > \phi^*$ in Fig. 174), heating the system can only cause the particles to increase their Brownian motion while maintaining their time-averaged positions within the crystal lattice. Evidence for this comes from actual photomicrographs in which the images of particles of a single latex at different concentrations become more blurred with dilution, as seen in Fig. 179 [222]. The Brownian motion has been measured in detail as shown in Fig. 133.

9.2.1.2 Neutron scattering

Further evidence for increased Brownian motion as a function of the volume fraction comes from neutron scattering. The structure factor, $S(Q)$ in Eq. 75 (Section 5.2.6) gives information about the correlations of motions and positions among particles via the correlation function $g(r)$. In plots of $S(Q)$ as a function of Q the position and breadth of the peaks, such as those shown in Fig. 180, tell us about the interparticle spacing (peak position) and degree of Brownian motion (peak breadth). At constant ϕ, above the Krieger point, increasing the salt concentration (Fig. 180b) causes a broadening of the peaks without change in position (at least up to $[NaCl] = 10^{-3}$ M), interpreted as causing no change in interparticle distances but an increase in

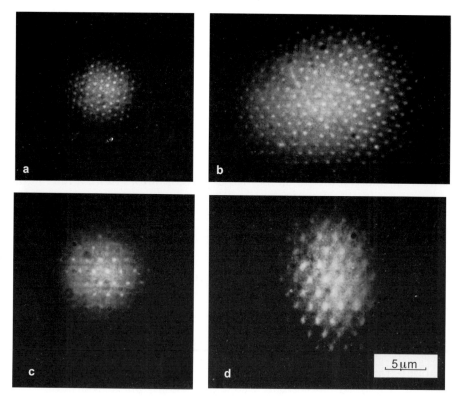

Fig. 179. Photomicrographs of PS colloid at: (a) 4%; (b) 1.5%; (c) 0.55%; and (d) 0.4% solids. Diameter = 341 nm [222].

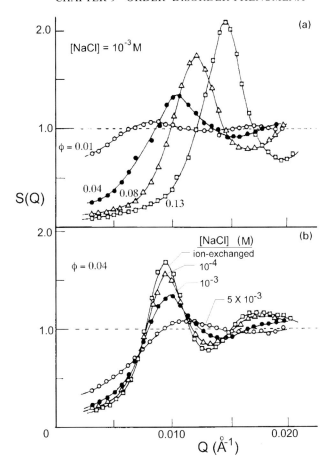

Fig. 180. Neutron scattering curves for a PS latex, $D_0 = 310$ nm, at various volume fractions (a), and salt concentrations (b) [223].

thermal motion [223]. The increase in salt concentration causes the contraction of the diffuse electrical layer, apparently thereby weakening repulsions and allowing greater mobility.

> Ottewill and coworkers concluded that the charged particles behave as 'soft spheres.'

On the other hand, an increase in volume fraction at constant ionic strength, as in Fig. 180a, shifts the peaks towards higher Q (shorter interparticle spacing) and sharpens them. Use of Eqs 75–78 converts these curves to their corresponding radial distributions, which have been shown in Fig. 54 for $\phi = 0.01$, 0.04 and 0.13 for the same set of experiments. At $\phi = 0.01$ the system has very little structure, behaving like a gas, whereas at the highest volume fraction there are strong correlations, and the system is highly ordered, as discussed earlier in Section 5.2.6. From the slopes of these curves,

Ottewill and coworkers concluded that the particles behave as 'soft spheres,' with a DLVO pair potential up to a certain interparticle distance at which the potential rises to infinity.

> In concentrated systems, a particle interacts with others and diffuses in a collective mode.

There is a direct relationship between the Fickian diffusion coefficient, D, of a particle and its thermal motion, according to Einstein's Law of Diffusion:

$$D = \frac{kT}{f} \qquad (163)$$

where kT is the thermal energy of a particle and f is its viscous resistance. Stokes showed that for hard spheres $f = 6\pi\eta a$, where η is the viscosity of the medium and a is the radius (cf. Eq. 142). Dynamic light scattering can be used to obtain values for D, according to Eq. 68 (Section 5.1.6):

$$g(\tau) = 1 + \exp(-2DQ^2\tau) \qquad (68)$$

in which $g(\tau)$ is the autocorrelation function and Q is the scattering vector. By choosing different values of τ, the correlation delay time, one can obtain a short-time and a long-time self diffusion coefficient. The former measures D over only a few Brownian random steps, whereas the latter is over a sufficiently long time that, in concentrated systems, the particles interact with others and diffuse in a collective mode. At very low volume fractions the free diffusion, D_0, of the particles should equal that calculated by the Stokes–Einstein equation. Ottewill and Williams used a small amount of a PMMA latex stabilized with polyhydroxystearate (PHS) as tracer particles in a PVAc (polyvinyl acetate)/PHS colloid in an organic medium which was refractive-index-matched to the PVAc [224]. Both particles were the same size with the same stabilizer, but light scattering could only 'see' the PMMA particles. The measured diffusion coefficients are shown in Fig. 181. The long-time values of D/D_0 decrease with increasing volume fraction until one reaches the Alder freezing point at $\phi_f = 0.5$, at which no further collective diffusion takes place. The particles are essentially 'locked' into their crystal lattice positions. The short-time diffusion continues since it measures motions of particles even within a crystal lattice like those in Fig. 179.

9.2.1.3 Diffusing wave spectroscopy

A more versatile method of obtaining diffusion coefficients in various time domains as a function of the physical conditions of the colloid relies on multiple backscattering of laser light from relatively concentrated dispersions [225]. This is known as diffusing wave spectroscopy (DWS), and treats the

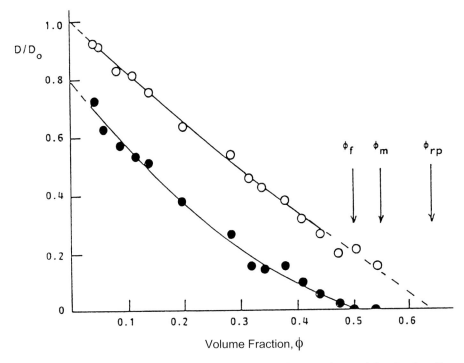

Fig. 181. Normalized self-diffusion coefficients of PMMA particles in decalin as a function of volume fraction. −○−, short-time values; −●−, long-time values [224].

path of light as it gets scattered from particle to particle before exiting as a diffusion problem. The various wave interferences which result are analyzed with an autocorrelator. In the past we have considered light scattering only from colloids so dilute that no 'multiple scattering' of this kind was important. With DWS advantage is taken of the multiple scattering to glean more information about the system, including the full spectrum of diffusional relaxation times and hydrodynamic interactions between particles (in still more concentrated systems). There is no requirement for refractive index matching, so that it works well in aqueous systems under a variety of ionic strengths, particle sizes and volume fractions. From the slope of the time-dependence of the autocorrelation function one can calculate the so-called 'q-averaged' diffusion coefficient, $\langle D(q)q^2 \rangle$. Some results of the normalized, inverse values of this parameter as a function of volume fraction, ϕ, and ionic strength, [I], expressed as molarity of KCl, are shown in Fig. 182.

> With DWS advantage is taken of the multiple scattering to glean more information about the system.

The effects of changes in ionic strength and concentration are readily apparent and the resultant changes in D are dramatic when the volume fraction of polymer is high enough. Clearly a decrease in salt concentration

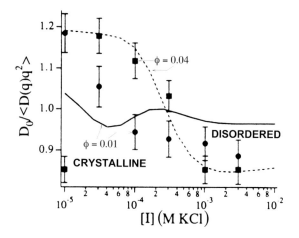

Fig. 182. Inverse dimensionless diffusion coefficients of PS anionic, aqueous latex. Points are experimental; curves are theoretical [225].

leads to ordering when $\phi = 0.04$ but not when $\phi = 0.01$; and the corresponding expansion of the diffuse part of the electrical double layer leads to restricted diffusion.

9.2.1.4 Crystalline and glassy states

If the nonaqueous colloid represented in Fig. 181 is brought to a concentration above the melting point, ϕ_m, then diffusion of particles over sufficiently long distances to find lattice positions becomes extremely slow, and the system tends to reside in a disordered, 'glassy' state. Above the random packing concentration, ϕ_{rp}, essentially all diffusion stops, and the colloid remains permanently glassy. Beautiful examples of these various states in this PMMA/PHS-PVAc/PHS system are shown in Plate 5. Because the refractive index of the particles is almost matched by that of the medium, these colloids are nearly transparent. It is still possible to nucleate crystallization from the glassy state, as evidenced in vial B in Plate 5. In an uncharged, nonaqueous dispersion the order–disorder transitions occur only at concentrations at or near the Alder values because of their short-range interparticle interactions, in contrast to electrostatically stabilized colloids, as shown in Fig. 139.

These principles have been put together in an elegant and beautiful fashion by Hachisu to make synthetic black opals (personal communication). He first made a monodisperse, lightly crosslinked polystyrene latex and transferred it into methyl methacrylate monomer (Section 7.7.2). This nonaqueous dispersion in the form of flat sheets, approximately 2 mm thick, was allowed to crystallize over several months at constant temperature. Then the MMA monomer was polymerized to 'freeze in' the large colloidal crystals which had formed. The sheets were cut into small, gem-sized pieces and set into gold mounts, like that shown in Plate 1. Professional jewelers characterized this

9.2.2 Reversible crystallization and melting

Under conditions where repulsions cannot dominate, e.g. where $\phi' < \phi < \phi^*$, the particles must be in a free energy minimum something like that shown in Fig. 132. A little heating could randomize the structure if the well were shallow enough. Hiltner and Krieger have shown that this is exactly what happens under these conditions, i.e. reversible melting of colloidal crystals occurs over a very narrow temperature range, as shown in Fig. 183.

A 1 K rise in temperature was seen to wipe out the opalescence, and cooling the same amount restored it, indicating that ϕ' increased with temperature. However, ϕ^* was unchanged upon heating within the experimental limits of these rather delicate systems. From the interparticle spacing data like those in Fig. 174, there must be an expansion in volume of the colloid upon melting, which leads to the conclusion that an increase in pressure would tend to reverse the process, causing crystallization. The parallels with molecular systems are striking, but must not be taken too far, given the different nature of the pair interaction potentials from those among molecules.

9.3 Bicomponent colloidal crystals

What happens when two, pure monodisperse polymer colloids, A and B, of different particle sizes, are mixed under conditions which should lead to

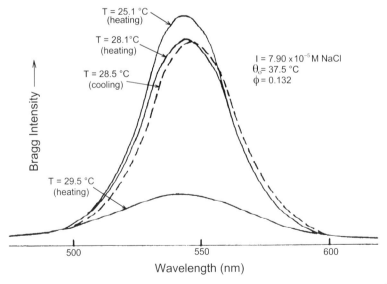

Fig. 183. Reversible melting of a PS colloid in the two-phase region, $\phi' < \phi < \phi^*$ [217].

crystallization? Does one act as an 'impurity' to hinder the crystallization of the other? Can they form an alloy with separate domains of crystals of each type? Can they form crystalline compounds, such as AB_2, AB_5 or AB_{13}? Can one construct a phase diagram for such bicomponent systems? The answer turns out to be yes for each of these questions.

9.3.1 Direct microscopic observation

Yoshimura and Hachisu were the first to observe that binary mixtures of aqueous, charge-stabilized PS colloids could form mixed crystals [226]. They used a specially designed metallurgical light microscope (Fig. 184) to examine samples held on microscope slides beneath cover slips. They allowed the samples to crystallize over tens to several tens of hours, and subsequently followed further structure development.

A pinhole below the sample was used to restrict the incoming light to a very small area in order to avoid scattering from neighboring areas of the colloid. Samples of about 10% solids at $[KCl] = 10^{-5}$ M were placed in the cell and allowed to crystallize. Typically islands of crystals in a 'sea' of disordered colloid were first observed. An example is shown in Fig. 185 of the 'pinhole' image of a mixed crystal from 550 nm and 310 nm anionically charged PS latexes. In Fig. 185a each large particle is surrounded by six smaller ones, marked by black dots in one location; in Fig. 185b the large

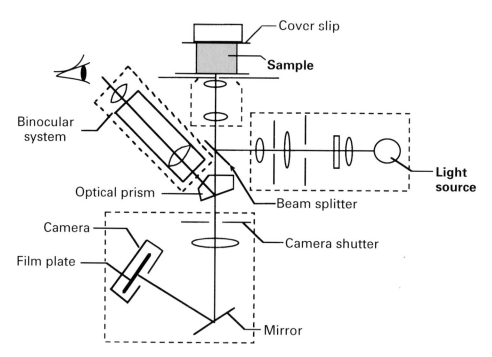

Fig. 184. Inverted 'metallurgical' microscope used for studying order–disorder phenomena [227].

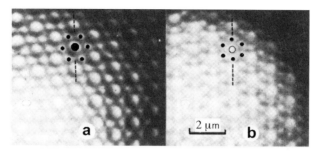

Fig. 185. Crystal of binary mixture of 550 nm and 310 nm PS particles in colloidal crystal. (a) First plane; (b) 0.4 μm above first plane [226].

particles are slightly out of focus and a small one can be seen between each large particle.

A diagram of this arrangement is shown in Fig. 186, which points to the conclusion that this structure has the formula of AB_5, similar to that of the intermetallic compound $CaCu_5$. At different ratios of large to small sized particles and different particle sizes, AB_2 and AB_{13} type crystals have been observed as well.

9.3.2 Binary phase diagram

Others have found these same structures in binary latex systems. In fact Bartlett, Ottewill and Pusey [228] have constructed a phase diagram for the binary PS/PHS nonaqueous colloidal system where $D_A = 642$ nm and $D_B = 372$ nm, so that $D_A/D_B = 0.58$. The medium was a decalin/carbon disulfide mixture like that used for the colloids shown in Plate 4. They kept the number ratio of particles, n_A/n_B, constant while changing the total volume

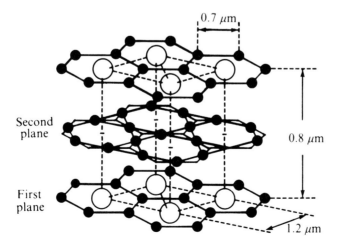

Fig. 186. Schematic drawing of the AB_5 structure (S. Hachisu, personal communication).

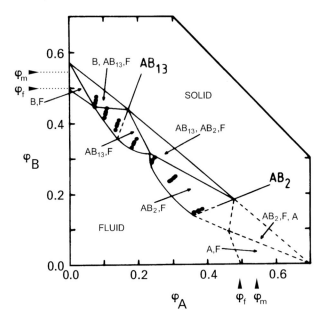

Fig. 187. Phase diagram of binary nonaqueous PS colloid in decalin/CS$_2$ mixture [228].

fraction to produce the lines radiating from the origin shown in the phase diagram in Fig. 187. The filled circles are experimental points, and ϕ_f and ϕ_m are the freezing and melting concentrations of the pure components, respectively. Bartlett and company had earlier shown that segregation of particles according to size could occur by sedimentation at a rate faster than the crystallization. To avoid this, they simulated zero gravity conditions by gently rotating their samples at a rate of about 1 revolution per day, fast enough to offset the calculated average rate of sedimentation and slow enough to minimize shear forces.

They investigated the structures of these systems by means of 'powder light diffraction,' using visible laser light and applying the well-known methods of X-ray powder diffraction to fluid dispersions of randomly oriented crystallites. An example of a plot of diffracted light intensity as a function of the scattering vector, Q, is given in Fig. 188. The calculated scattering pattern for AB$_2$-type crystals is given by the lines at the top of Fig. 188. There is almost perfect correspondence between theory and experiment.

Bartlett and coworkers also allowed samples to evaporate to dryness very slowly over several months, shadowed the surfaces with gold, and then looked at them with the scanning electron microscope. The AB$_2$ structure is clearly seen in Fig. 189 where alternating layers of large and small particles exist. The structure is built up from a hexagonally close packed layer of large particles, in the trigonal interstices of which are laid down a layer of small particles. This is simply repeated in alternating layers to form the long range order, as shown in the top part of Fig. 190.

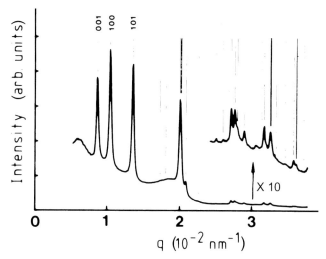

Fig. 188. Light powder diffraction pattern for PS/PHS colloid in decalin/CS$_2$. Lines above are theoretical for AB$_2$ crystals [228].

If that structure is rotated about 45° to the plane of the page, as shown in the lower part of Fig. 190, one obtains the arrangement seen in the SEM micrograph of Fig. 189. These structures, incidentally, could not be formed from a mixture with the exact ratio of $n_B/n_A = 2$, but rather there had to be a considerable excess of the B-type particles. Even more fascinating was the finding that a small, 7% change in particle size of only one of the colloids, to $D_B = 398$ nm from 372 nm, prevented the formation of AB$_2$-type crystals, even after many months incubation and with number ratios varying from $2 < n_B/n_A < 10$! Also surprising was the fact that AB$_{13}$-type structures formed more readily than the AB$_2$-type, and that the latter crystallized most rapidly when $n_B/n_A = 6$.

These nonaqueous colloids were prepared in such a manner as to avoid the formation of charge on the particles, so that they behave as classical hard

Fig. 189. Scanning electron micrograph of binary PS AB$_2$-type crystal [228].

Fig. 190. Ball models of AB$_2$ structure.

sphere systems, as we have seen earlier. Bartlett, Ottewill and Pusey have concluded, in agreement with Hachisu and coworkers, that in these high-volume-fraction systems the crystallization is entropy-driven, undergoing an Alder-type transition.

9.4 Applications of colloidal crystal arrays

9.4.1 Diffraction gratings

9.4.1.1 Liquid arrays

Figure 173 shows that the bandwidth of a Bragg diffraction peak of a monodisperse colloid can be very narrow at low volume fractions and low ionic strengths. Thus narrow bandpass filters can be constructed for transmission mode, since light which is diffracted is not transmitted. The modified Bragg equation, Eq. 158, indicates that the peak is a function of both wavelength, λ_0, and angle, θ. These features are illustrated in Fig. 191 in which the narrow Bragg peak, shown in absorbance mode, shifts from 497 nm at 90° to 493 nm at 80° [229]. Liquid colloid crystal arrays (CCAs) with a cell thickness of <0.5 mm have been measured to have a transmission of less than $10^{-8}\%$ at the Bragg peak while transmitting more than 80% at other wavelengths on either side.

> CCAs promise the possibility of Raman spectrometers with greatly improved performance at reduced cost.

It turns out that these are just the characteristics required for a Raman spectrometer where the excitation frequency along with the so-called Rayleigh

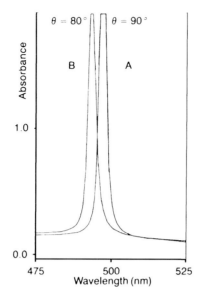

Fig. 191. Bragg diffraction spectra at two angles [229].

scattering ('so-called' because it is *quasi*-elastic scattering, whereas Rayleigh scattering is purely elastic), must be rejected so that the Raman lines can be seen. The net result is the possibility of Raman spectrometers with improved performance at reduced cost compared with current instruments: better signal-to-noise ratios, simplified construction and detection limits approaching the theoretical values [229].

9.4.1.2 Gel arrays

Liquid CCAs when used in devices for instrumentation have the disadvantage that they are liquid and subject to disordering by agitation or shock as well as from ionic impurities, such as those leached from glass over time. To overcome these problems Asher and coworkers devised a technique for locking the crystalline array in a gel matrix. They added to the crystallized colloid water-soluble monomers to make the aqueous phase have the following composition: 50% *N*-vinyl pyrrolidone, 17% acrylamide, 4.5% methylene-bis-acrylamide and 1% benzoin methyl ether photoinitiator. Irradiation with ultraviolet light caused polymerization of these monomers to produce an elastic gel without disturbing the order of the CCA, as evidenced by the diffraction spectra in Fig. 192 in which the peaks for the liquid and gelled CCAs are practically superimposable [230]. The polymer in this case was polyheptafluorobutyl methacrylate (PHFBMA), 155 nm particles, crystallized into a b c c array with the 1 1 0 plane parallel to the glass wall of the cell.

The choice of the perfluorinated polymer was made because it has a low refractive index of 1.393, close to that of water. Small additions of dimethyl

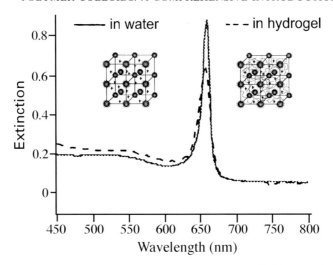

Fig. 192. Transmission spectra of 155 nm PHFBMA colloid crystal in liquid and gel [230].

sulfoxide (DMSO) to the medium lead to refractive index matching. The reasons for doing this become evident in the next section.

9.4.2 Photothermal nanosecond light-switching devices

If the refractive-index-matched PHFBMA crystalline colloid described in the previous section contains covalently bound dye molecules it can absorb light, but not diffract. The absorbed energy is dissipated into heat, such that the temperature of the particles rises upon illumination. When that happens, the refractive index of the polymer decreases and the system diffracts at the Bragg wavelength, cutting off any incident radiation at that frequency. Thus with two lasers, one at the λ_{max} for the dye to 'pump' the colloid to produce a step-jump in temperature, and the other at the Bragg wavelength to 'probe' the CCA, Asher and colleagues demonstrated both theoretically and experimentally that fast switching of the probe beam could be effected. An example of the apparatus used is shown in Fig. 193 in which a yttrium-aluminum-garnet (YAG) laser is used to pump both the dye laser and to create the photothermal heating in the CCA. Its frequency is tuned to the absorbance of the dye in the particles. The dye laser is tuned to the frequency of the Bragg diffraction peak at the detection angle.

> These effects can occur over a time period as short as 3 ns.

These effects can occur over a time period as short as 3 ns, so that one has the ability to switch light on and off on an extremely rapid time scale [231]. An example of results which can be expected is shown in Fig. 194. The pump beam (A) was directed at the CCA cell for 5 ns. The transmittance reached a

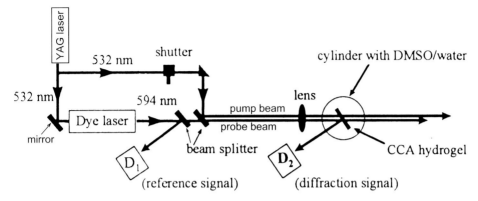

Fig. 193. Nanosecond laser set-up for CCA-based, nonlinear photothermal light switching [230].

minimum at this time, almost zero. The particles then rapidly cool, so that after about 20 ns the system has returned to its original state.

9.4.3 Electric field effects

Tomita and van de Ven found that beautiful color changes due to diffraction occur when an external electrical field is applied to colloidal crystals [232]. For example, a PS anionic latex with particle radius of 64 nm at 10% volume concentration is green when viewed normally (at 90° to the cell face). When a −1.5 V field was applied across a cell with a gap of 1 mm, the color changed to a 'reddish pink,' and when the polarity was reversed to +1.5 V, the colloid turned blue!

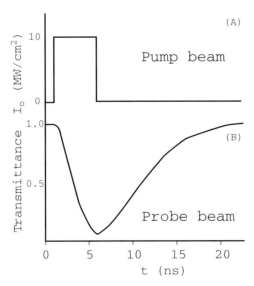

Fig. 194. Response characteristics of a CCA resulting from a 5 ns incident pump beam pulse [231].

In such an experiment there is a balance of forces at equilibrium: the electrophoretic force due to the external field and the 'colloidal' force due to diffusion resulting from the creation of a concentration gradient by the externally applied field. These are illustrated schematically in Fig. 195. When the particles are negatively charged they will be compressed towards the positive electrode such that the interlayer spacing, d, is reduced near the positive cell surface. When the polarity is reversed, d will consequently be increased.

The motion of the particles is described by the convective diffusion equation:

$$\frac{\partial c}{\partial t} = \frac{\partial}{\partial x}\left(D\frac{\partial c}{\partial x}\right) - \frac{\partial}{\partial x}(uc) \qquad (164)$$

where c is the particle concentration, t is time, x, the distance from the positive electrode, D, the diffusion coefficient and u, the electrophoretic velocity. The colloidal force, F_{col}, and the electrophoretic force, F_{ext}, act in opposite directions to move the particles with the velocity u against their viscous resistance, f:

$$F_{ext} + F_{col} = fu \qquad (165)$$

Combination of these two equations and integration leads to Eq. 166:

$$\frac{Df}{c}\frac{dc}{dx} = F_{ext} + F_{col} \qquad (166)$$

The relationship between f and D is given by the Einstein diffusion equation (Eq. 163), and the concentration, c, is related to the particle radius, a, and the volume fraction, ϕ, by:

$$c = \frac{3\phi}{4\pi a^3}$$

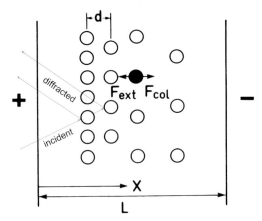

Fig. 195. Bragg diffraction from a CCA under the influence of an external electrical field [233].

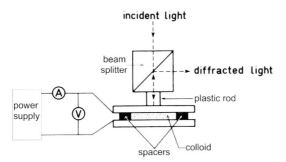

Fig. 196. Apparatus for studying electrochromic effects in colloids [233].

This leads to the simple dependence of the concentration gradient, $d\phi/dx$, upon the two forces:

$$\frac{kT}{\phi}\frac{d\phi}{dx} = F_{ext} + F_{col} \quad (167)$$

The problem now becomes one of expressing the two forces in terms of experimentally available parameters. Tomita and van de Ven have chosen the theory of Wiersema, Loeb and Overbeek, a predecessor to the theory of O'Brien and White (reference 200, Section 8.2.2.1), to obtain the electrophoretic force [233], which takes the form:

$$F_{ext} = 6\pi a \varepsilon \zeta E g(\kappa, \zeta) \quad (168)$$

where ζ is the zeta potential, E, the applied electrical field, and $g(\kappa, \zeta)$ is a function for the relaxation effect. The colloidal force is derived from the DLVO theory for spheres (Eq. 105, Section 7.2.3.1. Because there is a concentration gradient of particles, as shown in Fig. 195, the pair potentials among the particles are unbalanced. This leads to the net force, F_{col}, given by Tomita and van de Ven as:

$$F_{col} = 2\pi\varepsilon\zeta^2\kappa a f(\kappa a, \phi, d\phi/dx) \quad (169)$$

where f is a function of the variables in parentheses in Eq. 169. The force can be derived from the repulsive potential in Eq. 105 by differentiation with respect to x. Use of Eq. 169 assumes that the attractive interparticle interactions are negligible.

Solving these equations in terms of experimental variables, e.g. ionic strength (through κ) and volume fraction, leads to plots of d, the interparticle spacing, as a function of the position in the cell, x. The spacing, in turn gives the color, in terms of the peak wavelength λ_A, via the Bragg diffraction equation, Eq. 158 (using d instead of D for the spacing to avoid confusion with the diffusion coefficient). A comparison of some experimental results (filled triangles), obtained with the apparatus shown in Fig. 196, with those calculated (cross-hatched areas) are shown in Fig. 197.

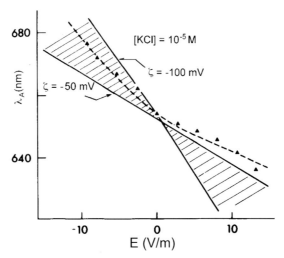

Fig. 197. Bragg peak wavelength as a function of applied electrical field for a PS anionic colloidal crystal [233].

The agreement of the experimental points with calculated values is excellent, but only for negative potentials. Also at higher volume fractions the agreement tends to be worse. This is partly due to the approximative nature of the equations used, but this theory nevertheless clearly establishes trends of behavior, their approximate magnitude and their causes.

> Beautifully colorful effects would seem to have applications where broad areas of color could be changed at will.

These beautifully colorful effects would seem to have practical applications where broad areas of color could be changed at will. A couple of practical considerations must be taken into account, however: (1) because the changes occur as a result of electrophoretic migration, the time required to reach steady state may be on the order of 200 s [233], and (2) because of the diffusive nature of the particles, sharp image boundaries apparently cannot be obtained [232]. More important from a scientific perspective is the possibility that further study along these lines could lead to a more quantitative understanding of the electrophoretic and diffusive forces involved.

9.4.4 Elastic gel colloidal crystalline arrays

The gelled colloidal crystals described earlier have some interesting properties worthy of mention. Once the latex particles are fixed in the polyvinyl pyrrolidone/acrylamide/methylene-bis-acrylamide (PVP/PAm/BAM) gel, they can no longer diffuse about. Subsequent substitution of the water by organic media will cause no changes in the nature of the crystal structure, but only its dimensions. This arises from the polymer–solvent interactions of the gel

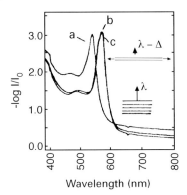

Fig. 198. Transmittance spectra of gelled colloid crystal [234].

polymer to expand or contract the gel matrix. The *relative* positions of the particles remain undisturbed. Addition of electrolytes, likewise, will have no effect on relative positions of the particles, although it may cause swelling or contraction of the gel matrix. These effects are observable in the diffraction behavior of the gelled colloidal crystalline arrays.

> Addition of electrolytes or organic solvents will have no effect on relative particle positions in gel CCAs.

An intriguing aspect of these gels is that they have mechano-optical properties, i.e. their Bragg diffraction peaks can be reversibly altered by simply stretching [234]. An example was prepared as follows: a polystyrene latex with particle size of 170 nm was deionized, allowed to crystallize, and then highly purified monomers introduced to give an aqueous phase composition of PVP/PAm/BAM = 50/6/1%. After photopolymerization the Bragg diffraction spectrum is essentially unchanged from that of its precursor latex, like the system shown in Fig. 192. The transmittance spectrum $-\log I/I_0$ versus wavelength — is shown in Fig. 198, where I and I_0 are the transmitted and incident beam intensities, respectively. When this elastic gel CCA is stretched in the plane of the film, the interparticle spacing in the normal direction is decreased, with a consequent decrease in the Bragg peak wavelength (peak b shifts to peak a). When the tension is released, the film reverts to its original dimensions and spectrum (peak c).

The number of practical applications for these systems and for various devices derived from them seem almost endless, and the reader is invited to find novel ways in which they might be used!

References

217. Krieger, I.M. and Hiltner, P.A. (1971) In *Polymer Colloids* (R.M. Fitch ed), p. 63, Plenum Press, New York.

218. Hachisu, S. and Takano, K. (1988). In *Ordering and Organization in Ionic Solutions* (N. Ise and I. Sogami eds), p. 376, World Scientific Publishing Co., Singapore.
219. Alder, B.J. and Wainright, T.E. (1962). *Phys. Rev.* **127**, 359.
220. Yoshino, S. (1993). *Polym. Int.* **30**, 541.
221. Fitch, R.M., Su, L.S. and Tsaur, S.L. (1990). In *Scientific Methods for the Study of Polymer Colloids and Their Applications* (F. Candau and R.H. Ottewill eds), p. 373, Kluwer, Dordrecht, Netherlands.
222. Kose, A., Ozaki, M., Takano, K., Kobayashi, Y. and Hachisu, S. (1973). *J. Colloid Interface Sci.* **44**, 330.
223. Ottewill, R.H. (1989). In *The Langmuir Lectures, Langmuir* **5**, 4.
224. Ottewill, R.H. and Williams, N.St.J. (1987). *Nature* **375**, 232.
225. Nilsen, S.J. and Gast, A.P. (1994). *J. Chem. Phys.* **101**, 4975.
226. Hachisu, S. and Yoshimura, S. (1980). *Nature* **283**, 188.
227. Kose, A., Ozaki, M., Takano, K., Kobayashi, Y. and Hachisu, S. (1973). *J. Colloid Interface Sci.* **44**, 330.
228. Bartlett, P., Ottewill, R.H. and Pusey, P.N. (1992). *Phys. Rev. Lett.* **68**, 3801.
229. Asher, S.A., Flaugh, P.L. and Washinger, G. (1986). *Spectroscopy* **1**, 26.
230. Asher, S.A. and Pan, G. (1997). 'Crystalline Colloidal Array Self Assembly: A Motif for Creating Mesoscopic Periodic Smart Materials,' in *Nanostructured Materials* (J. Fendler ed), NATO Workshop, Kluwer Publishing Co., Dordrecht.
231. Kesavamoorthy, R., Super, M.S. and Asher, S.A. (1992). *J. Appl. Phys.* **71**, 1116.
232. Tomita, M. and van de Ven, T.G.M. (1984). *J. Optical Soc. Amer. A* **I**, 317.
233. Tomita, M. and van de Ven, T.G.M. (1985). *J. Phys. Chem.* **89**, 1291.
234. Asher, S.A., Holtz, J., Liu, L. and Wu, Z. (1994). *J. Amer. Chem. Soc.* **116**, 4997.

Chapter 10

Rheology of Polymer Colloids

10.1. Introduction

The word rheology comes from two ancient Greek words, 'rhein,' to flow and 'logos,' word or speech. The latter has come to include the study of a subject. So rheology is the study of the flow of materials, solids, liquids and gases. It is such a huge subject, with so much written about it, that we must restrict ourselves – after a brief introduction – to just those aspects peculiar to polymer colloids.

A colloid flows when a force is applied to it. We shall start with a simple, continuous shear force, F, acting on an element of a fluid initially in the shape of a cube. The fluid is considered to be comprised of a stack of planar, horizontal sheets of area A which can slide over each other, as shown schematically in Fig. 199. The dots in Fig. 199 represent latex particles suspended randomly or arrayed as shown in a continuous medium. The base of the cube is fixed, while the top layer moves at the velocity u, so that there is a velocity gradient, du/dz (s^{-1}) produced, represented by the horizontal arrows of different vector lengths. Then the viscosity, η, is the force per unit area, F/A, required to produce a velocity gradient, du/dz, usually represented as the stress, σ, divided by the strain rate, $\dot{\gamma}$, as given in Eq. 170:

$$\eta \equiv \frac{F/A}{du/dz} = \frac{\sigma}{\dot{\gamma}} \qquad (170)$$

where η has the SI units of kg m^{-1} s^{-1} or pascal-seconds (Pa s). If cgs units are used, 1 g cm^{-1} s^{-1} (0.1 Pa s) is called a poise, named after Poiseuille who

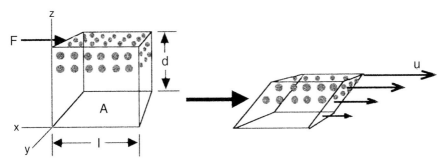

Fig. 199. Action of a shearing force, F, on a fluid with a fixed base of area A to produce a velocity gradient du/dz.

formulated the viscous flow of fluids through capillaries. The viscosity of water at room temperature, for example, is 0.01 poise or 1 centipoise (1 CP = 1 mPa s). If the viscous response of the fluid is time-independent, it is said to be 'Newtonian,' whereas if the viscosity changes with time at a given shear rate, it is said to be non-Newtonian.

If both sides of Eq. 170 are multiplied by du/dz, dimensional analysis can show that the viscosity also measures the energy dissipation rate per unit volume:

$$\frac{dE/dV}{dt} = \eta \left(\frac{du}{dz}\right)^2 \tag{170a}$$

10.2 Viscosity of hard sphere dispersions

> The viscosity of hard spheres at infinite dilution was shown by Einstein to have a very simple dependence on volume fraction.

The viscosity of a dispersion of hard spheres at infinite dilution in a fluid medium was shown by Einstein to have a very simple dependence on the volume fraction, ϕ, and the viscosity of the medium, η_0:

$$\eta_r \equiv \frac{\eta}{\eta_0} = 1 + 2.5\phi \tag{171}$$

where η_r is called the relative viscosity. The remarkable result is that the slope of 2.5 is independent of the particle size as long as the particles behave as hard spheres, defined as having negligible double layer and hydrodynamic interactions between particles. The 'intrinsic viscosity,' $[\eta]$, is defined as follows, with its numerical value given for hard spheres:

$$[\eta] \equiv \lim_{\phi \to 0} \left(\frac{\eta_r - 1}{\phi}\right) = 2.5 \tag{172}$$

At higher volume fractions interparticle interactions occur, so that Eq. 171 will have higher-order terms:

$$\eta_r \equiv \frac{\eta}{\eta_0} = 1 + 2.5\phi + O(\phi^2) \tag{173}$$

where the O represents terms 'of the order of' as a result of binary interactions. Furthermore, one expects that at some very high concentration of hard spheres, no flow at all can take place. This is known as the 'packing fraction,' $\phi = p$, and these ideas are captured in Fig. 200.

It is conventional to plot the relative viscosity as a function of shear rate. Experimentally Krieger found that for many polymer colloids an analogous plot of relative viscosity as a function of volume fraction, according to Eq. 173, gives an intercept of 1.0, an initial slope of 2.7 (not exactly 2.5 probably because of the presence of a few aggregates) and packing fractions

CHAPTER 10 RHEOLOGY OF POLYMER COLLOIDS

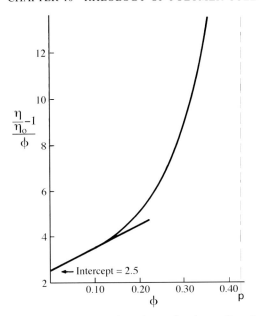

Fig. 200. Reduced viscosity as a function of volume fraction (schematic) [235].

which depend upon the shear rate [235]. He found both low shear and high shear limiting values of p, as shown in Fig. 201. From these experimental results he formulated a 'rheological equation of state' for rigid sphere dispersions in steady flow as follows:

$$\eta_r = \eta_{r_{hi}} + \frac{\eta_{r_0} - \eta_{r_{hi}}}{(1 + 0.431)|\sigma_r|} \qquad (174)$$

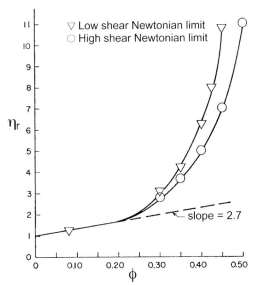

Fig. 201. Experimental (points) and theoretical (curves) showing relative viscosities of polymer colloids [235].

where σ_r is the reduced shear stress, defined below. From these experiments, it was found that the curves in Fig. 201 could be fit with the following parameters:

$$\left.\begin{array}{l} \eta_{r_{hi}} = (1 - 1.47\phi)^{-1.82} \\ \eta_{r_0} = (1 - 1.75\phi)^{-1.50} \\ \sigma_r \equiv \dfrac{\sigma a^3}{kT} \end{array}\right\} \quad (175)$$

The fact that p depends upon the shear stress indicates that these colloids are not Newtonian.

10.2.1 Principle of corresponding rheological states

Krieger developed a theory for the rheology of dispersions of spheres at all concentrations up to p using a principle of corresponding rheological states and by employing reduced variables. He noted that the viscosity of a dispersion of hard spheres was a function of nine variables:

$$\eta = f_1(\dot{\gamma}, t, \eta_0, \rho_0, a, n, \rho_p, kT) \quad (176)$$

where η_0 and ρ_0 are the viscosity and density of the medium (considered to be continuous relative to the size of the spheres), t is time, n is the number concentration of particles, and ρ_p is the polymer density.

All nine of these variables are expressible in terms of the basic variables of mass, length and time, so that they may be reformulated into $9 - 3 = 6$ dimensionless groups:

Relative viscosity	$\eta_r \equiv \dfrac{\eta}{\eta_0}$
Volume fraction	$\phi = \frac{4}{3}\pi a^3 n$
Reduced shear stress	$\sigma_r \equiv \dfrac{\sigma a^3}{kT}$
Relative time	$t_r = t/t^*$
where	$t^* \equiv \dfrac{\eta_0 a^3}{kT}$
Relative density	$\rho_r = \rho_p/\rho_0$
Internal Reynolds Number	$R_i \equiv \dfrac{a^2 \dot{\gamma} \rho_0}{\eta_0}$

The Internal Reynolds number is the ratio of inertial forces to viscous forces. When it is low, the inertial forces can safely be neglected.

CHAPTER 10 RHEOLOGY OF POLYMER COLLOIDS

Then the *relative* viscosity becomes:

$$\eta_r = f_2(\phi, \sigma_r, t_r, \rho_r, R_i) \tag{177}$$

The system can be simplified still further by restricting the experimental conditions to steady flow, neutrally buoyant particles and laminar flow, so that t_r approaches infinity, ρ_r is unity and R_i is zero, respectively. Now the relative viscosity is simply a function of two reduced (dimensionless) variables:

$$\eta_r = f(\phi, \sigma_r) \tag{178}$$

This suggests that a plot of relative viscosity as a function of reduced shear stress should be the same for all *hard sphere* colloids at a given volume fraction. For aqueous systems this means that sufficient neutral electrolyte must be added to suppress most of the electrical double layer effects without actually destabilizing the colloids, at least during the time of the measurements (Section 7.2.3.1). Typical results are shown in Fig. 202. The three smallest particle size colloids all fall on the same curve, whereas those larger than 500 nm in diameter do not. The explanation appears to be that above this size and at these ionic strengths the pair potentials, such as those in Figs 130 or 132, contain a secondary minimum leading to some aggregation. These would require an additional force, F_e, to break up, leading to higher viscosities.

Krieger emphasized that the particle size, a, which is cubed in the calculation of the shear stress, must be that of the core particle *plus* any adsorbed ionic or steric stabilizer.

Some lightly crosslinked polystyrene colloids were transferred into organic solvents after their behavior in aqueous media was determined. This was to see if the theory for hard spheres were applicable to all media of various viscosities and dielectric constants. Characteristic of all systems, aqueous and organic, are the curves in Fig. 203 for PS spheres in benzyl alcohol [235]. At

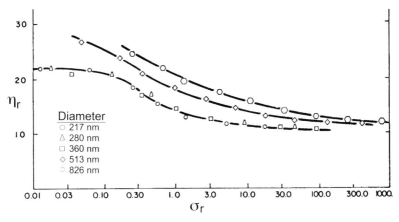

Fig. 202. Relative viscosity vs reduced shear stress for several aqueous PS latexes at $\phi = 0.50$ [235].

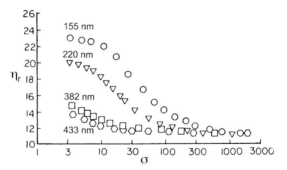

Fig. 203. Relative viscosity vs shear stress in benzyl alcohol. PS colloids of various particle sizes.

low shear stresses the systems are non-Newtonian and have relative viscosities, η_{r_0}, which depend strongly on particle size, whereas all curves converge to the same high shear-stress limiting viscosity, $\eta_{r_{hi}}$.

> The principle of corresponding rheological states applies over a concentration range from at least $\phi = 0.1$ to 0.5.

When these data, along with those shown in Fig. 202 (diameters less than 500 nm), are plotted as a function of *reduced* shear stress, they all fall on the same master curve as shown in Fig. 204. The data cover a 100-fold range in shear stress and a 400-fold variation in t^*! The brilliance of this achievement is especially compelling in light of the fact that the principle of corresponding rheological states applies over a concentration range from at least $\phi = 0.1$ to 0.5, not all of which has been shown here.

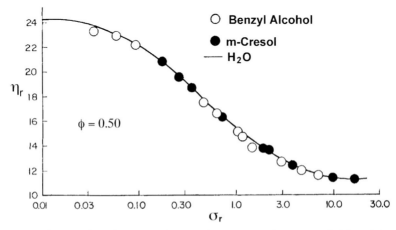

Fig. 204. Relative viscosity vs reduced shear stress for PS colloids of four sizes in various media [235].

10.2.2 The Krieger–Dougherty equation

An equation to describe the behavior of hard sphere systems under all of these conditions was derived by Dougherty and Krieger, based upon the earlier work of Mooney and others [236]. To model the behavior shown in Fig. 204, i.e. the dependence of the relative viscosity on the reduced shear stress, Krieger and Dougherty reasoned that when two particles in a shear field approach each other because of their Brownian motion, hydrodynamic forces cause them to tumble as a doublet until they move to different shear planes, when they will separate. This causes dissipation of energy among shear planes, and is responsible for non-Newtonian behavior. At greater shear stresses the Brownian diffusion to form pairs becomes less important compared to the shearing forces, and the relative viscosity will be lower. On the basis of these ideas they derived the following equation:

$$\eta_r = \eta_{r_{hi}} + \left(\frac{\eta_{r_0} - \eta_{r_{hi}}}{1 + b\sigma_r} \right) \quad (179)$$

where b is a fitting parameter which must be the same for all systems of hard spheres. This equation not only applies to the data in Fig. 204, but also to a large number of experiments conducted on PMMA colloids *sterically* stabilized in silicone oils of various viscosities and at several volume fractions. These are plotted in Fig. 205 with the corresponding theoretical curve [237]. Suffice it to say that the fit with theory is excellent except at high volume fractions, the largest particle sizes, high viscosities of the medium, and at high shear rates where the data slope upward. This is due to particle crowding, or a 'snowplow' effect (dilatancy) and is not comprehended by the theory.

Again, as in the other examples cited above, it is imperative to include the thickness of the stabilizer layer in calculating the particle size, since it determines the volume fraction according to Eq. 159.

Krieger and Dougherty also noted that the viscosity at finite concentrations could be divided into two parts: the first was that of a starting dispersion with

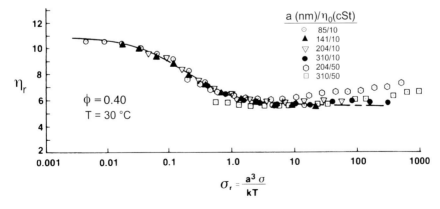

Fig. 205. Theoretical (Eq. 179, solid curve) and experimental data for sterically stabilized PMMA NADs in silicone oils [237].

a volume fraction ϕ_1 and the second, due to an incremental addition of more spheres to give a volume fraction ϕ_{12}. Then

$$\eta_r(\phi_1 + \phi_2) = \eta_r(\phi_2)\eta_r(\phi_{12})$$

whose solution is:

$$\eta_r = \left(1 - \frac{\phi}{p}\right)^{-[\eta]p} \qquad (180)$$

This is generally referred to as the Krieger–Dougherty equation, where p is the packing fraction and $[\eta]$ is the intrinsic viscosity, as defined in Eq. 172. If the system is crystalline $p = 0.74$ for a face-centered cubic (fcc) structure and $p = 0.68$ for a body-centered cubic (bcc) structure. For random close packing $p \cong 0.63$.

The theoretical curves in Fig. 201 were calculated from this equation using $[\eta] = 2.67$ and $p_0 = 0.57$ and $p_{hi} = 0.68$ for the zero-shear and high-shear packing limits, respectively. Additional correlation of many more data is shown in Fig. 206 in which polystyrene particles in water and silica particles in cyclohexane of particle diameters ranging from 49 nm to 433 nm all fall on the same theoretical curves! These data, from different laboratories, were correlated by using $[\eta]p = 2$ and $p = 0.63$ for the low shear, and 0.71 for the high shear limits [238, 239].

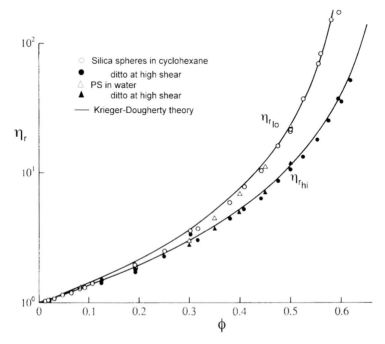

Fig. 206. Hard sphere limiting viscosities of PS and silica colloids at high and low shear [235, 238 (silica), 239 (curves)].

Russel and Sperry rearranged Eq. 180 to obtain the plot shown in Fig. 207 with viscosity data taken from the same authors as above along with results on glass beads with particle diameters up to 105 000 nm suspended in Aroclor taken by Lewis and Nielsen [240, 241]. Again the correlation is excellent for $\eta_{r_{hi}}$ over a great variety of systems. Extrapolation to infinite viscosity gives the value of p to be 0.70. The lower curve in Fig. 207 is based upon data taken with an oscillatory viscometer at high frequency and low amplitude, to be discussed below.

10.2.3 The four limiting viscosities

A generally accepted measure of the importance of the shear-induced force, $6\pi\eta_0 a^2 \dot{\gamma}$, relative to that resulting from Brownian motion, kT/a, is the dimensionless Peclet number:

$$Pe \equiv \frac{6\pi\eta_0 a^3 \dot{\gamma}}{kT} \qquad (181)$$

At very low values of Pe, Brownian motion dominates as the restoring force, and the system behaves as if it were at equilibrium. This is the condition at the zero-shear limit under steady flow and at low amplitude, high oscillatory frequencies. Under the latter conditions an elastic storage modulus also may be observed due to the Brownian restoring force. At $Pe \gg 1$ the shearing forces dominate, inducing an ordered, dynamic structure in which there are shear planes sliding over each other, almost literally as shown in Fig. 199. At even higher shear rates, represented as $Pe = \infty$, hydrodynamic turbulence

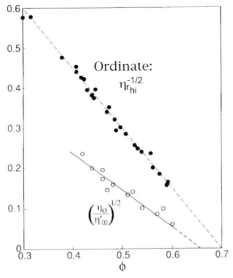

Fig. 207. High shear limiting viscosities of various hard sphere polymer colloids [240, 241].

leads to disordering not unlike that in the system at equilibrium, but now with a 'snowplow' effect leading to high viscosities. The degree of ordering, whether induced by shear, Brownian motion or crystallization, can be expressed in terms of an equilibrium-radial distribution function, $g(r)$, or a nonequilibrium-pair distribution function, $p_2(r)$ (Eqs 76 and 78), whose difference from $g(r)$ is measured by a factor $f(r)$.

> At very low values of Pe, Brownian motion dominates as the restoring force, and the system behaves as if it were at equilibrium.

The four limiting viscosities are defined by the Peclet number limits (high or low) and method of measurement, i.e. whether under continuous shear or in a low amplitude, oscillatory mode at low or high frequency, Ω. These are summarized, according to Russel and Sperry, in Table 10.1. The primes on the symbols represent oscillatory experiments. S^H and S^B are interparticle dipoles which arise from hydrodynamic and Brownian forces, respectively.

The expansion of the Einstein equation, Eq. 173, for each of these viscosity regimes is given below.

10.2.3.1 Low frequency, low shear regime

$$\eta_{r_{lo}} = 1 + 2.5\phi + 2.5\phi^2 + 7.5\phi^2 \int_2^\infty r^2 \langle J \rangle_2 g \, dr + 0.225\phi^2 \int_2^\infty r^2 \langle W \rangle_2 g f \, dr \quad (182)$$

where $r = R/a$ and R is the center-to-center radial distance, and $\langle J \rangle$ and $\langle W \rangle$ are average hydrodynamic and Brownian coefficients associated with the corresponding dipoles (subscript 2 indicating pair interactions) S^H and S^B in Table 10.1.

10.2.3.2 High frequency oscillatory regime

The viscosity here is derived solely from the hydrodynamic interaction of the system otherwise at equilibrium, and therefore given by the first four terms in

TABLE 10.1
The four limiting viscosities for hard spheres [240]

	Pe	Ω	S^H	S^B	$p_2 - g$	Microstructure
η'_∞	$\ll 1$	$\gg 1$	+	0	0	Brownian isotropic
$\eta_{r_{lo}}$	$\ll 1$	$\ll 1$	+	+	$\cong Pe$	Slightly perturbed
$\eta_{r_{hi}}$	$\gg 1$	0	+	0	$\cong 1$	2D slip planes
η_{hyd}	∞	0	+	0	$\cong 1$	Hydrodynamic isotropic
G'	$\ll 1$	$\gg 1$	+	0	$\cong Pe/\Omega$	Brownian isotropic

Eq. 182:

$$\frac{\eta'_\infty}{\eta_0} = 1 + 2.5\phi + 2.5\phi^2 + 7.5\phi^2 \int_2^\infty r^2 \langle J \rangle_2 g \, dr \quad (183)$$

whereas the shear modulus is determined by Brownian dipoles, and is thus a function of the last term in Eq. 182:

$$\frac{a^3 G'_\infty}{kT} = \frac{0.225}{3\pi} \phi^2 \int_2^\infty r^2 \langle W \rangle_2 g f \, dr \quad (184)$$

10.2.3.3 High shear limiting viscosities at $Pe \gg 1$ and $Pe = \infty$

The difference between steady flow, high shear and extremely high shear viscosities lies in the microstructure of the colloid under these conditions, and is therefore reflected in the pair distribution functions, p_2^∞, and p_2^{hyd}, respectively:

$$\eta_{r_{\text{hi}}} = 1 + 2.5\phi + 2.5\phi^2 + \phi^2 \int \langle \Delta S^H \rangle_2 p_2^\infty \, dr \quad (185)$$

and

$$\frac{\eta_{\text{hyd}}}{\eta_0} = 1 + 2.5\phi + 2.5\phi^2 + \phi^2 \int \langle \Delta S^H \rangle_2 p_2^{\text{hyd}} \, dr \quad (186)$$

The dependence of the two kinds of forces, Brownian and hydrodynamic, on the Peclet number and the corresponding dependence of the four limiting hard sphere viscosities is graphically illustrated in Fig. 208. Here the various

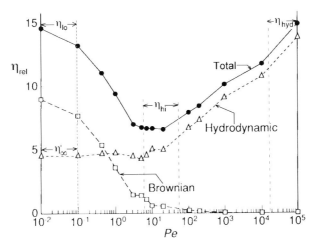

Fig. 208. Simulations of contributions to limiting viscosities of Brownian and hydrodynamic forces [242].

viscosity regimes are approximately positioned on theoretically modeled curves calculated by Phung [242].

Comparisons of a large number of experimental data for hard spheres by Russel and Sperry show the relative magnitudes of the four limiting viscosities as a function of volume fraction (Fig. 209). The curves in this figure are not theoretical, but accurate correlations of experimental data, with points shown only for the top curve.

10.3 Gelled dispersions

In the series of experiments on PMMA colloids in silicone oils described earlier in Section 10.2.2, when the molecular weight of the silicone oil was 11 000 dalton and its kinematic viscosity was 200 cSt (centistokes; $1\,\text{cSt} = 10^{-6}\,\text{m}^2\,\text{s}^{-1}$) ($\eta_0 = 181$ cP), the behavior of the colloids was entirely different [237]. They were gels at all volume fractions down to 10% and exhibited extremely pronounced shear-thinning behavior, as seen in Fig. 210. That this was not due to the viscosity of the fluid medium was shown by taking the 'fluid' dispersions to low temperatures where the viscosity of the 50 cSt oil was almost equal to, or considerably higher than, that of the 200 cSt oil. Even at $-44°C$ where the 50 cSt oil had a viscosity of 446 cP, the colloid at 20% solids remained fluid, whereas that in the 200 cSt oil was a gel at room temperature.

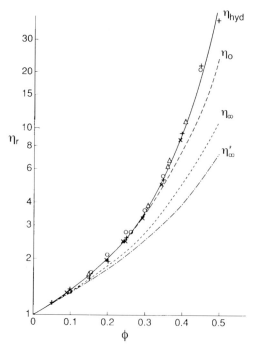

Fig. 209. Four limiting hard sphere experimental viscosities vs volume fraction [240].

The shear thinning, incidentally, often required as long as 10 to 20 *minutes* of steady shear before a constant relative viscosity was attained. Here, as in Fig. 203, all of the colloids approach the same viscosity at high shear rates, indicating that any structure formation is broken down at high shear rates.

The gelled dispersions, as the name implies, also displayed a yield stress, σ_y, in which no flow occurs until a certain critical force is applied. This is seen as the asymptotic rise in viscosity at very low shear rates in Fig. 210. Such behavior has been modeled by Casson's equation in which the square root of the shear stress is given as a function of the square root of the shear rate, in contrast to Eq. 170 which can be recast as

$$\sigma = \eta \dot{\gamma} \qquad (170)$$

Casson's equation is

$$\left. \begin{array}{ll} \sigma^{1/2} = \sigma_y^{1/2} + \eta_{hi}^{1/2} \dot{\gamma}^{1/2} & \text{for } \sigma > \sigma_y \\ \dot{\gamma} = 0 & \text{for } \sigma < \sigma_y \end{array} \right\} \qquad (187)$$

where σ_y is the yield stress and η_{hi} is the high shear limiting viscosity. Examples of the data plotted according to Casson's equation are shown in Fig. 211. The intercepts extrapolated to zero shear rate clearly show the yield stresses.

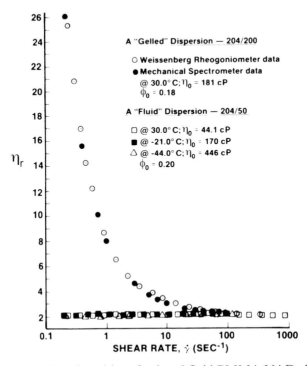

Fig. 210. Relative viscosities of gel and fluid PMMA NADs [237].

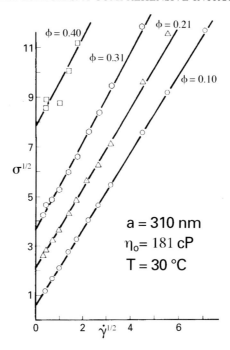

Fig. 211. Gel PMMA/NAD colloid rheology. Casson equation plot [237].

The explanation given by Choi and Krieger for this kind of behavior is that the molecules of the medium are so large that in occupying the spaces between particles in concentrated dispersions, the molecules, both those of the steric stabilizing layer and those of the medium, are compressed. This leads to what has been called a 'steric-elastic' force, which is larger at elevated temperatures. This *increases* the viscosity at higher temperatures, as shown in Fig. 212, even though the viscosity of the medium, η_0, *decreases* as the temperature is raised. Again, at high enough shear rates these effects tend to disappear.

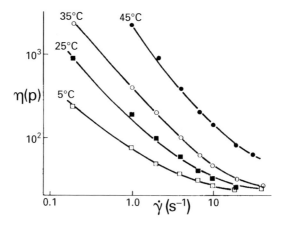

Fig. 212. Temperature-dependence of gel NAD viscosity for large-molecule media [237].

10.4 Electroviscous effects

Aqueous, ionically stabilized polymer colloids can have a strong dependence of viscosity on inert electrolyte concentration. Considering the extent of electrical double layer expansion upon deionization, this may not be surprising. At low shear stresses the particles may be expected to have an effective radius, a_{eff}, larger than that of the polymer particle itself, and that this will translate into a higher viscosity. This phenomenon is generally referred to as the 'secondary electroviscous effect.' The experimental data are dramatic, as shown in Fig. 213 in which the relative viscosity is plotted as a function of the reduced shear stress [243]. The latex was a polystyrene colloid at volume fraction, $\phi = 0.40$, with a radius, $a = 110$ nm, and which was made from persulfate initiator, so that it had $-$OH, $-$COOH and $-$SO$_4^-$ surface groups. The surface charge density of $-$SO$_4^-$ groups was determined by conductometric titration to be $0.95\,\mu\text{C cm}^{-2}$. Using HCl as the added inert electrolyte would suppress the ionization of surface carboxyl groups, so that the charge on the particles would be due solely to the strong acid groups.

It appears from the data in Fig. 213 that once again the relative viscosities all approach the same high-shear limit. The asymptotic rise in η_r as the shear stress decreases indicates that at $[\text{HCl}] < 1.88 \times 10^{-2}\,\text{M}$ there exists a yield stress, probably due to crystallization of the system. Brilliant opalescence is often observed in these highly viscous or gelled colloids.

A most remarkable result is the independence of the salt effect on the valence of the counterions. Monovalent, divalent, trivalent and tetravalent cations all had the same effect on the relative viscosity at the same *equivalent* concentrations (1:1/2:1/3:1/4 *molar* ratios, respectively), as seen in Fig. 214.

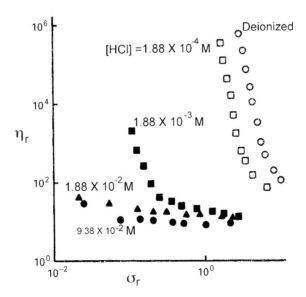

Fig. 213. Relative viscosity as a function of reduced shear stress at various HCl concentrations. Anionic PS latex. $\phi = 0.40$, $a = 110$ nm, $\sigma_0 = 0.95\,\mu\text{C cm}^{-2}$ [242].

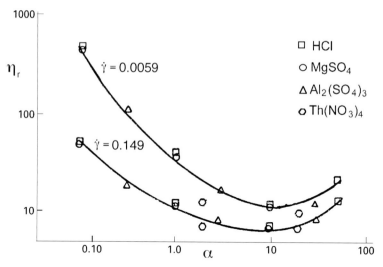

Fig. 214. Relative viscosity as a function of electrolyte equivalent ratio, α, in the presence of salts of various valences [242].

The variable α in Fig. 214 is the ratio of the number of charges on the counterions in solution to the number of ion charges on the particles. So it appears that the second electroviscous effect is regulated by the neutralization of charge rather than by compression of the diffuse part of the electrical double layer. And this dependence on ionic charge is of the same nature at widely different shear rates. The ordinate scales in both Figs 213 and 214 are logarithmic over several orders of magnitude, indicating how very large these electroviscous effects can be. More recently the theory of Russel and Benzing, which uses a self-consistent field model of a unit cell immersed in the surrounding material of average colloidal properties, has as a key parameter the ratio Q/N in which Q and N are the dimensionless surface charge and ion concentration, respectively [244].

A theory due to Smoluchowski suggests that another electroviscous effect (the 'primary') arises from the stripping away of ions in the diffuse part of the double layer under shear. There is as a result a conduction of ions from their equilibrium positions and an associated excess dissipation of energy proportional to the square of the conduction current, I^2, and inversely proportional to the ion conductivity, λ:

$$\frac{\eta_r - 1}{\phi} = 2.5 + \frac{2.5}{\lambda \eta_0 a^2} \left(\frac{\varepsilon I}{2\pi} \right)^2$$

This equation, which was given without derivation [245], is now believed to represent a relatively minor effect compared to that due to expansion of the double layer (the 'secondary') presented above. There is also a 'tertiary' electroviscous effect arising from the expansion of polyelectroytes, either bound to particles or in solution, upon changes in pH or ionic strength.

10.5 Effects of ordering on rheology

10.5.1 The shear modulus

Colloids which exhibit crystalline order due to long-range electrostatic interactions generally have an *elastic* response to small stresses. The crystal is distorted under the applied stress and relaxes back to its original condition upon release of the stress, i.e. it displays a storage modulus, G. Below a critical stress, the yield stress (Section 10.3), there appears to be no flow. In fact at high enough concentrations some latexes upon being inverted are capable of supporting their own weight for many months! The measurement of such tiny stresses and strains is described later in this chapter (Section 10.7.4). At higher stresses, of course, viscous flow will occur. Such systems are said to be viscoelastic.

> Crystalline colloids generally have an *elastic* response to small stresses.

10.5.1.1 Buscall, Goodwin, Hawkins and Ottewill (BGHO) theory

To understand more quantitatively the behavior of ordered polymer colloids, the approximative theory of Buscall, Goodwin, Hawkins and Ottewill (BGHO) [246] is given here. It involves taking a crystalline array of charged latex particles and determining the restoring force created when the particles are displaced a very small distance from their equilibrium positions. The geometry involved is shown in Fig. 215 in which one particle is at the origin

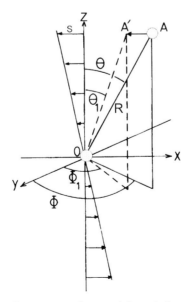

Fig. 215. Geometry of a particle pair in a stress field.

and the other at point A. A strain displacement S in the x–z plane leads to a movement of the upper particle from A to A'.

The theory assumes that the restoring force arises entirely from the pair interaction potential for spheres, and postulates that it is solely repulsive under the conditions of their experiments. The relationship used is similar to Eq. 106, employing assumptions of low surface potentials:

$$V_R \approx \frac{\varepsilon a^2 \psi_d^2}{R} e^{-\kappa H_0} \tag{188}$$

Then the total potential energy, E_{total}, of interaction for a given particle is the sum of these pair potentials over all n of its nearest neighbors:

$$E_{\text{total}} = \tfrac{1}{2} \sum_1^n (V_R)_i \tag{189}$$

The restoring force acting along the shear-path is determined from the component in the x-direction (Fig. 199) of the pair potential:

$$F_x = R \frac{\partial^2 V_R}{\partial R^2} \left(\frac{\mathrm{d}s}{\mathrm{d}z}\right) \sin^2 \theta \cos \theta \sin^2 \phi \tag{190}$$

where R is the center-to-center distance, $\mathrm{d}s/\mathrm{d}z$ is the strain in the x–z plane, and θ and ϕ are the two angles in spherical coordinates. The stress is the force per unit area (Eq. 170), which in this case is the area projected onto the x–y plane by the particle pair in question. Geometrical analysis and summing over all nearest neighbors gives:

$$\sigma = \frac{3np}{32} \frac{\partial^2 V_R}{\partial R^2} \left(\frac{\mathrm{d}s}{\mathrm{d}z}\right) \tag{191}$$

The shear modulus is defined as the shear stress divided by the strain gradient:

$$G_0^{\text{th}} = \frac{\sigma}{\mathrm{d}s/\mathrm{d}z} = \frac{\alpha}{R} \frac{\partial^2 V_R}{\partial R^2} \tag{192}$$

where $\alpha = 3np/32$, and G_0^{th} is the theoretical high frequency, low amplitude limiting modulus in an oscillatory experiment. For fcc and hexagonally close packed arrays $\alpha = 0.833$ while for bcc it is 0.510. BGHO have chosen the fcc geometry for their theoretical modeling, and have shown that bcc results are not as different as may be thought because of the cube root dependence of R on p (Eq. 159).

TABLE 10.2
Latex L70 [246]

Experimental	Assumed	Calculated from experiment
$a = 34.3\,\text{nm}$	$p = 0.74$	$\psi_d = 50\,\text{mV}$
$\kappa a = 2.48$	$\alpha = 0.833$	
$[\text{NaCl}] = 5 \times 10^{-4}\,\text{M}$		

Differentiation of Eq. 188 and substitution into Eq. 192, along with an approximation of $\kappa < 3$ or $\kappa > 10$ gives, respectively Eqs 193 and 194:

$$G_0^{\text{th}} = \alpha \varepsilon a^2 \psi_d^2 \left(\frac{\kappa^2 R^2 + 2\kappa R + 2}{R^4} \right) e^{-\kappa H_0} \quad \text{for} \quad \kappa < 3 \qquad (193)$$

$$G_0^{\text{th}} = \frac{\alpha \varepsilon a \psi_d^2}{R} \left(\frac{\kappa^2 e^{-\kappa H_0}}{(1 + e^{-\kappa H_0})^2} \right) \quad \text{for} \quad \kappa > 10 \qquad (194)$$

10.5.1.2 Comparison of experiments with theory

A plot of experimental results for a polystyrene latex designated L70 with the properties listed in Table 10.2 against those calculated by Eq. 193 is shown in Fig. 216.

The ionic strength of the colloid was fixed not by adding a calculated amount of salt, but by dialysis against an NaCl solution of the desired concentration. This cleverly avoids any assumptions about the relative

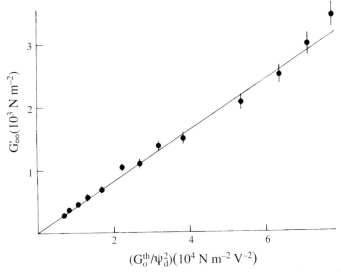

Fig. 216. Correlation plot of elastic modulus of aqueous PS crystalline latex at various volume fractions [246].

contributions of added ions and counterions to the ionic strength at all volume fractions. The correlation with this theory is excellent up to $\phi = 0.3$ where $G'_\infty \cong 3000\,\mathrm{N\,m^{-2}}$, beyond which many-body interactions are expected to become important.

10.5.1.3 Particle size effects

There is a large effect on the modulus due to particle size, as seen in Fig. 217, resulting from at least two factors: (1) the spacing between particles is greater for larger particles at a given volume fraction for a given array geometry, and (2) in this case, at least, the surface potential generally increased with larger size.

The latter effect is not surprising given the method of synthesis of these PS colloids, in which larger sizes are created by increasing the ionic strength of the emulsion polymerization medium to induce a greater degree of coagulation during particle formation (Section 2.2.2). This in turn would leave a greater charge density on the particles in order to overcome greater van der Waals attractions due to the larger size. And higher charge density leads to higher surface potentials according to the Gouy–Chapman theory (Section 7.2.2.1):

$$\sigma_d = \frac{\varepsilon k T}{v e} \sinh\left(\frac{v e \psi_d}{2kT}\right) \tag{195}$$

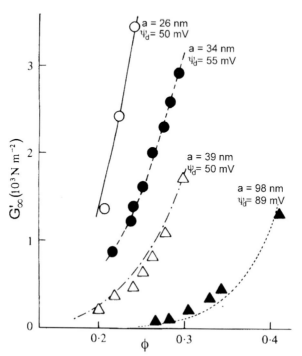

Fig. 217. High-frequency shear modulus as a function of volume fraction and particle size [246].

CHAPTER 10 RHEOLOGY OF POLYMER COLLOIDS

where σ_d is the charge density at the Stern layer surface and ψ_d is the corresponding potential.

10.5.2 The viscosity

The viscosity of these ordered, ionic polystyrene colloids was found to be highly dependent on volume fraction. This in turn was determined to be a result of the rapid increase in the pair potentials as particles are forced closer together with an increase in concentration (Fig. 218). The effect of an increase in volume fraction on viscosity is many orders of magnitude larger than that on the modulus, as shown in Fig. 217. The left-hand ordinate scale in this case is proportional to the relative viscosity divided by the interparticle distance, R (where C is a constant). This latter is proportional to the coefficient for viscous drag, f, for spheres in an array:

$$f \propto \frac{\eta_{lo}}{R\eta_0}$$

The Stokes coefficient for spheres at infinite dilution is given by Eq. 81 with the substitution of f_0 for ϕ to avoid confusion with volume fraction:

$$f_0 = 6\pi\eta a$$

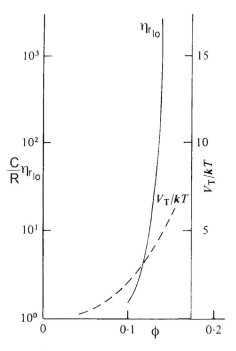

Fig. 218. Relative viscosity and pair potential vs volume fraction for latex L70 [246].

so that the ratio of the experimental coefficient to the Stokes value may be expressed as

$$\frac{f}{f_0} = \frac{C}{R}\frac{\eta_{lo}}{\eta_0} = \frac{C}{R}\eta_{r_{lo}} \qquad (196)$$

which is the ordinate in Fig. 218.

10.5.3 Creep compliance

A measurement of great practical importance is the creep compliance, J, in which a constant stress is applied to the system and the strain is measured as a function of time:

$$J(t) = \frac{\varepsilon(t)}{\sigma} \qquad (197)$$

which is dimensionally the reciprocal of the modulus. A viscoelastic colloid can be described quite precisely by a rather simple mechanical analog comprised of springs (elastic) and dashpots (viscous) which is shown in Fig. 219. The creep compliance of such a system is given by:

$$J(t) = \frac{1}{G'_\infty}\left[2 - \exp\left(-\frac{tG'_\infty}{\eta_{lo}}\right)\right] + \frac{1}{\eta_{lo}} \qquad (198)$$

BGHO investigated the creep of their latex L70 at three closely spaced volume fractions, i.e. $\phi = 0.138$, 0.141 and 0.177, and obtained three very different

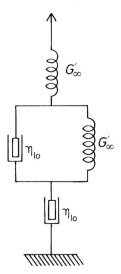

Fig. 219. Mechanical analog of a colloid.

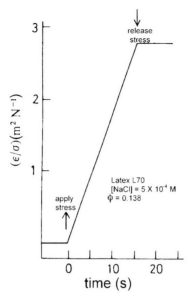

Fig. 220. Creep compliance for viscous anionic colloid [246].

types of behavior which are shown in Figs 220, 221 and 222. At the lowest concentration Newtonian viscosity (slope independent of time) was observed (Fig. 220), with no elastic component observable as long as the applied stress was less than $0.3\,\mathrm{N\,m^{-2}}$.

At the slightly higher concentration of $\phi = 0.141$ the behavior is much more complex, exhibiting both viscous and elastic components, as seen in Fig. 221. There is an initial elastic response (vertical rise in J) upon the application of the shear force, followed by a linear viscous flow. When the applied stress is

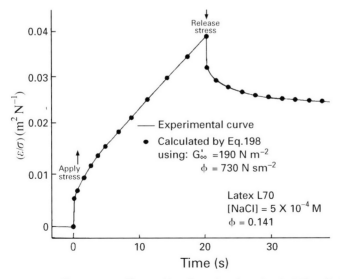

Fig. 221. Creep compliance for viscoelastic anionic PS colloid [246].

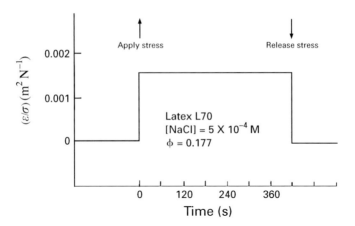

Fig. 222. Creep compliance of elastic anionic PS colloid [246].

released, the elastic part is rapidly recovered. At still higher concentration the creep response becomes purely elastic, at least at the very small applied stresses involved in these experiments. This is shown in Fig. 222 in which there is a time-independent strain as long as the stress is applied. Immediate and complete recovery of the original dimensions is obtained upon release of the stress. All of these experiments were carried out at extremely low Peclet numbers.

10.6 Rheology of nonspherical particles

10.6.1 Permanent aggregates and multilobed particles

Aggregates of spherical particles can form non-spherical macroparticles. In their tumbling in a shear field such particles will sweep out, or entrap, a spherical volume larger than their actual size, thereby behaving as particles with an effective volume and volume fraction larger than otherwise (Fig. 223). Lewis and Nielsen formed aggregates of glass beads by sintering them, and sorted them according to size by sieving. They then measured their viscosities in Aroclor at $Pe \cong \infty$ as a function of the number, N, of primary beads in an aggregate [241]. Their results are shown in Fig. 224 in which it is clear that at a given volume fraction a higher aggregate number gives higher viscosity.

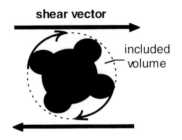

Fig. 223. Hydrodynamic volume of a multilobed particle.

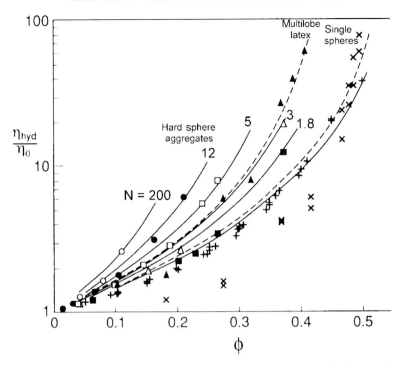

Fig. 224. Ultrahigh shear viscosities of permanent aggregates of glass spheres and a multilobed PS latex [241].

Latex particles can be multilobed either through controlled aggregation or by synthesis of multidomain structures (see Fig. 1 and Section 2.3). These also display the same characteristics of higher viscosities at a given volume fraction (also shown in Fig. 224), and in fact have been commercialized for latex paints. The curves drawn in Fig. 224 are fitted to the Krieger–Dougherty equation with appropriate values of p ranging from 0.28 for $N = (40 - 300)$ to $p = 0.52$ for $N = 1.8$.

10.6.2 Shear-dependent aggregates

Loose aggregates of particles can be formed through secondary minimum interactions or through bridging flocculation upon the addition of adsorbing polymer (Fig. 42). These aggregates, or 'floccules,' can be broken up to a greater or lesser extent in a shear field depending upon the magnitude of the shear stress. Often, because of the long-range structures built up by flocculation the systems will exhibit a modulus of elasticity as well. Typically high viscosity and modulus are observed at low shear stresses and very pronounced shear-thinning at high shear. A common example is gelled latex paint which does not flow until it is subjected to the high shear forces under a brush as it is spread onto a surface.

A very common steric stabilizer and thickening agent is so-called polyvinyl alcohol, which ordinarily is produced by the *partial* hydrolysis of polyvinyl acetate, PVAc. The hydrolysis is autocatalytic in the sense that when an $-\text{OH}$ group is formed it aids the hydrolysis of a neighboring acetate group, with the result that one obtains a blocky microstructure somewhat like that shown in Fig. 135. There will be hydrophilic blocks of polyvinyl alcohol, PVA, and blocks of unhydrolysed, hydrophobic PVAc. If the molecular weight of this material is high enough, and if the surface coverage is incomplete, then bridging flocculation is highly likely. The degree of bridging will depend upon the relative and absolute concentrations of latex and block copolymer and the manner of mixing.

As an example, the work of Tadros [247] on the rheology of a polystyrene colloid ($a = 92$ nm) with varying degrees of surface coverage by a commercial 'polyvinyl alcohol,' PVA-b-PVAc, is shown in Figs 225 and 226. The PVA-b-PVAc was Alcotex 88/10 from Revertex Ltd. and had a molecular weight $M_w = 45\,000$ and 12% acetate groups. Creep compliance measurements gave viscoelastic curves similar to that in Fig. 221, with values of J up to about $42\,\text{m}^2\,\text{N}^{-1}$ over a period of about 45 min. From the magnitudes and slopes of these curves the zero-shear viscosity, η_{lo}, and the shear modulus, G_∞, could be calculated. The yield stress, τ, was obtained from a series of experiments with varying shear stress. All of these are plotted as a function of surface coverage, Γ, in Fig. 225. It is clear from these curves that all three rheological characteristics decrease with increasing surface coverage by the polyvinyl alcohol. At low surface coverage there are unoccupied bare spots on the particle surface so that the block copolymer can attach itself to more than one particle and thus effect bridging. This leads to large floccules which trap much solvent medium and thus have high viscosity, modulus and yield stresses.

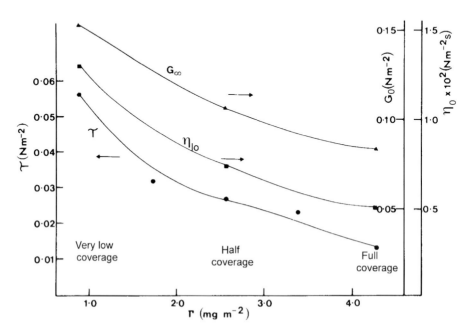

Fig. 225. Rheology of a PS colloid with varying surface coverage of PVA-b-PVAc [247].

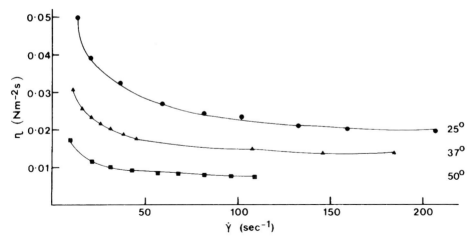

Fig. 226. Viscosity vs shear rate for PS colloid at full surface coverage by PVA-b-PVAc [246].

Even at full coverage the colloid is still viscoelastic (it has positive τ and G_∞ as well as η_{lo} in Fig. 225) so that it must be somewhat flocculated. It is highly shear-thinning at room temperature, as can be seen in Fig. 226. The viscosity approaches infinity at low shear rates, reflecting the yield value. At higher temperatures the floccules apparently break up to some extent because of their thermal agitation, so that this effect is not so great.

A simple theory which treats the floccules as equivalent hard spheres which sweep out a larger volume fraction than the particles themselves provides a semi-quantitative estimate of the various parameters involved. The yield stress, τ, must give a measure of the strength of internal pair potentials, E_s, in the flocs, where the latter is a function only of the bridging interaction, since the van der Waals forces are relatively insignificant. If we define the hydrodynamic volume fraction as

$$\phi_H \equiv \phi\left(1 + \frac{\delta}{a}\right)^3 \qquad (199)$$

where $(a + \delta)$ is the equivalent hydrodynamic radius of a particle, then according to Tadros, the pair potential is

$$E_s = \frac{\tau \pi^2 (a + \delta)^3}{3\phi_H^2} \qquad (200)$$

Some values of E_s as a function of electrolyte concentration calculated from the data in Fig. 225 and other experiments are given in Table 10.3.

Since the PS particles as synthesized are negatively charged by surface sulfate groups, the electrical double layer repulsions tend to offset the bridging tendency of the surface polyvinyl alcohol chains extending out into the

TABLE 10.3
Pair potentials for a flocculated PS/PVA-b-PVAc colloid as a function of sodium sulfate concentration [246]

[Na_2SO_4]	τ (N m^{-2})	E_s/kT	ϕ_H/ϕ
0.25	6.8	47.7	1.63
0.26	16.1	114	—
0.28	21.3	150	1.84
0.32	39.0	275	1.9

medium. Addition of divalent sulfate counterions has a large effect on the range of these electrostatic forces (cf. Section 7.2, Figs 125 and 130), allowing the bridging interaction to become stronger.

Tadros used a precursor of the Krieger–Dougherty equation, the Mooney equation, to obtain estimates of the floc volume fraction from his viscosity data:

$$\eta_r = \exp\left(\frac{2.5\phi_H}{1 - (\phi_H/p)}\right) \tag{201}$$

The Einstein value of 2.5 is used here instead of the intrinsic viscosity, presumably because obtaining the latter by a series of dilutions and extrapolation to zero concentration would give a meaningless result. It is the interparticle interactions which are of interest here, and they are lost upon dilution. Values of ϕ_H relative to the true volume fraction, ϕ, are also given in Table 10.3.

This 'elastic floc model' can provide further estimates of floccule and system properties, such as the number of interparticle bonds per floc, floc strength and the force required to break a floc doublet. But the approximations required are great, so that the exercise is hardly worth the effort.

10.7 Rheological measurements

As stated at the outset of this chapter, the discussion of rheology in this text is limited primarily to special features of polymer colloids. The reader is referred to general texts for broader treatments of rheology [248–250].

10.7.1 Steady flow methods

There are two generally used geometries to obtain a constant shear gradient throughout the sample being measured: the couette and the cone-and-plate. Both require one part to move under the applied shear stress, and the other to be stationary – the rotor and the stator. The latter is mounted on a torsion

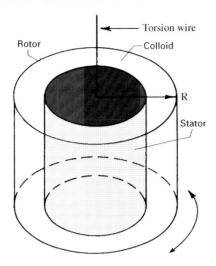

Fig. 227. Couette viscometer (schematic).

wire or some kind of strain gauge to measure the force. Schematic diagrams of these two devices are shown in Figs 227 and 228.

The cone-and-plate geometry automatically compensates for the difference in linear velocities at different radii by increasing the vertical distance between rotor and stator. This provides for a constant shear *gradient* throughout the sample. In both types the rotor is constantly moving. The rotational speed can be varied at will, ordinarily by means of a computer program, so that a full range of shear rates may be available.

10.7.2 Oscillatory methods

Oscillatory methods can employ the same geometry as the steady flow ones, but with the difference that the rotor is caused to reverse motion sinusoidally at the desired frequency. The amplitude of rotation is an additional variable.

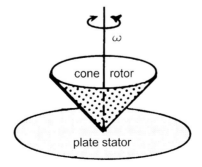

Fig. 228. Cone-and-plate viscometer (schematic).

10.7.3 Creep compliance

Again, in this technique the geometries can be the same as those above. The rotor is caused to move a fixed distance to provide a given stress, and then maintained at that position while the strain is measured as a function of time.

10.7.4 Pulse method

The pulse method is different from those above and deserves additional discussion because of its novel and unique design and capabilities. In it a small wave deformation is propagated through the sample from a wave generator, and the resulting disturbance is detected at a given, variable distance away. The apparatus devised and built by Buscall and coworkers is shown in Fig. 229. The driving pulse is imposed on the lower disk by its attached LiCl transducer at a frequency, ω, of approximately 200 Hz. The rotational amplitude of the disk was 10^{-4} radian or 6×10^{-3} degree, an almost imperceptible nudge! The propagated wave is detected by the upper

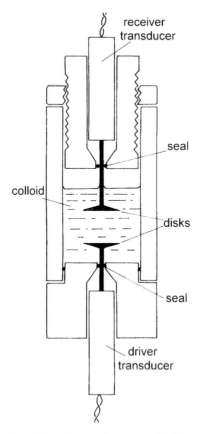

Fig. 229. Pulse rheometer [246].

disk/transducer and recorded on an oscilloscope. The velocity, v, of the wave through the colloid is determined, and this is related directly to the modulus and viscosity according to Eqs 202 and 203.

$$G' = \frac{v^2 \rho (1 - r^2)}{(1 + r^2)^2} \qquad (202)$$

$$\eta' = \frac{2\rho v^2}{\omega (1 + r^2)^2} \qquad (203)$$

where ρ is the density of the colloid.

The experimental results of Buscall, Goodwin, Hawkins and Ottewill, given above in Section 10.5.1, were largely obtained by means of this pulse rheometer.

Instrumentation with a wide range of capabilities is available commercially. Among the current major manufacturers are: Anton Paar GmbH, Graz, Austria; Bohlin Reologi AB, Sjobo, Sweden; ReoLogica Instruments AB, Lund, Sweden; and Rheometric Scientific, Piscataway, NJ, USA. All have home pages on the Internet.

10.8 Practical applications

The various kinds of rheological behavior all have names, most of which have been identified earlier in this chapter. They are presented schematically in Fig. 230. Extrapolation of the linear portion of the pseudoplastic curve to the σ-axis (not shown) gives an intercept often referred to as the yield point. Such systems in practical applications frequently behave as if they had a true yield stress, but careful measurements show that at very low stresses, they still flow. Shear-thinning behavior, in which loose structures are broken down upon application of stress, have time-dependent curves, i.e. they exhibit hysteresis, as shown in Fig. 231. As the shear stress is increased, more floccules are broken apart and the system behaves more like that of individual particles. However, upon subsequent decrease of the applied stress, floccules start to reform at a rate which may or may not be matched by the rate of change in the applied stress. This is what leads to the hysteresis.

> Shear-thinning: as the shear stress is increased, more floccules are broken apart.

The ratio of the yield value to the viscosity is called in industry the 'shortness factor,' SF. The viscosity is measured at the shear rate of the industrial application. For example, high speed printing can lead to the formation of fluid ink filaments in the printing roller nip, a phenomenon which shows a strong correlation with the SF. High SF correlates with shorter filaments,

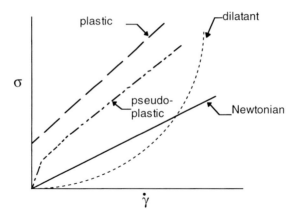

Fig. 230. Various rheological behaviors.

which in turn means less misting of the ink. However, ink surface roughness increases at higher shortness factors, because with high yield values there is less flow-out, or levelling, under the force of gravity immediately after application to the substrate. This leads to lower gloss. These considerations also apply to roller-applied latex paints.

There is an almost endless variety of products involving polymer colloids, either by themselves or mixed with pigments, fillers, polymers, surfactants, electrolytes and solvents. Environmental regulations are forcing industries to formulate water-borne systems to replace those which traditionally have used organic solvents. Besides printing inks, there are latex clear varnishes including floor finishes and photographic film, and pigmented paints and coatings of many kinds. Latexes are mixed with cement and lime to form

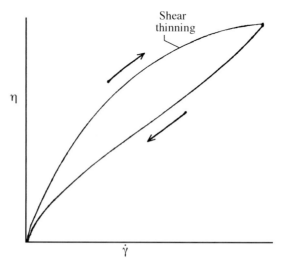

Fig. 231. Shear-thinning or thixotropic rheology.

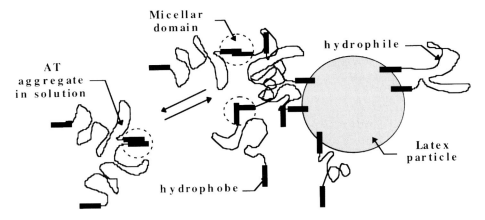

Fig. 232. Interaction of associative thickeners with a latex particle.

rubber-reinforced concrete and stucco. Adhesives and putties are now widely latex-based.

Organic solvents are sometimes added in small amounts to film-forming colloids to enhance particle coalescence by temporarily softening the particles (cf. Chapter 11). This can change the dielectric constant of the medium sufficiently to affect the pair potentials, and thereby the rheology. Ionic surfactants are often required to aid in pigment dispersion. To the extent that they also adsorb onto the polymer particles, they will affect interparticle interactions. The same can be said for the addition of salts. Many pigments and surfactants contain significant amounts of electrolytes as impurities. Sometimes salts are deliberately added to affect the rheology of a system.

Added polymers can have a profound effect on the rheological properties of a polymer colloid. Nonadsorbing polymers in small amounts simply act to thicken the medium, such that effectively η_0 is increased. In larger amounts they can cause 'depletion flocculation' by commandeering enough solvent so that the particles are forced into aggregation. This, of course, leads to large

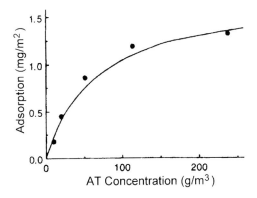

Fig. 233. Adsorption isotherm of AT onto acrylic latex. Curve is fitted using the Langmuir model [252].

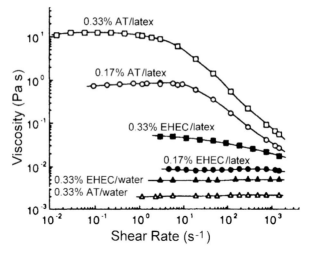

Fig. 234. Viscosity of solutions and acrylic latex with AT (empty symbols) and EHEC (filled symbols) [252].

increases in the viscosity and its dependence upon shear rate. Adsorbing polymeric additives, especially so-called 'associating polymers' or associative thickeners (ATs) usually bring about bridging flocculation even when added in small amounts. They are thus extremely powerful rheology modifiers [251]. ATs act because their hydrophobic moieties tend to aggregate in micellar clusters in solution as well as to adsorb onto latex particles, as shown schematically in Fig. 232. Adsorption of the hydrophobes appears to be Langmuirian, as shown in Fig. 233, according to the results of Huldén [252]. Viscosity is enhanced as a result of this clustering mechanism by as much as three orders of magnitude, as seen in Fig. 234. The micellar domains are held together with weak hydrophobic interactions, and these can be broken under shear, also observed in Fig. 234. In this set of experiments Huldén compares the viscosity/shear rate behavior of a pure AT in solution and in the presence of a commercial latex, Primal HG-44 against that of ethyl hydroxyethyl cellulose (EHEC). The AT contained C_{18} end groups and had a molecular weight of $31\,kg\,mol^{-1}$, while the EHEC had a molecular weight of $80\,kg\,mol^{-1}$ and is not associating.

The AT-containing systems clearly have much higher viscosity at low shear, but are much more shear-thinning than the EHEC systems. This is characteristic of many AT-thickened polymer colloids.

As an example, Table 10.4 shows the effect of a PEO/polyurethane hydrophobically modified nonionic associating polymer additive ('HEUR' – see below) on the rheology of an acrylic latex paint in comparison to the effect of a non-adsorbing polymer, hydroxyethyl cellulose, HEC. To achieve the same Stormer viscosity (approx. $100\,s^{-1}$ shear rate) less than one fifth of the associating polymer is required. At very high shear rates – the 'ICI' viscosity, approximately $10\,000\,s^{-1}$ shear rate (which obtain under a brush or roller) – the viscosity is much lower. And because there is negligible yield stress in the associating systems, the flow-out at essentially zero shear stress is excellent,

CHAPTER 10 RHEOLOGY OF POLYMER COLLOIDS

TABLE 10.4
Effect of added polymers on latex paint rheology [251]

Polymer	Mol. wt.	Polymer content (lb/gal)	Stormer viscosity (KU)	ICI viscosity (P)[a]	Leveling (10 is best)
HEC	850 000	3.34	81	0.43	2
HEUR	40 000	0.60	82	0.20	8

[a] 1 P (poise = 0.1 Pa s).

leading to high gloss. The relatively low molecular weight of the associating polymer to produce these high viscosities indicates that the effect is achieved by formation of an infinite network of loosely bridged particles in this colloidal system.

One can imagine that the addition of conventional surfactant molecules to an associatively thickened system would lead to a competition for adsorption sites between the two different hydrophobes among the micellar domains and the particle surface. If the adsorption energy of one type of hydrophobe is greater than that of its competitors, then it may displace the other. Many adsorption studies have shown that this is so, with great consequences for the

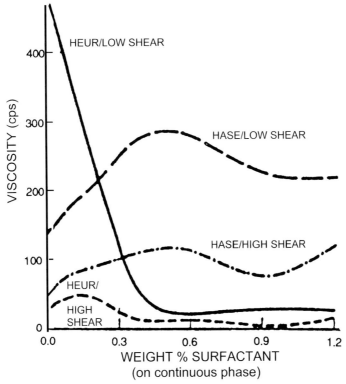

Fig. 235. Effect of added SDS on viscosity of AT/latex systems.

viscosity. Schaller and Sperry have reported on some of these, for example of two different kinds of ATs in competition with sodium dodecyl sulfate (SDS) [253] as shown in Fig. 235. The terms used are as follows:

- HEUR: *H*ydrophobically modified (poly) *E*thylene oxide/(poly)*UR*ethane.

- HASE: *H*ydrophobically modified *A*lkali-*S*oluble *E*mulsion – a latex with sufficient RCOOH content that it becomes a water-soluble AT upon increasing the pH with alkali.

It turns out that the HASE is apparently more strongly adsorbed to the latex particle surface, with the result that there is little change in viscosity upon addition of SDS, whereas there is a dramatic drop in low-shear viscosity in the HEUR/latex system, as seen in Fig. 235.

Rheology plays a crucial role in the performance of many, if not most, products: the brushing characteristics of latex paints, application speed of printing inks, the slumping behavior of putties, cements and stuccoes, and the way coatings in general flow out once they have been applied. So, although the subject is complex, its study is important to anyone who wishes to be successful in its practical applications.

The reader at this point can understand why Professor Irvin Krieger has said that, 'Rheology has much in common with theology. In both fields the most interesting phenomena are very difficult to reproduce, and defy rational explanation!'

References

235. Krieger, I.M. (1972). *Adv. Colloid Interface Sci.* **3**, 111.
236. Dougherty, T.J. (1959). Ph.D. Dissertation, Case Western Reserve University, Cleveland, OH; Krieger, I.M. (1967). In *Surfaces and Coatings Related to Paper and Wood* (R. Marchessault and C. Skaar eds), Syracuse University Press.
237. Choi, G.N. and Krieger, I.M. (1986). *J. Colloid Interface Sci.* **113**, 94.
238. de Kruif, C.G., van Iersel, E.M.F., Vrij, A. and Russel, W.B. (1985). *J. Chem. Phys.* **63**, 4717.
239. Russel, W.B., Saville, D.A. and Schowalter, W.R. (1995). *Colloidal Dispersions*, p. 468, Cambridge University Press, Cambridge.
240. Russel, W.B. and Sperry, P.R. (1994). *Prog. Organic Coatings* **23**, 305.
241. Lewis, T.B. and Nielsen, L.E. (1968). *Trans. Soc. Rheol.* **12**, 421.
242. Phung, T.K. (1993). Ph.D. Thesis, California Institute of Technology, Pasadena, CA.
243. Krieger, I.M. and Eguiluz, M. (1976). *Trans. Soc. Rheol.* **20**, 29.
244. Russel, W.B. and Benzing, D.W. (1981). *J. Colloid Interface Sci.* **83**, 163.
245. von Smoluchowski, M. (1916). *Kolloid Z.* **18**, 194.
246. Buscall, R., Goodwin, J.W., Hawkins, M.W. and Ottewill, R.H. (1982). *J. Chem. Soc., Faraday Trans. I* **78**, 2873–2889.
247. Tadros, Th.F. (1984). In *Polymer Adsorption and Dispersion Stability* (E.D. Goddard and B. Vincent eds), ACS Symposium Series 240, p. 411, American Chemical Society, Washington, DC.
248. Russel, W.B., Saville, D.A. and Schowalter, W.R. (1995). *Colloidal Dispersions*, Cambridge University Press, Cambridge (reprinted).
249. Ferry, J.D. (1980). *Viscoelastic Properties of Polymers*, John Wiley, New York.

250. Macosko, C.W. (1994). *Rheology Principles, Measurements and Applications*, VCH Publishers, New York.
251. Sperry, P.R., Thibeault, J.C. and Kostansek, E.C. (1987). *Adv. Org. Coatings Sci. Tech.* **9**, 1.
252. Huldén, M. (1994). *Colloids and Surfaces A: Physicochem. Eng. Aspects* **88**, 207.
253. Schaller, E.J. and Sperry, P.R. (1992). In *Handbook of Coatings Additives* (L.J. Calbo ed), Vol. 2, p. 105, Marcel Dekker, New York.

Chapter 11

Film Formation

11.1 Introduction

The formation of films from latexes is the most important practical application of polymer colloids because of its relevance to the paint and coatings industries, which include graphic arts (primarily printing inks), adhesives, caulks, varnishes, floor finishes, etc. Making a film is simplicity itself: one spreads liquid latex onto the substrate, allows the medium to evaporate, and voilà! a clear, continuous film remains. By mixing pigment into the colloid prior to application, opaque, white or colored films are formed. Other adjuvants are often added to affect wetting of the substrate, corrosion protection of steel, coalescence of the particles and cost.

11.1.1 The basic processes of film formation

The process of film formation can be broken down into eight steps, some of which may not always occur. As always, water is the chosen medium here, but it is understood that an organic medium also could be used.

(1) A wet film is spread upon a substrate.
(2) Water initially evaporates at the constant rate at which pure water evaporates. There may be some cooling of the film because of the latent heat of vaporization (Stage I, Fig. 236). This usually continues until the volume fraction of solids reaches 0.60 to 0.75.
(3) Often the particles form a regular array, fcc or bcc, depending upon the ionic strength, particle size uniformity, viscosity, etc. (see Chapter 9).
(4) The evaporation rate slows down as particles are crowded and occupy surface sites (Stage II, Fig. 236).
(5) A continuous film is first formed usually at the surface (it 'skins over'), so that further evaporation occurs by diffusion of water through this skin, first through interstitial channels and finally through the continuous polymer film (Stage III, Fig. 236).
(6) Particles after coming into contact start to deform into polyhedral structures.
(7) Water and nonvolatile, water-soluble components are squeezed into interstitial or interfacial regions.
(8) Polymer segmental interdiffusion may occur across the former particle boundaries.

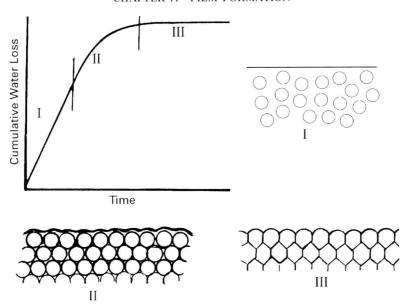

Fig. 236. Three stages of water loss from a latex film [254].

11.2 Experimental results

11.2.1 The minimum film forming temperature (MFT)

The MFT is perhaps the most common experimental technique used to characterize the film-forming properties of polymer colloids. The apparatus is simple: it is composed of a metal bar heated at one end and cooled at the other to form a linear temperature gradient. When a liquid latex film is spread upon the bar and the water evaporates, there will be a narrow zone below which the film is white and 'mud-cracked' and above which the film is clear and coalesced. The transition temperature is the MFT which is close to, but usually higher than, the glass transition temperature, T_g, of the polymer.

Most researchers have found a small but statistically significant dependence of the MFT on particle size. A typical example is shown in Fig. 237 in which terpolymers of MMA/BA/MAA all had the same composition, but varied in particle size and polydispersity. Each experimental point represents the average of five measurements. The entire ordinate scale covers only 5°C. It is interesting to note that in the polydisperse colloids it was the number-average particle size which correlated well with the MFT for the monodisperse sample, whereas the weight-average size did not [255].

It is clear from many studies that water plays an important role in film formation. When one reaches Stage II (in Fig. 236) the surface is curved between particles such that the surface tension of the water tends to force particles together. The ambient conditions of drying may thus play a role in the efficiency of particle compaction during the latter stages of film formation. Sperry and coworkers have shown not only that this is so, at least for fairly

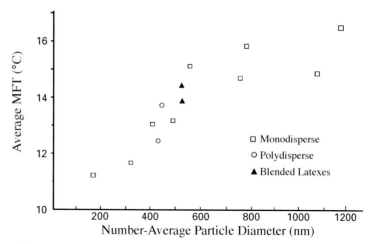

Fig. 237. Minimum film temperature as a function of latex particle size for PMMA/BA/MAA latexes [255].

hydrophilic polymers, but that the MFT is also time-dependent [256]. They combined two methods of casting the films with two environmental conditions to have four different test protocols as given in Table 11.1

Thus they had tests under DD, DH, WD, and WH conditions. They also employed the ASTM (American Society for Testing Materials) test, which would be something like a 'WW' condition because the humidity was almost saturated during the drying stage which therefore lasted about twice as long as their 'H' condition.

Examples of their experimental results are shown in Figs 238 and 239. For a hydrophobic polymer such as PEA/S/AA (47/52/1 wt%) there was essentially no difference among all the four test conditions, and the MFT gradually became lower over long periods of time, being linear with the logarithm of time (Fig. 238). The latter behavior is predicted by the Williams–Landel–Ferry (WLF) equation which states that there is a time/temperature superposition of the glass transition temperature, T_g. In Fig. 238 T_g is shown as that taken under standard conditions with a differential scanning calorimeter, i.e. relatively rapidly over just a few minutes. The drying curves extapolated to zero time correspond closely to T_g. The ASTM result, however, suggests that under extreme conditions of relative humidity the MFT will be somewhat lower.

That water can play a significant role is borne out by Sperry's results on hydrophilic polymers, as seen in Fig. 239. It is clear in this case that when the

TABLE 11.1
Test protocols

Casting of film		Test environment	
D	Dried below MFT; then bar heated	D	Rapid drying in dry air flow
W	Cast wet onto preheated bar	H	Open to room: 65% RH, 25°C

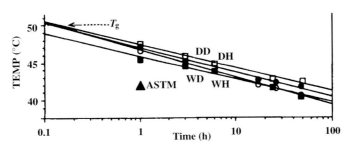

Fig. 238. MFT vs time for EA/S hydrophobic latex under various conditions of casting and drying [256].

latex is cast onto a preheated bar the MFT is more than 10 degrees lower regardless of the ambient drying conditions, WD and WH. Extrapolation to zero time suggests that T_g has been lowered because of plasticization by water, a well-known phenomenon. At longer times the water diffuses out of the film, so that it approaches the MFT of the dry latex in the DD and DH curves.

The combination of all of these results suggests that water and the viscoelastic properties of the polymer both play a role in film formation, and that in at least some circumstances the two are inseparable. Further evidence comes from a detailed examination of the morphology of films by electron microscopy.

11.3 Electron microscopy

The studies described below were done on ideal systems in which the particles were monodisperse; in some cases cleaned; and in all cases they formed ordered arrays. A hint that surface tension forces must be important came early when electron micrographs of polystyrene particles dried at room temperature appeared to be fused together like those shown in Fig. 240 [257]. The

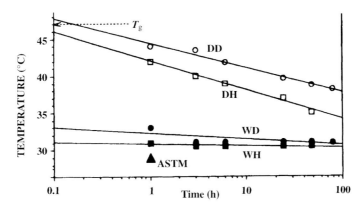

Fig. 239. MFT vs time for a PEA/MMA hydrophilic colloid under various conditions of casting and drying [256].

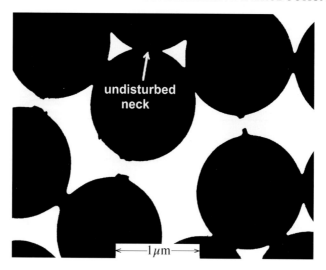

Fig. 240. TEM of 1.3 μm polystyrene particles dried in air at room temperature [257].

T_g of PS is about 104°C, so that these particles have undergone viscous flow at a temperature some 80°C below T_g! The effect is most readily seen between particles that have been pulled apart during the TEM preparation. Careful measurements indicated that the 'necks' between particles were not due to interfacial, water-soluble polymer, but the polystyrene itself. In Fig. 240 it can be seen that where the particles have not been disturbed there is fusion, with an extremely small radius of surface curvature where the particles are joined. It is the interfacial tension forces in this region of high curvature that Vanderhoff and colleagues argue are the source of the large force required to deform the particles. More on this in the theoretical section below. If this is so, then the driving force for coalescence should be a function of the degree of curvature, i.e. the particle radius.

The undisturbed neck in Fig. 240 suggests that either the radius of the circle of contact (a in Fig. 241) or the half angle of contact (θ in Fig. 241) may be used as a measure of the degree of coalescence of particles. Eckersley and Rudin took for their model studies of this problem a series of polymer colloids of different particle sizes but the same composition (PMMA/BA/MAA = 49.5/49.5/1.0 wt%) (see also Fig. 237). The samples were spread and dried under identical conditions, and then their surface structures were examined by scanning electron microscopy (SEM). Their results, in Table 11.2, show a dependence of the radius of contact on particle size, to be expected, since a should increase with r for a given degree of coalescence. But there is also a dependence of the half angle of contact upon particle radius.

Some caution must be used in interpreting these data, however, because the colloids were not purified, so that the amount of interstitial surfactant (commercial sodium dodecyl benzene sulfonate) would be a function of the particle size.

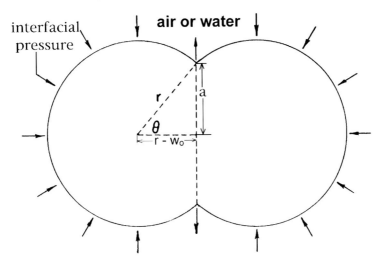

Fig. 241. Geometry of extent of coalescence of two colloidal particles.

The extent of coalescence as a function of time was also investigated by Vanderhoff, who used SEM to show that the degree of crosslinking of the polymer within the particles had a huge effect on the rate, presumably because the viscosity of the polymer was greatly affected by this [258].

The internal structure of films derived from model polymer colloids has been beautifully revealed by Sosnowski, Winnik and coworkers, who took air-dried, uncoalesced films of purified, monodisperse PMMA and PS latexes and gently compressed them for 5 min in a press a few degrees above T_g [259]. The resulting films were clear, but brittle because little diffusion of polymer could occur under these conditions. Film clarity indicated that all interstitial void volume due to the evaporation of water was eliminated, so that the particles must have become deformed into polyhedra such as those drawn in Fig. 242. Fcc (closest) packing would lead to the formation of dodecahedra (12 sided), and bcc packing would give tetrakaidecahedra (14 sided).

When such films were subjected to freeze fracture by plunging them into a very cold fluid such as liquid ethane and then breaking them apart, the

TABLE 11.2
Radii and angles of contact between particles in PMMA/BA/MAA films dried overnight and determined by SEM [255]

Radius, r (nm)	Contact radius, a (nm)	θ (°)
286	210	47.4
394	290	47.4
408	320	51.7
563	380	42.5
617	395	39.8

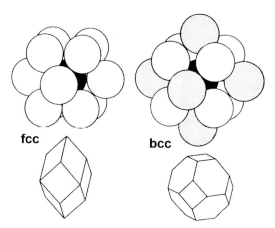

Fig. 242. Polyhedral structures derived from fcc and bcc packing of spherical particles.

internal structure was revealed. Evaporation of gold onto the surface – a standard way to eliminate charge build-up – and subsequent scanning electron microscopy gave the stunning results, samples of which are shown in Figs 243 and 244. In the former a polystyrene colloid had formed an fcc array which was then compressed. The fracture occurred along a (1 1 1) plane. Interestingly, there is a region of disorder in which a cell is missing, known in crystallography as a Schottky defect, indicated on the micrograph.

Figure 244 is an FFTEM micrograph of a PMMA dried colloid which has been lightly compressed. These particles assumed a bcc structure, a more open configuration than that of an fcc array. Minor imperfections can be observed, most of which are artifacts introduced during preparation of the sample. It must be remembered that these examples were made by the application of an external force, and may not represent structures which are formed spontaneously. One question which remains, for example, is what happens to the water-soluble components of polymer colloids in film formation when they are present, as they always are in commercial products? Do they get trapped in microdomains within the film? Do they remain at the interface between particles, or do they migrate to the interfaces, air/film and substrate/film? The answer, from many investigations on practical systems, is probably 'all of the above,' depending upon the details of composition and conditions under which the films are formed.

> Surfactant migrates to both film interfaces indicating that serious problems may arise.

Many commercial latexes, for example, are made by copolymerization of acrylic acid (AA) or methacrylic acid (MAA) neutralized to pH of about 7.0 to enhance stability. The result is a hydrophilic stabilizer surface layer which is immobile. Polyvinyl acetate rather easily hydrolyzes during and after emulsion

Fig. 243. Polyhedral structure formed from monodisperse PS Colloid. Diameter = 1250 nm. FFTEM. Micrograph 40 500× [259].

polymerization, forming surface polyvinyl alcohol which acts in the same manner. Staining studies have shown that this material retains its structure in the final film as a fine network occupying the interfaces among the polyhedra [258, 260]. Other studies have shown that surfactant migrates to both film interfaces, indicating that serious problems may arise with respect to adhesion, especially wet adhesion, to the substrate, as well as adhesion of a second coat of paint, for instance [261].

11.4 Polymer diffusion between particles

Once particles have come into intimate contact, it is reasonable to expect that, because of their thermal motion, polymer chain segments will reptate across

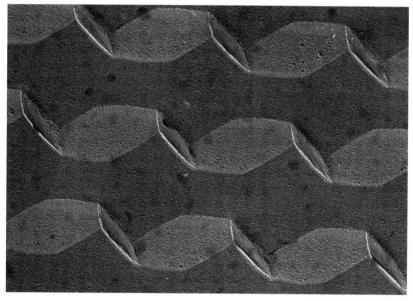

Fig. 244. FFTEM micrograph of a film derived from a PMMA colloid. Diameter = 810 nm. 40 500× [259].

the interparticle interface in both directions. This may be especially so for small particles where it has been shown that the average chain dimensions within the particle are constrained to be less than those for a free chain in the bulk. The problem is analogous to that of the healing of microcracks in damaged polymers, which has been studied extensively.

Small angle neutron scattering (SANS) is an excellent tool for this purpose (see Section 5.2) because it has the needed resolution for objects of molecular

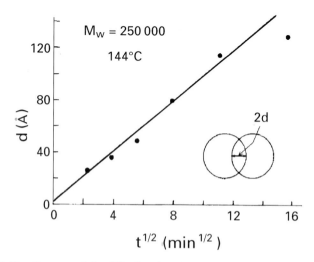

Fig. 245. Interparticle diffusion in PS film determined by SANS [262].

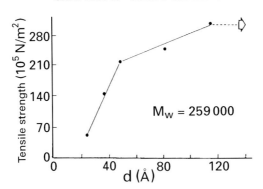

Fig. 246. Tensile strength as a function of interdiffusion depth, d in PS film [262].

size. Yoo, Sperling, Glinka and Klein took a mixture of two PS colloids, one made from normal, protonic styrene, and the other, from deuterated styrene [262]. This was dried to a film and subsequently pressed at 110°C to form a coalesced film in which almost no segmental diffusion had occurred. This film was then heated at 144°C for varying times and investigated by SANS. The average interpenetration depth, d, could be calculated and was plotted as a function of the square root of time as shown in Fig. 245. This time-dependence can be rationalized on the basis of the de Gennes model for segmental reptation employed by Kline and Wool [263].

One would expect that the strength of a film will be enhanced by such 'stitching' together, and the results clearly indicate that this is the case: when the film tensile strength is measured, there is an eight-fold improvement, as shown in Fig. 246. There appear to be three regions, two with positive slopes and a final one with zero slope, although the data – which are difficult to obtain, since a nuclear reactor is required for starters – are rather skimpy. The first transition occurs at a penetration distance of 45–50 Å, which is close to the radius of gyration of the critical molecular weight for polystyrene entanglements. Full tensile strength is reached at the second transition when the penetration is c. 110–120 Å, corresponding to the theoretical weight-average radius of gyration for a perfectly random PS chain of this molecular weight. Thereafter the tensile strength is fully developed and further diffusion cannot improve it.

Although interdiffusion of chain segments does not imply any net flow, there is considerable evidence that in many systems with high polymer viscosity some flow continues for days, weeks and even months. For example, a polystyrene/butadiene (67/33) latex dried in air for 33 days showed a gradual change from a pebbly surface after 1 day to one with hardly any discernible features at 33 days, as seen in Fig. 247. Incidentally, the same film dried for the same time in a nitrogen atmosphere showed no such morphological surface change [261]! The rationale for this remarkable result is that a butadiene-containing copolymer is crosslinked because of its residual unsaturation from pendant vinyl double bonds. These crosslinks inhibit viscous flow, preventing coalescence to any significant degree. However, in the

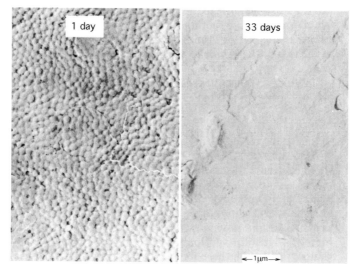

Fig. 247. Surface morphology of a PS/B film as a function of time in air [261].

presence of oxygen the crosslinks are rather easily oxidized and subsequently broken.

11.5 Theories of film formation

11.5.1 The role of water

> Water plays an important role in film formation.

The experimental evidence is strong that water plays an important role in film formation in most cases, and in every case – even with the most hydrophobic polymers – if a continuous film without 'mud-cracks' is desired. It is the water/air surface tension which acts to bring all the particles into close proximity and to hold them there as the process of evaporation and coalescence goes forward. This can be envisioned by considering just two particles in a droplet of water as the water evaporates. When the only water left is the small amount between the particles, and when it has a concave surface, the water/air surface tension will act in the direction of the heavy arrows at the aqueous surface in Fig. 248 to force the particles together.

Vanderhoff and coworkers have argued that the first step in contact is the touching of the Stern layers and that a certain minimum force is required to 'rupture' this layer. But consideration of the dynamic aspects of the electrical double layer and the variation in its dimensions with electrolyte concentration (Section 7.2.2) would suggest that it is not such a discrete entity and thus not subject to a catastrophic event such as rupture. On the other hand, if the surface contains strongly adsorbed or bound polymeric steric stabilizer, then rupture may not occur at all. Evidence for this exists.

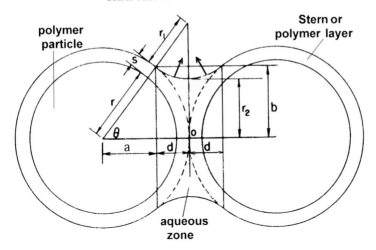

Fig. 248. Schematic of two particles with an interstitial aqueous zone [257].

11.5.2 The role of the polymer interface

This water/air interfacial tension can act until the film 'skins over,' after which there is only polymer in contact with air at the surface, while the remaining water is in contact solely with the internal polymer. Another look at Fig. 240 indicates that the juncture between the partially coalesced particles has an extremely small radius of curvature, designated as r_1 in Fig. 249. If we let P_{ext} be the external pressure, then the differential pressure acting to force the particles together is a function of the three radii of curvature, r_1, a and r_0, the last now representing the original particle radius. Then, according to Young and LaPlace, the coalescing pressure is given by Eq. 204:

$$P \equiv P_2 - P_1 = \gamma\left(\frac{1}{r_1} - \frac{1}{a} + \frac{2}{r_0}\right) \qquad (204)$$

where γ is the either the air/water (Fig. 248) or the polymer/water (Fig. 249) interfacial tension, and $P_2 - P_{\text{ext}} = 2\gamma/r_0$. This difference in pressures between the two regions within a particle acts until coalescence is complete, or when the force is insufficient to overcome the viscoelastic resistance, G^*, of the polymer. For the situation in Fig. 249, since $r_1 \ll a, r_0$, at least in the early stages of coalescence i.e. Eq. 204 can be approximated by

$$P \cong \frac{\gamma_{\text{air/water}}}{r_1} \cong \frac{\gamma_{\text{polymer/water}}}{r_1} \qquad (205)$$

As can be seen from Fig. 240 the value of r_1 must be extremely small, so that even though the interfacial tension is small – probably on the order of 10–30 mN m^{-1} for either interface – the driving pressure can be several

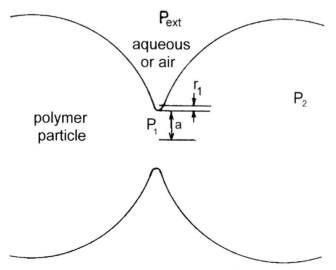

Fig. 249. Schematic of two particles after some coalescence in contact with air or water [257].

hundred atmospheres [257]. For example, if $\gamma = 10\,\mathrm{mN\,m^{-1}}$, $r_0 = 50\,\mathrm{nm}$ and $r_1 = 25\,\mathrm{\AA}$, then $P = 580\,\mathrm{psi}$ ($\approx 4.0\,\mathrm{MPa}$). This is quite enough to cause 'cold flow' of normal plastics. The shapes of the curves derived from Eq. 204 are shown in Fig. 250, in which the pressure, P, is plotted against the polymer volume fraction in a water droplet as a function of the original particle radius, r_0, using the geometry in Fig. 248. This behavior – essentially a step-function in the pressure – is usually what is observed experimentally, i.e. that when the water is gone, coalescence is complete, at least to the naked eye.

A problem arises, however, when one considers the tensile or adhesive strength of the water. Water cavitates, i.e. it forms microbubbles of vapor under high degrees of tension, such as during ultrasonic irradiation. These voids are undoubtedly nucleated by ions, cosmic radiation, dust particles, and surface or other impurities. So far no one has measured the tensile strengths of the aqueous phases in polymer colloids. The highest value found experimentally for pure water under ideal conditions is 277 atm, or about 3800 psi ($\sim 26\,\mathrm{MPa}$).

Once the particles have been brought into contact and forced to deform and flow, the geometry in Fig. 249 obtains, such that the polymer/water interfacial tension can operate. Calculations again show the possibility for large forces, such as those shown in Fig. 251. In this case the value of r_1 is kept constant arbitrarily, as it *appears* in many electron micrographs, although this is an assumption because of the impossibility of resolving and measuring such a small radius.

Because both pressures in Eq. 205 are about the same magnitude, they evidently operate simultaneously as long as there is water present. The model is made more realistic by including more than one nearest neighbor, keeping the total polymer volume constant, and by imposing the restriction that film contraction as water evaporates can only occur in the direction normal to the

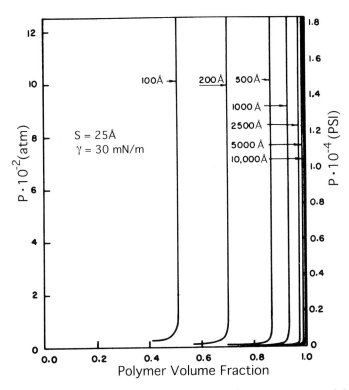

Fig. 250. Pressure due to water/air surface tension between two particles in a water droplet vs. particle size [257].

plane of the film. Further, it must be remembered that once the film skins over, there is only the polymer/aqueous interface that is present within the film until all the interstitial water has left (there still may be water dissolved by the polymer acting to plasticize it). At that point all the polyhedra are fully formed. If not, there must exist voids filled with water vapor which may 'heal' with time, and these of course will depend upon the polymer/air (or water vapor) interfacial tension for the compressive force.

11.5.3 The role of the viscoelastic properties of the polymer

Given that the forces acting on the particles are sufficient to bring about coalescence, the *rate* at which this occurs depends on the viscosity of the polymer. There is no complete theory for the rate of film formation because of the complexities involved, but much can be learned about the magnitudes of the factors influencing the process and which parameters are important.

Most theories start with the equation of Frenkel, which provides the dependence of the radius of the circle of contact, a, on the viscosity of the polymer and the time [264]. It is based upon the geometry of two particles already in contact shown in Fig. 241, and relies on the radius of curvature of the

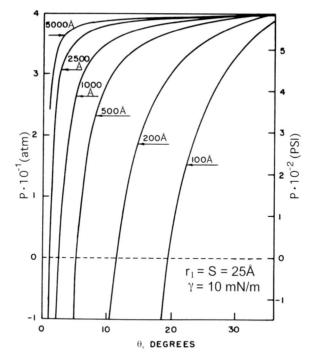

Fig. 251. Pressure between two particles driven by the polymer/water interfacial tension, γ [257].

particles themselves, r, and the polymer/medium (aqueous or air) interfacial tension, γ:

$$a = \sqrt{\frac{3\gamma rt}{2\pi\eta}} \quad (206)$$

Eckersley and Rudin (E & A) employed this equation to apply to their experimental SEM results (Table 11.2), using the polymer/aqueous value for γ calculated for *pure* PMMA (27 mN m^{-1}). This value is no doubt unrealistically high because of adsorbed surfactant which was present in their system.

E & A used Eq. 206 along with one analogous to that of Vanderhoff and colleagues which recognizes the role of the high curvature in the interstices among particles. They employed the equivalent radius of curvature, r', of the aqueous medium in the 'throat' among three particles in a hexagonal array as shown in Fig. 252, instead of the r_1 of Fig. 249. They also assumed that the surface tension against air, $\gamma_{w/a}$ of the aqueous phase was 30 mN m^{-1} because it would contain considerable concentrations of surface-active solutes. The Young–LaPlace pressure for this capillary column would then be

$$P_c = \frac{2\gamma_{w/a}}{r'} \quad (207)$$

CHAPTER 11 FILM FORMATION

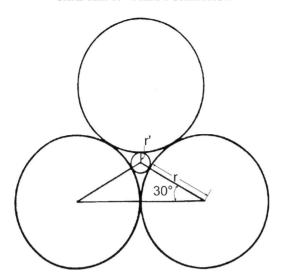

Fig. 252. Geometry of equivalent interstitial capillary column.

By trigonometry the value of r' can be related to the particle radius $(r/(r + r') = \cos 30°)$ to give Eq. 207 in terms of the particle radius:

$$P_c = \frac{12.9 \gamma_{w/a}}{r} \tag{208}$$

E & A had to employ some assumptions concerning the area, A, over which this pressure is applied in order to obtain the capillary force, F_c ($F = P \times A$). A was taken as a regular hexagon with the edge length of $[2(r - w_0)/3]^{1/2}$. This gave Eq 209:

$$F_c = \sqrt{2}(r - w_0)^2 P_c \tag{209}$$

where $r - w_0$ is the distance between the particle center and the circle of contact (Fig. 241).

The radius of contact, a, can then be calculated as a function of time and the polymer compliance, J_c (related to the viscosity – see Section 10.5.3):

$$a = \left(\frac{3 J_c(t) F_c}{16} \right)^{1/3} \tag{210}$$

Substitution of Eqs 208 and 209 into Eq. 210 gives

$$a = [3.42 J_c(t) \gamma_{w/a} (r - w_0)^2]^{1/3} \tag{211}$$

There is a direct geometrical relationship between $(r - w_0)$ and the extent of coalescence for a given particle size. This, along with an assumption of fcc

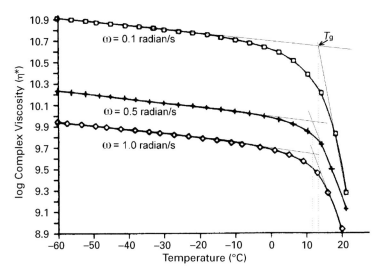

Fig. 253. Dependence of viscosity of bulk PMMA/BA/MAA vs. temperature and shear rate [255].

closest packing leads to the time-dependence of a in terms of experimentally determined parameters:

$$a = [2.80 J_c(t') \gamma_{w/a} r^2]^{1/3} \quad (212)$$

where t' is the time at which all water has been expelled, and is taken as the limit for this process.

The compliance, $J(t)$, was obtained from their data for the bulk viscosity of their PMMA/BA/MAA polymer. The latter results are shown in Fig. 253 in which the viscosity was measured in a mechanical spectrometer as a function of applied frequency and temperature. The maximum strain was 0.09%. Just as a matter of interest the intersection of the extrapolated linear slopes below and above T_g has been taken as a measure of T_g. The dependence of T_g on the rate of deformation can be seen. Assuming that under these conditions they had purely viscous flow, the compliance is inversely related to the viscosity (Eq. 198).

Thus from the data in Fig. 253 the value of J could be found for any given temperature at one of the three rates of oscillation in the dynamic spectrometer. Eckersley and Rudin used the sum of Eqs 206 and 212 to calculate theoretical values for the radius of contact under two limiting conditions: a low estimate, based on a 1 h drying time, t, and a high estimate based on a 2 h drying time. These are plotted, along with the experimental data for the contact radius given in Table 11.2, and are shown in Fig. 254.

The experimental points all fall within the limits of the two estimates, with a definite trend towards the upper one. Although this would appear to be a triumph of the theory, so many assumptions had to be made along the way

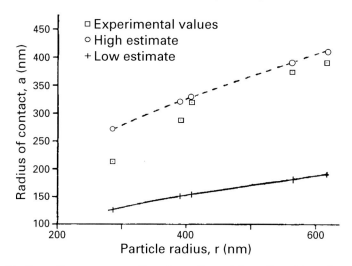

Fig. 254. Radius of contact as a function of particle radius. Experimental data and two theoretical models [255].

that overconfidence is not warranted. It does seem clear, however, that the basic tenets of the theory are correct.

11.5.4 Application to practical systems

There are several considerations which make the theory outlined above difficult to apply to commercially practical systems:

(1) Most commercial polymer colloids are polydisperse and therefore cannot form regular fcc or bcc arrays. This may actually be an advantage in that the smaller particles can fill the interstitial voids among the larger ones. In fact some people have deliberately synthesized bi- and tri-modal particle size distributions calculated so that the smaller sizes exactly fit the voids formed by the next larger sizes.

(2) Most practical systems contain surface-active species such as surfactants, amphiphilic polymers added to control rheology, and water-soluble oligomers with ionic end groups, comprising the lowest molecular weight 'tail' of the molecular weight distribution. These tend to concentrate in interstitial voids and at the upper and lower interfaces of the film. Wet strength and wet adhesion are thereby adversely affected.

(3) The theory tacitly assumes that the stability of the colloid is maintained throughout the process of film formation, i.e. no aggregation occurs. If flocs or hard aggregates are formed as the film concentrates during evaporation, there will be many voids left in the film because of the fractal nature of these structures. In most practical systems this is unlikely because of the relatively high levels of stabilizers employed. However, in high surface-tension formulations, where surfactants are largely absent from the medium, this could be a problem.

(4) The presence of pigment particles will disrupt the ordering of the polymer particles, and they of course cannot coalesce.

> If aggregates are formed as the film concentrates, there will be voids left in the film because of their fractal nature.

Other practical considerations include wetting, adhesion to the substrate, rheology during application, durability and appearance.

11.5.4.1 Wetting

When a latex is applied to a surface it is expected to spread evenly and not break up into small globules. This will happen for most latex systems when they are placed on substrates of sufficiently high surface free energy, such as wood, paper, oxidized old paint and metal. However, on low surface-energy-substrates such as polyolefin films or plastics, wetting and spreading is poor. High concentrations of surfactants can be added, but these will cause other problems mentioned above.

There are several ways around this difficulty. One is to use a volatile surfactant or co-surfactant such as an alcohol. The latter can act in conjunction with an ionic surfactant to bring the surface tension practically to zero. Another commonly used way is to pretreat the surface to increase its free energy, such as by means of a corona discharge which activates the molecules in the air or other gas environment. The active ions and free radicals formed then attack the surface, say of a polyolefin film, to form $-COOH$, $-OH$, $-CHO$, $-NH_2$, etc. groups. Like all good thermodynamic systems, of course, these groups will tend to diffuse into the bulk of the material, thereby lowering the surface free energy. So one has only a limited time after corona treatment to coat the substrate. Obviously, glassy substrates in which diffusion is extremely slow will remain activated for longer times.

The spreading coefficient, $S_{A/B}$, is a measure of the ability of a liquid, A to spread over the surface of another liquid or a solid, B (see Section 2.3.1.1):

$$S_{A/B} \equiv \gamma_A - \gamma_B - \gamma_{A/B} \qquad (213)$$

where γ_A, γ_B and $\gamma_{A/B}$ are the corresponding interfacial tensions. When $S_{A/B}$ is positive the liquid will spread across the substrate, increasing its own surface area in the process. Thus if B is the latex composition and A is the substrate, then the formulator wants to maximize γ_A and minimize γ_B and $\gamma_{A/B}$ to achieve spreading. Good spreading does not ensure good adhesion to the substrate, however.

11.5.4.2 Adhesion

Adhesion is really a problem of wetting of the substrate by the *polymer* (as opposed to the latex). If there is good contact between coating and substrate

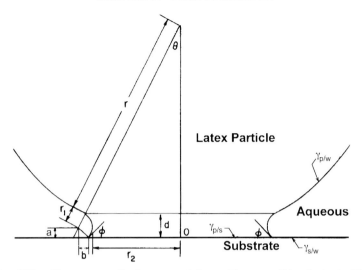

Fig. 255. Geometry of a latex particle wetting a solid substrate [257].

at the molecular level, then van der Waals forces are sufficiently large to create a bond which is almost as strong as *co*hesive bonds. The problem is to achieve reasonably perfect wetting with nothing in the interface which may interfere, such as surface-active materials. Such are the challenges to the formulator: removal of all surface-active species leads to higher surface tension which means poorer wetting. The Gods can be cruel!

The situation is analogous to that of the coalescence of latex particles, in that interfacial forces must act in regions of high curvature to deform the spheres so that they spread over, and wet the substrate. The geometry is shown in Fig. 255. Calculations analogous to those for coalescence lead to forces of about the same order of magnitude, assuming that it is the polymer/aqueous interfacial tension, $\gamma_{p/w}$ at work and that the spreading coefficient for the polymer over the substrate is positive. This last is in most cases impossible to achieve if surfactant can migrate to the interface, since it will be strongly adsorbed. To achieve good adhesion, then, the polymer must displace the surfactant, which if it can be achieved at all, must be a very slow process.

There are ways to dramatically improve adhesion:

(1) The value of $\gamma_{p/w}$ should be as low as possible. Therefore one should put on the particle surface functional groups which are readily solvated by water (see Section 6.2). The energy of solvation lowers the surface free energy – as long as water is present. This, of course, usually means greater sensitivity to moisture of the derived film. Since only the surface may be involved, a shell of polar material on a hydrophobic core will perform well in a film in the presence of moisture, at least after some interdiffusion between particles has taken place.

(2) The use of 'adhesion promoters,' which are usually polar monomers, can be effective. They may operate as described in the preceding paragraph, but there is evidence that specific interactions with the substrate may be involved. For example, small amounts of comonomers

containing groups such as −COOH, cyclic ureide, and even −OH and −CHO will form hydrogen bonds with cellulosic materials, while carboxyl groups may react with metals to form insoluble salts.

$$\begin{array}{c} R1 \\ O \diagdown | \diagup O \\ HN \quad NH \\ \diagdown | \diagup \\ O \end{array}$$

A cyclic ureide

(3) Every solid material has impurities on the surface which can interfere with adhesion. All solids adsorb gases from the air; metals may contain mill oils; all may have finger prints. Thus, cleaning the material without the use of surfactants prior to coating can improve adhesion.

11.5.4.3 Rheology

The flow behavior of polymer colloids and their derived compositions can make or break its success in use. The basic aspects of their rheology have been discussed in Chapter 10. Latex compositions of all kinds – paints, varnishes, adhesives, decorative and protective coatings, etc. – all have different rheological requirements during filling of containers in the factory, application, film formation and final development of film properties. For example, a gelled latex house paint must flow into the can, set up to a gel in the can, flow readily under a brush or roller, flow out onto the substrate to form a smooth finish, coalesce to a continuous film, and develop at least minimal film strength so that it doesn't peel off when someone goes to clean it or accidentally hits it!

On the other hand, a printing ink is applied from rollers to paper at extremely high speeds representing enormous shear rates in the η_{hyd} range (see Fig. 208). Once on the paper, the ink must penetrate the pores and spread over the fibers, adhere to them and dry in a very short time so that the printed paper can be rolled up or stacked in sheets. The dried ink must not flow onto the back of the paper above it on the roll or in the stack, a phenomenon known as 'blocking.' Addition of a water-insoluble wax or other material which will float to the film surface and provide a barrier to blocking has been found on occasion to solve this problem.

> Environmental laws, however, will provide little choice: waterborne systems are the wave of the future.

Water as a medium has the great advantage that it is environmentally nonpolluting. A major disadvantage is that it has a high latent heat of vaporization compared with most organic liquids. It is thus slower to dry and it requires more heat energy. Environmental laws, however, will provide little choice: water-borne systems are the wave of the future.*

11.5.5 Appearance

Appearance involves gloss, color, opacity ('hiding power'), transparency and refractive index.

Gloss. Gloss is generally measured by the intensity of light reflected at a given angle, although glossy *appearance* involves psychological factors which are considerably more complex. For example, a clear varnish may be required to show the wood grain underneath. Light scattering from imperfections in the polymeric film (e.g. interstitial surfactants or water-soluble polymers or voids) would produce a slight haze *perceived* as poor gloss. A polymer film of high refractive index has a high reflectivity which is seen as high gloss.

A smooth surface is essential to high gloss. This requires that particle coalescence during film formation is complete. The data in Table 11.2 indicate that small particles coalesce more completely than large ones at a given time and temperature. Even at the same degree of coalescence, small particles will produce a smoother surface in the sense that the hills and valleys will be smaller.

Color. Pigments are mixed with polymer colloids to form colored films. Titanium dioxide, the best white pigment, is a colorless, crystalline material in the bulk. It becomes white upon subdivision into micrometer-sized particles. Thus its whiteness is a result of the scattering of light. Because it has a high refractive index, it continues to scatter light when dispersed in a polymeric film. Refractive index difference and particle size determine the intensity of backscattering according to Eq. 62 (Section 5.1.2).

The intensity of color of pigments which absorb certain wavelengths of light is also a function of the complex scattering – Mie scattering – as given by Eq. 63. Even carbon black is not a pure black, but somewhat reddish or brownish, depending upon its degree of dispersion. Pigments come commercially as aggregated particles, and must be completely disaggregated by grinding in the presence of surfactants in order to achieve maximum tinctorial strength. When pigments are ground to a very fine particle size of below 1 μm, their scattering intensity is minimal and their absorbance is maximal.

> Even carbon black is not a pure black, but reddish or brownish, depending upon its degree of dispersion.

An ideally pigmented film will have the pigment particles randomly dispersed without any aggregation. This means that there should be no forcing of

---

*Powder coatings are also nonpolluting, but really do not compete well with latex-based systems. The driving forces to bring particles into close order and to keep them compacted during film formation are absent. Very low molecular weights are required in powder coatings so as to have sufficiently low viscosity for coalescence.

the pigment into interstitial voids during film formation. Furthermore, to prevent void formation, the polymer particles should wet and spread over the pigment particles.

Opacity and transparency. Opacity, as we have seen, is achieved through a high degree of backscattering of light or through a large extinction coefficient for light absorption. So where opaque films are required, large particle sizes will be preferred. The latex particles themselves can add to the light scattering if they contain more than one phase with a large refractive index difference (Section 2.3). As an example, particles with voids in them have been made [265] and are offered commercially.* Such latexes are lower in density than those without voids, offering some savings in requiring fewer kilograms per liter, but these are offset to some extent by a higher manufacturing cost. Additional cost savings comes in the use of a smaller amount of expensive TiO_2, for example, to achieve the same hiding as in a conventional latex paint.

Good transparency requires a minimum of haziness in the film derived from a polymer colloid. With large particle sizes this is almost impossible to achieve because of the large interstitial voids formed as water evaporates. Upon film formation these will likely contain water-soluble components or vacuoles which scatter light. Small particle size colloids (less than c. 100 nm diameter), even if they have the same percentage of material in the interstices, will have much less haze, because the scattering centers are smaller. These will likely be in the Rayleigh region where the scattering intensity is proportional to the sixth power of the particle size, according to Eq. 61. Of course, avoidance of any interstitial material is always the best solution when it is possible and economically feasible.

Another technique, mentioned earlier, for improving not only transparency, but also rheology and the protective properties of the derived film is to conduct the emulsion polymerization in such a manner that the particle size distribution is bi- or tri-modal, with the smaller sizes calculated just to fit the interstitial voids of the next larger ones, as shown in Fig. 256. The drawing in the figure is, of course, a two-dimensional representation of a three-dimensional arrangement. The remaining voids are so small as to have minimal light scattering intensity.

The protective properties of the film also are enhanced by this technique because diffusion of gases and moisture through the film should be reduced from that of a film derived from a monodisperse colloid: the diffusion path is more tortuous and the void/interstitial volume containing water-sensitive material is reduced.

The rheololgy is 'improved' in the sense that the value of the packing limit, p, can be significantly higher than that of a monodisperse colloid (Section 10.2). This means that products in which a very high volume fraction of polymer is desired, such as a caulk or putty, can be obtained without impossibly high viscosities.

---

* Ropaque^{TM}, Rohm & Haas Co., Philadelphia, PA, USA.

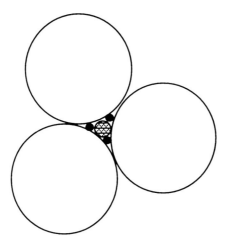

Fig. 256. Trimodal size distribution to fill interstitial voids.

11.5.5.1 Durability

Durability implies that the desired physical properties are retained after long periods of use. For a house paint this means that the protective and decorative qualities are not destroyed by weathering under ultraviolet radiation and moisture, oxidation by air and attack by molds and mildew. For an adhesive this may mean withstanding repeated mechanical distortion without adhesive or cohesive failure under a variety of environmental conditions from bookbinding to furniture to aircraft parts. For cement and stucco it means holding crystals of limestone and calcium phosphate together under high heat, humidity and physical abuse. For a floor polish it means retention of gloss and protection of the substrate under repeated abrasion and impaction by hard heels of shoes.

It is impossible to generalize. The author hopes that the reader will have gleaned enough from the contents of this book to at least have a good start at solving his or her particular problems. Good luck!

References

254. Vanderhoff, J.W., Bradford, E.B. and Carrington, W.K. (1973). *J. Polym. Sci., Symposium No. 41*, p. 155.
255. Eckersley, S.T. and Rudin, A. (1990). *J. Coatings Tech.* **62**, 89.
256. Sperry, P.R., Snyder, B.S., O'Dowd, M.L. and Lesko, P.M. (1994). *Langmuir* **10**, 2619.
257. Vanderhoff, J.W., Tarkowski, H.L., Jenkins, M.C. and Bradford, E.B. (1966). *J. Macromol. Chem.* **1**, 361.
258. Vanderhoff, J.W. (1977). *Polymer News* **3**, 194.
259. Sosnowski, S., Li, L., Winnik, M.A., Clubb, B. and Shivers, R. (1994). *J. Polym. Sci., Part B, Polym. Phys.* **32**, 2499

260. Joanicot, M., Wong, K., Maquet, J., Chevalier, Y., Pichot, C., Graillat, C., Lindner, P., Rios, L. and Cabanne, B. (1990). *Prog. Colloid Polym. Sci.* **81**, 1.
261. Bradford E.B., and Vanderhoff, J.W. (1972). *J. Macromol. Sci. – Phys.* **B-6**, 671.
262. Yoo, J.N., Sperling, L.H., Glinka, C.J. and Klein, A. (1990). *Macromolecules* **23**, 3963.
263. Kline, D.B. and Wool, R.P. (1988). *Polym. Eng. Sci.* **28**, 52.
264. Frenkel, J. (1943). *J. Phys. (USSR)* **9**, 385.
265. Okubo, M., Ichikawa, K. and Fujimura, M. (1991). *Colloid Polym. Sci.* **269**, 1257.

Index

ABS (acrylonitrile/butadiene/styrene) 41, 45
acid-catalyzed hydrolyses of small molecules 160–1
acrylamide inverse microemulsion polymerization 66–7
acrylamide microemulsion polymerization 67
acrylic acid (AA) 106, 320
acrylic latex 309, 310
activation energy 202
adhesion 332–4
adhesion promoters 333
adhesives 71, 145, 309
admittance 122, 123
adsorbed layer 106
adsorption isotherms 131–2, 309
adsorption of polymers by SdFFF 113
aggregates
   rheology of 300–4
   shear-dependent 301–4
aggregation number 16
AIBN (azobis-iso-butyronitrile) 69
Alder-Wainwright transition 254
Alexander-Napper-Gilbert theory 39–41
aliphatic hydrocarbon 63–4
amine-containing colloids 158–9
amphoteric colloid 242
angular dependence of light scattering intensity 76–8
angular dependence of neutron scattering intensity 87, 90
anionic polymerization 62–3
antibody conjugation 165
apparent ionic strength 124
aqueous phase reactions 31
aqueous polymer colloids 115
associating polymers (ATs) 310
atomic force microscopy (AFM) 98–100
attractive forces 255
autocatalyzed surface hydrolysis 157
autocorrelation function 84, 244
automotive finishes 69–70
Avogadro's number 30

batch emulsion polymerization 49–50

batch reactor 50
bead polymerization 24
bentonite clay dispersions 127
benzyl alcohol 281, 282
benzyl chloride groups 158–9
Bessel functions 37, 38
BGHO theory 293–5, 298, 307
bicomponent colliodal crystals 263–8
binary mixtures 264, 265
binary phase diagram 265–8
binodal phase separation 42–4
biomedical applications 164–71
biomolecule binding 164
bis(2-hydroxyethyl) terephthalate 63
block copolymers 193
blocking 334
Bragg angle 250
Bragg diffraction 90, 250–2, 268, 269, 272, 275
Bragg peak wavelength 274
Brownian coefficient 286
Brownian collision 11, 18
Brownian diffusion 283
Brownian motion 17, 84, 92, 105, 231, 256–63, 285, 286
Brownian particles 100
buoyant density 104
butadiene-containing copolymer 323
butyl acrylate (BA) 21

cancer therapy 167–8
capillary hydrodynamic fractionation (CHDF) 100–4
ε-caprolactam 63
carboxyls 116
Casson's equation 289
cationic colloids 147
centrifugation and washing 132–3
cetyl trimethyl ammonium (CTA) 141
chain-growth mechanism 63
chain transfer-derived surface groups 155–6
chair diagram 35
charged particles, interactions between 180–92

chemical modification of surface groups 156–9
chemotherapy 170–1
cleaning of polymer colloids 121, 128–33
clinical diagnostics 168–9
coagulation 11, 12, 28–9, 168, 204–7, 213, 219–20, 241
  kinetics 202–16
coagulation rates 16–18, 210–11
coagulum 57
coalescence 319, 325, 326, 333
coalescing pressure 325
coatings 70–1
  rheology 334–5
coherent neutron scattering length density 86
coherent scattering length 85–6
Cole-Cole distribution constant 122
collision frequency 203
collision rate 204
colloidal crystal arrays (CCAs) 271, 272
  applications 268–75
color of pigments 335
comonomer 52
comonomer-derived functional groups 151–5
compliance 329, 330
compositional drift 50
compression cell 199
condensation polymers 62–3
cone-and-plate viscometer 305
confocal laser scanning microscopy 92–3
contamination 121, 129, 132
continuous emulsion polymerization 53–4
continuous hollow fiber dialysis 130
continuous stirred tank reactors (CSTRs) 4, 53–7
contrast matching technique 88–9
controlled heterocoagulation 218–19
copolymer equation 50
core/shell particle morphologies 88–9, 118
corona treatment 332
correlation function 83
couette viscometer 305
counterion condensation 125–6, 179, 242
counterions 84, 124, 129, 230, 304
creep compliance 298–300, 306
critical chain length 15
critical coagulation temperature (CCT) 195
critical micelle concentration (CMC) 17, 21, 22, 40, 64
crystalline colloids 293
crystalline state 262
crystallization 263
crystals 265–7

Debye-Grans region 76
Debye-Hückel approximation 176, 178

Debye-Hückel osmotic pressure 190
Debye length 178, 231
decarboxylation 161–2
degree of polymerization 7
Derjaguin approximation 183, 197
desorption constant 37
destabilization 224
dialysis 130–1
DIANA 169
dibutyl zinc 63
dielectric increment 239–40, 242
dielectric probe 248
dielectric spectrometer 246–8
dielectric spectroscopy 120–8, 239–43, 245–8
diethyl aluminum chloride/titanium trichloride catalyst 62
$1,1'$-diethyl carbocyanine 142
diffraction gratings 268–70
diffusing wave spectroscopy (DWS) 260–2
diffusion 321–4, 322
diffusion coefficients 82, 84, 203, 262, 272
diffusion theory 203
dimethyl sulfoxide (DMSO) 270
disk centrifuge photosedimentometer (DCP) 108
dispersion polymerization 64
1,4-divinyl benzene (DVB) 221–2
DLVO theory 17, 18, 101, 180, 188, 190–2, 212, 255, 256, 273
DNA 169–70
DNA-binding proteins 169
DNA fragments 168–9
Doppler shift 243–4
double layer effects 127
double layer relaxations 120–1
drug delivery 171
durability 337
dye laser 270
dye partition 141–2
dynamic light scattering 82–5, 243
dynamic mechanical properties 118

effective degree of dissociation 212
Einstein's Law of Diffusion 231, 260, 272, 286
elastic floc model 304
elastic gel colloidal crystalline arrays 274–5
elastic modulus 295
elastic response 293
electric field effects 271–4
electrical double layer (EDL) 121, 176–9, 194, 228, 256
electrical FFF 112
electrical potential between two plates 180
electrically charged hydrophobic colloids, kinetic stability 174
electrochromic effects 273

electrokinetic instrumentation 243–8
electrokinetic potentials 229
electrokinetics 228–48
electrolyte concentration 206
electron microscopy 94–8
  film formation 317–21
electron scattering for chemical analysis (ESCA) 135
electroneutrality 176
electro-osmotic flow 228
electrophoresis 229–39
electrophoretic force 272
electrophoretic light scattering (ELS) 243–5
electrophoretic migration 274
electrophoretic mobility 230, 232, 234, 245
  of highly charged polymer colloids 236
  of low charge polymer colloids 235–6
  profiles 238
electrostatic interactions 190
electrostatically stabilized colloids 262
electrosteric stabilization 201–2, 217
electroviscous effects 291
emulsion polymerization 1, 5–47
  applications 48–58
  in supercritical fluids 68–9
  intervals during 28
  two-stage 118–20
encapsulants for pigment particles 71
equilibrium three-phase systems 252–3
equivalent hydrodynamic radius 83
equivalent interstitial capillary column 329
ethyl acrylate (EA) 21
2-ethylhexyl acrylate (2-EHA) 21
ethyl hydroxyethyl cellulose (EHEC) 310
externally catalyzed surface hydrolysis 158

fast atom bombardment-mass spectrometry (FAB-MS) 139
fibers 218
Fickian diffusion coefficient 82, 260
Fick's first law of diffusion 203
Field flow fractionation 110–114
film formation 314–38
  application to practical systems 331–2
  basic processes 314
  electron microscopy 317–21
  internal structure 319
  minimum temperature (MFT) 315–17
  role of polymer interface 325–7
  role of water 316–17, 324
  stages of water loss 314–15
  tensile strength in 323
  theories 324
film-forming latexes 4
fingerprints 239
Fitch-Tsai equation 12

Fitch-Tsai theory 9–12
flocculants 71–2
flocculation 205
floccules 307
  rheology of 301, 303
Flory-Huggins interaction parameter 27, 51, 68, 197, 220
Flory-Huggins theory 195
Flory-Krigbaum free energy 196
flow cytometer 167
flow FFF 112
foreign particles 25
Fourth Law of thermodynamics 128
Fraunhofer scattering 215
free induction decay (FID) 119
free radicals 16, 29, 31, 145
  transfer reactions 134
freeze fracture 319
freezing behavior of colloidal crystals 257–63
Frenkel equation 327
frequency-dependent total impedance 121
Fuchs stability factor 17
Fuchs stability ratio 206, 208, 211
functional groups 145–59
fusion 11

gel arrays 269–70
gelled dispersions 288–90
Gibbs free energy 194, 195
Gibbs-Helmholtz equation 194
glass transition temperature 316
glassy state 262
gloss 335
glycidyl methacrylate 164
Gouy-Chapman theory 296

hairy latex 240–3
Hamaker constant 174, 175, 182, 183, 189, 211, 212
Hamaker function 175, 185–8
Hansen-Ugelstad theory 13–15
hard sphere model 253–5
hard spheres 278–88, 303
HASE 312
heat treatment 241–3
heterogeneous catalysis 160–4
heterogeneous nucleation 21–5
HEUR 310, 312
higher order Tyndall spectra (HOTS) 80–1
HMHEC 108
Hofmeister (or lyotropic) series 208
homogeneous nucleation 7–21
homogeneous start (HOST) method 109
Hückel equation 232

Hückel theory 230
HUFT theory 25, 147
hydrodynamic exclusion chromatography (HEC) 100–4
hydrodynamic interactions 207–8
hydrodynamic radii 83, 103
hydrolysis 162–3, 301–2
  of surface groups 156–8
hydrolyzed acrylate latex particles 115
hydrophilic colloids 317
hydrophilic polymers 316
hydroxyethyl cellulose (HEC) 310

immunofluorescent cell separation 166–7
immunomagnetic cell separation 166
immunospecific cell separation 166–8
industrial production 4
initiation 29
initiator concentration 34, 35
initiator-derived functional groups 146
interface chemistry 145–72
interfacial area 2
interfacial tension 27
interparticle distance 90
interparticle spacing 253
intrinsic viscosity for hard spheres 278
inverse emulsion polymerization 64–8
inverse microemulsion polymerization 50, 65–8
o-iodoso-benzoate (IBA) 162
ion effects 208–9
ion-exchange 125–6, 128
ion scattering spectroscopy (ISS) 138–40
ion scattering spectroscopy/secondary ion mass spectrometry (ISS/SIMS) 133, 138–40
ionic strength 124, 184, 189, 252, 261
ionization 125
ionization potential 135

Kelvin, Lord 27
Kelvin effect 20
kinetic stability 173
  of electrically charged hydrophobic colloids 174
  of electrically neutral hydrophobic colloids 192–201
Krieger-Dougherty equation 283–5, 304
Krieger point 257

laboratory apparatus 48–9
Langmuir model 309
Langmuir-Ise-Sogami (LIS) Gibbs potential 190

Langmuir-Ise-Sogami (LIS) theory 189–90, 192, 212
latex, defined 1
latex paints 145, 216–17, 308, 311
latex stability 173–27
Legendre polynomials 79
Lifshitz-Hamaker function 187
Lifshitz-Parsegian-Ninham (LPN) theory 185–8, 256
lift-hyperlayer FFF 112
light scattering 73–85
  dynamic 82–5
  integrated method 209–13
  Lorenz-Mie theory 78–9
  Mie theory 78–9
  Rayleigh theory 73–6
  Rayleigh-Debye theory 76–8
line start (LIST) technique 109
liquid colloid crystal arrays (CCAs) 268–9
long-range electrostatic attractions 93
Lorenz-Mie equations 108
Lorenz-Mie theory of light scattering 78–9
lyophilic colloids 174, 220–5
  applications 225
lyophobic colloids 173, 174

matrix materials 4–5
mechanical properties 45–6
medical diagnostics 168–9
melting 263
  behavior of colloidal crystals 257–63
mercaptan surface groups 159
META 219
methacrylic acid (MAA) 320
methyl methacrylate (MMA) 14, 15, 21, 30, 55, 60, 61, 262
  emulsion polymerization 10, 17, 56
methylene bisacrylamide (BAM) 222
micellar nucleation 21–4
microelectrophoresis 235–6, 239
microemulsion nucleation 21–4
microgels 221–2
microscopy 91
  metallurgical 264–5
microtomy 94–5
Mie theory of light scattering 78–9
miniemulsion polymerization 24
minimum film forming temperature (MFT) 315–17
MMA/BA/MAA 315
mobility spectrum 243–4
molding resins 217–18
molecular weight distribution 331
monodisperse polymer colloids 145, 250, 263
monodisperse systems 3, 12
monomer composition 53

monomer concentration 30
monomer droplets 24
monomer feed 52
monomer partition 26
monomer-starved emulsion polymerization 52
Mooney equation 304
Morton equation 27, 51
motional averaging 114–15
multilobed particles 300–1
multiphase copolymers 41–5
multistage emulsion polymerization 52–3

Navier-Stokes equation 231
negative staining technique 28
neutron scattering 85–91, 258
 experiments 87–8
Newtonian system 277
6-nitrobenzisoxazole-3-carboxylate (NBC) 161
NMR peak assignments 117–18
NMR spectra
 of aqueous polymer colloids 115–18
 temperature-dependence 116
NMR spectroscopy 114–20, 140
nonaqueous dispersions (NADs) 59–72
 applications 69–72
 general characteristics 59
 methods of preparation 63–4
nonaqueous emulsion polymerization 60–4
nonionic initiators 147
nonionic surfactants 129
nonNewtonian behavior 283
nonNewtonian system 277
nonradical polymerization 61–3
nonspherical particles, rheology of 300–4
Norrish-Tromsdorff effect 28, 29, 61
nuclear magnetic resonance (NMR) spectroscopy see NMR

O'Brien and White theory 232
oligomeric free radicals 14
oligoradical capture 19
onion skin structure 53
opacity 336
optical microscopy 91
order-disorder equilibrium 192
order-disorder phenomena 127, 250–76
ordering effects on rheology 293
organic solvents 309
organosols 225
oscillatory methods 305
O'Toole equation 38
oxidizing agents 150

pair potentials 294, 297
particle distribution 18
particle formation 6–26, 40, 60–1
particle growth 26–41, 61, 67
particle interactions 196–201
particle morphologies 2, 41–5
particle nucleation 13, 67
particle number 15, 16, 39
particle pair 293
particle separation 102, 111
particle size 12, 18, 20, 34, 35, 38, 39, 127, 154–5, 222, 296–7
particle size distributions (PSDs) 12, 215
particle swelling kinetics and equilibrium 26–8
PBMA 99, 100
peak width 114–15
Peclet number 285–7, 300
permittivity 127
persulfate-initiated systems 146–7
persulfate-initiator reactions 156
phase growth 44–5
phase separation 41–5
photoinitiators 66
photomultiplier (PMT) detector 82
photon correlation spectroscopy (PCS) 82, 240
photothermal nanosecond light-switching devices 270
Pinacyanol 142
Planck's constant 85
plastisols 225
plug flow 54
PMMA 18, 25, 43, 44, 63, 69, 71, 111, 119, 120, 132, 197, 199, 200, 225, 243, 261, 283, 288–90, 319, 320, 322, 328
PMMA/BA/MAA 316, 319, 330
Poisson-Boltzmann distribution 178, 228
Poisson-Boltzmann-Nernst-Planck equation 231
polyacrylamide (PAM) 59, 64, 71, 243
polyacrylic acid (PAA) 225
polybutadiene rubber 45
polybutene 127
poly(1,1-dihydro perfluoro-octyl)acrylate 69
polydispersity effects 81
polyelectrolytes 201
polyethylene glycol (PEG) 103, 146, 223
polyethylene oxide (PEO) 170, 218
polyethylene terephthalate (PET) 63–4
polyhedral structures 320
polyheptafluorobutyl methacrylate (PHFBMA) 269–70
polyhydroxy stearic acid (PHA) 197, 199
polyhydroxystearate (PHS) 260
poly(lactide-co-glycolide) latexes 171

polylauryl methacrylate 63
polymer alloys 45–6
polymer colloids
 characterization 73–144
 chemistry of 3
 defined 1
 general descriptions 1–2
 physics of 3
polymerase chain reaction (PCR) technique 168
polymeric stabilizers 192–5
polymethyl acrylate (PMA) 157, 253
polymethyl methacrylate see PMMA
polysilicates 243
polystyrene 45, 78, 87, 90, 102, 103, 123, 132, 137, 141, 151, 208, 209, 212, 239, 258, 295, 302, 317, 318, 319
polystyrene/acrylonitrile (PS/AN) copolymer 41
polystyrene seed latex 26
polyvinyl acetate (PVAc) 18, 260, 302, 320
polyvinyl alcohol 301, 302, 303
polyvinyl chloride (PVC) 59, 225
poly(vinyl chloride-co-lauryl methacrylate) 63
polyvinyl toluene 236
potassium polyphosphotungstic acid (KPT) 95
potassium stearate 140
powder light diffraction 266
power feed 52
preformed seed particles 57
propagation 29–30
propyl acrylate (PA) 21
protective colloids 192
pulse method 306–7
purification 121, 129
putties 309
pyrrolidone/acrylamide/methylene-bis-acrylamide (PVP/PAm/BAM) gel 274–5

quasi-elastic light scattering (QELS) 82
quaternary ammonium groups 158–9

radial distribution function 91
radical capture 18–21
radical entry 29
radical exit 31
radical lifetime 34–6
Raman spectrometer 268–9
Randel circuit 121, 122
rate of capture of radicals 37
rate of generation of radicals 37
Rayleigh-Debye region 76

Rayleigh-Debye theory of light scattering 76–8
Rayleigh-Gans-Debye theory 209
Rayleigh jet technology 167
Rayleigh ratio 75
Rayleigh scattering 268–9
Rayleigh scattering intensity 75
Rayleigh theory of light scattering 16, 73–6
recursion equation 32
redox reactions at particle surface 159
Reerink-Overbeek theory 211
refractive index matching 262, 270
relative viscosity 281, 282
relaxation measurements 118
repulsive free energy 180–1
residence times 54, 55
retardation 175
reticuloendothelial system (RES) 170
reversible coagulation 219–20
Reynolds number 54, 100, 280
rheological equation of state 279
rheological measurements 304–7
rheological states, principle of corresponding 280, 282
rheology
 effects of ordering 293
 in coatings 334–5
 kinds of behavior 307
 of nonspherical particles 300–4
 of polymer colloids 277–313
Ricatti-Bessel functions 79
RNAs 170
rubberized concrete and stucco 218

salt concentration 211, 261
scanning electron microscopy (SEM) 97–8, 166, 267, 318, 319, 328
scanning transmission electron microscopy (STEM) 98
Schottky defect 320
secondary ion mass spectrometry (SIMS) 138–40
sedimentation 104–14, 114
sedimentation coefficient 106, 107
sedimentation field flow fractionation (SdFFF or SdF^3) 110
sedimentation rate 114
sedimentation rate equation 108
sedimentation velocity 105, 107
seeded emulsion polymerizations 25
semi-continuous emulsion polymerization 50–3
serum replacement 131–2
shadowing 95–6
shear dependence 126
shear modulus 293, 294, 296

shear rate 303
shear stress 280–3, 291, 307
shear-thinning 307–8
shearing force 278
shortness factor (SF) 307
single particle optical sizing (SPOS) 213–15
small angle neutron scattering (SANS) 87–8, 322
Smith-Ewart kinetics 31–6
Smith-Ewart recursion formula 61, 64, 69
Smith-Ewart-Roe equation 23
Smitham-Evans-Napper (SEN) theory 196–7
Smoluchowski equation 232
Smoluchowski fast coagulation rate 18
Smoluchowski theory 230
snowplow effect 283
sodium 10-mercapto-1-decane sulfonate (SMDSo) 155
sodium dodecyl sulfate (SDS) 17, 18, 21, 52, 132, 311, 312
sodium dodecyl sulfonate 139
sodium-sulfodecyl styryl ether (SSDSE) 154
sodium styrene sulfonate 65
Sogami potential 191
Solartron frequency response analyzer (FRA) 245
solution polymerization kinetics 35
solvent transfer through semi-permeable membrane 224–5
solvents 223–4
  mixing with 224
specific ion effects 126
spinodal decomposition 42–4
spreading coefficient 42, 332, 333
stabilizer precursor 63
staining techniques 94–5
steady flow methods 304–5
stearic acid 140
step-growth condensation polymerization 62–3
steric-elastic force 290
steric FFF 112, 113
steric stabilization 57, 69, 192, 197, 198, 201, 202, 218
Stern layer 126, 179, 191, 297, 324
Stern potential 211, 212, 229
Stokes coefficient 297–8
Stokes-Einstein equation 83, 204
Stokes equivalent spherical hydrodynamic diameter 109
Stokes law 260
structure factor 89–90, 258
styrene 14, 28, 219
styrene/acrylonitrile (S/AN) emulsion copolymerization 50

styrene/butadiene emulsion polymerization 54
styrene/butadiene rubber (SBR) 32
styrene/divinyl benzene copolymer 223
styrene emulsion polymerization 10
styrene/polystyrene/polymethylmethacrylate phase diagram 43
sulfonate derivatives of benzoyl peroxide (SMBP) 151
sulfonate surface groups 150–1
sulfonic acid 125, 126, 157, 161
sulfur 139
Sundberg theory and experiments 44
supercritical fluids 68–9
surface charge density 11, 127, 243
surface charges 191
surface chemistry 128–42
surface concentrations 3
surface coverage 302–3
surface group composition and concentration 133, 154–5
surface potential 191, 212
surface reactions 3
surfactant adsorption 140–1
surfactant concentration 11, 13, 56
surfactant-free PS colloids 146–51
surfactants 145
SURFER spline approximation algorithm 237
suspension polymerization 24, 64

tensile strength in film formation 323
termination 30–1, 34, 37
tetrahydrofuran (THF) 78
textile fiber coatings 71
thermal FFF 112
thermal motion 91
thermodynamic stability 173
thixotropic rheology 308
time-average intensity 84
time/temperature super-position 316
titration 125–6, 133–5
Torza and Mason theory and experiments 41–2
total free energy
  for flat plates 182–3
  for spheres 183
total repulsive potential energy between spheres 198
transmission electron microscopy (TEM) 94–7, 160, 318
transparency 336
transurfs 155–6
trimodal size distribution 337
tubular reactors 4, 53–4
Tyndall effect 73

Ugelstad-Hansen theory 37–9
ultracentrifugation 105–8
ultramicroscope 92

van der Waals attraction 101, 174–6, 190, 194, 296, 303, 333
vinyl acetate 55
vinyl benzyl chloride 158–9
4-vinyl pyridine (4-VP) 222
viscoelastic properties 327–31
viscoelastic resistance 325
viscoelasticity 299
viscosity 279, 281, 282, 285, 297–8, 303, 310, 311, 330
   definition 277
   hard sphere dispersions 278–88
   high frequency oscillatory regime 286–7
   high shear limiting 287
   low frequency, low shear regime 286
viscous drag coefficient 104
voids 93, 337

water-soluble monomers 64–5
wetting 332, 333
Williams-Landel-Ferry (WLF) equation 316

X-ray photoelectron spectroscopy (XPS) 133, 135–8
X-ray powder diffraction 266

yield stress 289
Yoshino quantized oscillator model 255–7
Young-LaPlace equation 325
Young-LaPlace pressure 328
yttrium-aluminum-garnet (YAG) laser 270

zero-one type kinetics 33
zeta potential 232–5, 243
Ziegler-Natta type reactions 62

DATE DUE

OhioLINK		
MAR 1 7 REC'D		
1985854		
#44217		
2/21/04		

WITHDRAWN

GAYLORD — PRINTED IN U.S.A.

SCI QD 549.2 .P64 F58 1997

Fitch, Robert McLellan, 1928-

Polymer colloids